T0342886

THE PENGUIN BOOK OF THE OCEAN

EDITED BY JAMES BRADLEY

James Bradley is the author of six novels, *Wrack*, *The Deep Field*, *The Resurrectionist*, *Clade*, *The Silent Invasion* and *Beauty's Sister*, and a book of poetry, *Paper Nautilus*. His books have won or been shortlisted for a number of major Australian and international literary awards and have been widely translated.

EDITED BY JAMES BRADLEY

THE PENGUIN BOOK OF THE OCEAN

HAMISH HAMILTON
an imprint of
PENGUIN BOOKS

HAMISH HAMILTON

Published by the Penguin Group
Penguin Group (Australia)
250 Camberwell Road, Camberwell, Victoria 3124, Australia
(a division of Pearson Australia Group Pty Ltd)
Penguin Group (USA) Inc.
375 Hudson Street, New York, New York 10014, USA
Penguin Group (Canada)
90 Eglinton Avenue East, Suite 700, Toronto, Canada ON M4P 2Y3
(a division of Pearson Penguin Canada Inc.)
Penguin Books Ltd
80 Strand, London WC2R 0RL, England
Penguin Ireland
25 St Stephen's Green, Dublin 2, Ireland
(a division of Penguin Books Ltd)
Penguin Books India Pvt Ltd
11 Community Centre, Panchsheel Park, New Delhi – 110 017, India
Penguin Group (NZ)
67 Apollo Drive, Rosedale, North Shore 0632, New Zealand
(a division of Pearson New Zealand Ltd)
Penguin Books (South Africa) (Pty) Ltd
24 Sturdee Avenue, Rosebank, Johannesburg 2196, South Africa

Penguin Books Ltd, Registered Offices: 80 Strand, London WC2R 0RL, England

First published by Penguin Group (Australia), 2010

Copyright in this collection and the Introduction © James Bradley 2010
Copyright © in individual stories remains with the authors

The moral right of the authors has been asserted

All rights reserved. Without limiting the rights under copyright reserved above, no part of this publication may be reproduced, stored in or introduced into a retrieval system, or transmitted, in any form or by any means (electronic, mechanical, photocopying, recording or otherwise), without the prior written permission of both the copyright owner and the above publisher of this book.

Cover design by Tony Palmer © Penguin Group (Australia)
Text design by Laura Thomas © Penguin Group (Australia)
Cover image by Katsushika Hokusai/Getty Images
Typeset in 12/16.5 pt Adobe Garamond by Post Pre-press Group, Brisbane, Queensland
Printed and bound in Australia by Griffin Press

National Library of Australia
Cataloguing-in-Publication data:

The Penguin book of the ocean / edited by James Bradley.
9781926428161 (pbk.)
Sea stories.
Bradley, James.

808.83932162

penguin.com.au

MIX
Paper from responsible sources
FSC® C009448

CONTENTS

INTRODUCTION

We will never know when humans first encountered the ocean. Nor is it easy to imagine what those ancient travellers made of it. Did they worship it? Fear it? Or is it possible they felt, as we so often do, something more like recognition, some sense its waters held an echo of their origins?

These are not questions we can answer. Yet perhaps even from the vantage point of the 21st century it is possible to imagine the manner in which that encounter altered them, altered us, to imagine the manner in which those first journeys outwards, away from land, enlarged our world, and with it our conception of who and what we might be.

Certainly this feeling of possibility, of passage into a new world echoes through our imagining of the ocean. Emily Dickinson describes the 'Exultation . . . The divine intoxication/Of the first league out from land', Melville the 'mystical vibration' of the first glimpse of the open ocean. Yet these feelings are only the outward echoes of a more profound relationship. The ocean has been one of the wellsprings of the human imagination for almost as long as we have existed. Its presence suffuses us, haunts us, echoing through our language, through our shared reservoir of myth and metaphor. We even bear its memory in the matter of our bodies, not just in the salt of our blood and tears and the water that fills our cells, but in our very DNA, its memory of a time when all life inhabited the warm oceans of the primeval Earth. In its immensity we see an echo of our own desire for transcendence, in its depths an echo of the vast and mysterious space of time, in its restlessness a reminder of the impermanence of all things, of the certainty of change, and loss.

In part this is a reflection of the manner in which the ocean has shaped our experience. Ever since those first travellers made their way onto the shores of Africa a hundred thousand years ago, the ocean has given us new horizons to cross, new possibilities to explore, even as the shifting cycles of its currents have shaped our environment, warming shores that would otherwise be cold, bringing rain, driving the great engine of the winds.

But it has also entered us in other, deeper ways. It is not accidental that when the historian Fernand Braudel sought to describe those cycles of history that exceed the human and stretch downwards, into the environmental, the geological, he reached for the marine metaphor of 'deep time', nor that Romain Rolland chose the term 'oceanic' feeling to describe the sensations of boundlessness and oneness with nature he believed were the birthplace of religious sentiment. Indeed, so pronounced is our tendency to reach for images of fluidity and submersion to describe our inner lives and the mysterious processes of creativity and creation that it is difficult not to wonder whether the association is somehow natural, less a habit of mind than something innate. But even if it is not, the association between water and dreams, time and the oceanic runs so deep it has become almost impossible to think of one without invoking the other.

This collection is an attempt to explore some of the complexities of this relationship. It is not, nor does it pretend to be a comprehensive survey; rather it is a personal selection of writing I believe has something to tell us about the ways we think about the ocean and, more particularly, the ways in which the ocean has shaped our imaginations, and by extension our selves.

Some of the pieces it contains will be familiar, others less so.

Yet my hope is that gathered together they will illuminate each other in new and hopefully interesting ways. Some of these connections and associations are explicit, as when Sebastian Junger quotes Ernest Shackleton's description of a wave so huge that in the darkness he mistook its white crest for a rift in the clouds, or in Emily Ballou's reimagining of the young Charles Darwin collecting specimens on the shores of South America. Others are more subtle: lines of influence and inspiration such as those that connect Owen Chase's terrifying account of the sinking of the whaleship *Essex* with Melville's masterpiece, *Moby-Dick*, or the manner in which Hakluyt's account of the horrible fate of the crew of the *Desire* can be heard echoing through the agitated, infernal textures of *The Rime of the Ancient Mariner*.

Yet all speak to the way in which our encounters with the ocean are also, perhaps inevitably, journeys inwards, explorations of aspects of our natures that might otherwise remain hidden.

Some of what is discovered in this process is good, even admirable. It is impossible to read Thor Heyerdahl's rapturous descriptions of the upwelling of life that surrounds the *Kon-Tiki* as she sails westwards, across the Pacific, without feeling something of the delight that echoes through them, or to hear William Beebe describe the sudden flash of life outside his bathysphere without sharing, however briefly, the same enlarging thrill of discovery that animates his words.

Something similar is true of the various works of natural history gathered here, from Matthew Fontaine Maury's extraordinary *The Physical Geography of the Sea and its Meteorology*, to more contemporary works such as Jennifer Ackerman's hymn to her strip of coast in upstate New York, or Deborah Cramer's expansive exploration of the interconnected cycles of wind and water in the Atlantic, all of which remind us of how little ultimately separates the observing eye of the scientist and the unifying vision of the poet.

Commensurately, our encounters with the ocean can be chastening, even discomfiting, a reminder of the manner in which those same desires to explore, to move beyond the known, and into the unknown, underpin our capacity for cruelty and greed. This uneasy understanding echoes through Derek Walcott's great song of loss and redemption, 'The Sea is History', but it is no less present in works such as Tim Winton's *Breath* or Kem Nunn's *The Dogs of Winter*, both of which explore the manner in which our urge to master the natural world distorts and corrupts.

Yet time and again we hear reminders of the ocean's capacity to take us out of ourselves. Like Slocum's description of the feelings of awe that overtook him as the *Spray* left the land behind, and of the manner in which his memory began to work 'with startling power', bringing visions and memories he had thought 'so long forgotten that they seemed to belong to another existence', the ocean is, in the most powerful sense, a reminder not just of what we are, but of what we might be.

These are, of course, thoughts written in the shadow of catastrophe. While sixty years ago Rachel Carson could write that 'man . . . cannot control or change the ocean as, in his brief tenancy of earth, he has subdued and plundered the continents', that confidence now seems misplaced. The ocean is in crisis, caught in a seemingly inexorable cycle of decline. Rising temperatures, increasing acidity, overfishing, all threaten the very life of the ocean.

Whether we will manage to avert the disaster we now face is unclear. What is clear is that the beginnings of a solution may also lie in the ocean, or more properly, in our capacity to grasp the meanings it offers us. For as Carson also reminds us, in:

> the artificial world of his cities and towns, [man] often forgets the true nature of his planet, and the long vistas of history, in which the

> existence of the race of men has occupied a mere moment of time. The sense of all these things comes to him most clearly in the course of a long ocean voyage, when he watches day after day the receding rim of the horizon, ridged and furrowed by waves; when at night he becomes aware of the earth's rotation as the stars pass overhead; or when, alone in this world of water and sky he feels the loneliness of his earth in space.

To understand the ocean, to glimpse its meaning is, in other words, to understand ourselves, and by extension our place in the larger order of things. As the philosopher Gaston Bachelard observes, 'To disappear into deep water or to disappear toward a far horizon, to become part of depth of infinity, such is the destiny of man that finds its image in the destiny of water.' Or, in the words of Thomas Farber, 'To name the qualities of even Earth's oceans . . . reveals our hungers. Takes us to the limits of our capacities. And beyond.'

And the earth was without form, and void; and darkness was upon the face of the deep.
And the Spirit of God moved upon the face of the waters.
And God said, Let there be light: and there was light.

Genesis 1:2–3

RACHEL CARSON

The Gray Beginnings

Beginnings are apt to be shadowy, and so it is with the beginnings of that great mother of life, the sea. Many people have debated how and when the earth got its ocean, and it is not surprising that their explanations do not always agree. For the plain and inescapable truth is that no one was there to see, and in the absence of eyewitness accounts there is bound to be a certain amount of disagreement. So if I tell here the story of how the young planet Earth acquired an ocean, it must be a story pieced together from many sources and containing whole chapters the details of which we can only imagine. The story is founded on the testimony of the earth's most ancient rocks, which were young when the earth was young; on other evidence written on the face of the earth's satellite, the moon; and on hints contained in the history of the sun and the whole universe of star-filled space. For although no man was there to witness this cosmic birth, the stars and moon and the rocks were there, and, indeed, had much to do with the fact that there is an ocean.

The events of which I write must have occurred somewhat more than 2 billion years ago. As nearly as science can tell, that is the approximate age of the earth, and the ocean must be very nearly as old. It is possible now to discover the age of the rocks that compose the crust of the earth by measuring the rate of decay of the radioactive materials they contain. The oldest rocks found anywhere on earth – in Manitoba – are about 2.3 billion years old. Allowing 100 million years or so for the cooling of the earth's materials to form a rocky crust, we arrive at the supposition that the tempestuous and violent events connected with our planet's birth occurred nearly 2.5 billion years ago. But this is only a minimum estimate, for rocks indicating an even greater age may be found at any time.*

The new earth, freshly torn from its parent sun, was a ball of whirling gases, intensely hot, rushing through the black spaces of the universe on a path and at a speed controlled by immense forces. Gradually the ball of flaming gases cooled. The gases began to liquefy, and Earth became a molten mass. The materials of this mass eventually became sorted out in a definite pattern: the heaviest in the center, the less heavy surrounding them, and the least heavy forming the outer rim. This is the pattern which persists today – a central

*Our concept of the age of the earth is constantly undergoing revision as older and older rocks are discovered and as methods of study are refined. The oldest rocks now known in North America are in the Canadian Shield area. Their precise age has not been determined, but some from Manitoba and Ontario are believed to have been formed about 3 billion years ago. Even older rocks have been discovered in the Karelia Peninsula in the U.S.S.R., and in South Africa. Geologists are generally of the opinion that present concepts of geologic time will be considerably lengthened in the future. Tentative adjustments of the length of the various periods have already been made and the age of the Cambrian has been pushed back 100 million years compared with the dating assigned to it a decade ago. It is in that immense and shadowy time that preceded the Cambrian, however, that the greatest uncertainty exists. This is the time of the pre-fossiliferous rocks. Whatever life may have inhabited the earth during that time has left few traces, although by indirect evidence we may infer that life existed in some abundance before its record was written in the rocks.

By studies of the rocks themselves geologists have established a few good benchmarks standing out in those vast stretches of time indicated on the chart as the Proterozoic and Archeozoic Eras. These indicate a billion-year age for the ancient Grenville Mountains of eastern North America. Where these rocks are exposed at the surface, as in Ontario, they contain large amounts of graphite, giving silent testimony to the abundance of plant life when these rocks were forming, for plants are a common source of carbon. An age-reading of 1 700 000 000 years has been obtained in the Penokean Mountains of Minnesota and Ontario, formerly known to geologists as the Killarney Mountains. The remains of these once lofty mountains are still to be seen as low, rolling hills. The discovery of even older rocks in Canada, Russia and Africa, dating back more than 3 billion years, suggests that the earth itself may have been formed about 4.5 billion years ago.

sphere of molten iron, very nearly as hot as it was 2 billion years ago, an intermediate sphere of semiplastic basalt, and a hard outer shell, relatively quite thin and composed of solid basalt and granite.

The outer shell of the young earth must have been a good many millions of years changing from the liquid to the solid state, and it is believed that, before this change was completed, an event of the greatest importance took place – the formation of the moon. The next time you stand on a beach at night, watching the moon's bright path across the water, and conscious of the moon-drawn tides, remember that the moon itself may have been born of a great tidal wave of earthly substance, torn off into space. And remember that if the moon was formed in this fashion, the event may have had much to do with shaping the ocean basins and the continents as we know them.

There were tides in the new earth, long before there was an ocean. In response to the pull of the sun the molten liquids of the earth's whole surface rose in tides that rolled unhindered around the globe and only gradually slackened and diminished as the earthly shell cooled, congealed, and hardened. Those who believe that the moon is a child of Earth say that during an early stage of the earth's development something happened that caused this rolling, viscid tide to gather speed and momentum and to rise to unimaginable heights. Apparently the force that created these greatest tides the earth has ever known was the force of resonance, for at this time the period of the solar tides had come to approach, then equal, the period of the free oscillation of the liquid earth. And so every sun tide was given increased momentum by the push of the earth's oscillation, and each of the twice-daily tides was larger than the one before it. Physicists have calculated that, after 500 years of such monstrous, steadily increasing tides, those on the side toward the sun became too high for stability, and a great wave was torn away and hurled into space.

But immediately, of course, the newly created satellite became subject to physical laws that sent it spinning in an orbit of its own about the earth. This is what we call the moon.

There are reasons for believing that this event took place after the earth's crust had become slightly hardened, instead of during its partly liquid state. There is to this day a great scar on the surface of the globe. This scar or depression holds the Pacific Ocean. According to some geophysicists, the floor of the Pacific is composed of basalt, the substance of the earth's middle layer, while all other oceans are floored with a thin layer of granite, which makes up most of the earth's outer layer. We immediately wonder what became of the Pacific's granite covering and the most convenient assumption is that it was torn away when the moon was formed. There is supporting evidence. The mean density of the moon is much less than that of the earth (3.3 compared with 5.5), suggesting that the moon took away none of the earth's heavy iron ore, but that it is composed only of the granite and some of the basalt of the outer layers.

The birth of the moon probably helped shape other regions of the world's oceans besides the Pacific. When part of the crust was torn away, strains must have been set up in the remaining granite envelope. Perhaps the granite mass cracked open on the side opposite the moon scar. Perhaps, as the earth spun on its axis and rushed on its orbit through space, the cracks widened and the masses of granite began to drift apart, moving over a tarry, slowly hardening layer of basalt. Gradually the outer portions of the basalt layer became solid and the wandering continents came to rest, frozen into place with oceans between them. In spite of theories to the contrary, the weight of geologic evidence seems to be that the locations of the major ocean basins and the major continental land masses are today much the same as they have been since a very early period of the earth's history.

But this is to anticipate the story, for when the moon was born there was no ocean. The gradually cooling earth was enveloped in heavy layers of cloud, which contained much of the water of the new planet. For a long time its surface was so hot that no moisture could fall without immediately being reconverted to steam. This dense, perpetually renewed cloud covering must have been thick enough that no rays of sunlight could penetrate it. And so the rough outlines of the continents and the empty ocean basins were sculptured out of the surface of the earth in darkness, in a Stygian world of heated rock and swirling clouds and gloom.

As soon as the earth's crust cooled enough, the rains began to fall. Never have there been such rains since that time. They fell continuously, day and night, days passing into months, into years, into centuries. They poured into the waiting ocean basins, or, falling upon the continental masses, drained away to become sea.

That primeval ocean, growing in bulk as the rains slowly filled its basins, must have been only faintly salt. But the falling rains were the symbol of the dissolution of the continents. From the moment the rain began to fall, the lands began to be worn away and carried to the sea. It is an endless, inexorable process that has never stopped – the dissolving of the rocks, the leaching out of their contained minerals, the carrying of the rock fragments and dissolved minerals to the ocean. And over the eons of time, the sea has grown ever more bitter with the salt of the continents.

In what manner the sea produced the mysterious and wonderful stuff called protoplasm we cannot say. In its warm, dimly lit waters the unknown conditions of temperature and pressure and saltiness must have been the critical ones for the creation of life from non-life. At any rate they produced the result that neither the alchemists with their crucibles nor modern scientists in their laboratories have been able to achieve.

Before the first living cell was created, there may have been many trials and failures. It seems probable that, within the warm saltiness of the primeval sea, certain organic substances were fashioned from carbon dioxide, sulphur, nitrogen, phosphorus, potassium, and calcium. Perhaps these were transition steps from which the complex molecules of protoplasm arose – molecules that somehow acquired the ability to reproduce themselves and begin the endless stream of life. But at present no one is wise enough to be sure.

Those first living things may have been simple micro-organisms rather like some of the bacteria we know today – mysterious borderline forms that were not quite plants, not quite animals, barely over the intangible line that separates the non-living from the living. It is doubtful that this first life possessed the substance chlorophyll, with which plants in sunlight transform lifeless chemicals into the living stuff of their tissues. Little sunshine could enter their dim world, penetrating the cloud banks from which fell the endless rains. Probably the sea's first children lived on the organic substances then present in the ocean waters, or, like the iron and sulphur bacteria that exist today, lived directly on inorganic food.

All the while the cloud cover was thinning, the darkness of the nights alternated with palely illumined days, and finally the sun for the first time shone through upon the sea. By this time some of the living things that floated in the sea must have developed the magic of chlorophyll. Now they were able to take the carbon dioxide of the air and the water of the sea and of these elements, in sunlight, build the organic substances they needed. So the first true plants came into being.

Another group of organisms, lacking the chlorophyll but needing organic food, found they could make a way of life for themselves by devouring the plants. So the first animals arose, and from that day to this, every animal in the world has followed the habit it learned in the ancient seas and depends, directly or through complex food chains, on the plants for food and life.

As the years passed, and the centuries, and the millions of years, the stream of life grew more and more complex. From simple, one-celled creatures, others that were aggregations of specialised cells arose, and then creatures with organs for feeding, digesting, breathing, reproducing. Sponges grew on the rocky bottom of the sea's edge and coral animals built their habitations in warm, clear waters. Jellyfish swam and drifted in the sea. Worms evolved, and starfish, and hard-shelled creatures with many-jointed legs, the arthropods. The plants, too, progressed, from the microscopic algae to branched and curiously fruiting seaweeds that swayed with the tides and were plucked from the coastal rocks by the surf and cast adrift.

During all this time the continents had no life. There was little to induce living things to come ashore, forsaking their all-providing, all-embracing mother sea. The lands must have been bleak and hostile beyond the power of words to describe. Imagine a whole continent of naked rock, across which no covering mantle of green had been drawn – a continent without soil, for there were no land plants to aid in its formation and bind it to the rocks with their roots. Imagine a land of stone, a silent land, except for the sound of the rains and winds that swept across it. For there was no living voice, and no living thing moved over the surface of the rocks.

Meanwhile, the gradual cooling of the planet, which had first given the earth its hard granite crust, was progressing into its deeper layers; and as the interior slowly cooled and contracted, it drew away from the outer shell. This shell, accommodating itself to the shrinking sphere within it, fell into folds and wrinkles – the earth's first mountain ranges.

Geologists tell us that there must have been at least two periods of mountain building (often called 'revolutions') in that dim period, so long ago that the rocks have no record of it, so long ago that the mountains themselves have long since been worn away. Then there came a third great period of upheaval and readjustment of the earth's crust, about a billion years ago, but of all its majestic mountains the only reminders today are the Laurentian hills of eastern Canada, and a great shield of granite over the flat country around Hudson Bay.

The epochs of mountain building only served to speed up the processes of erosion by which the continents were worn down and their crumbling rock and contained minerals returned to the sea. The uplifted masses of the mountains were prey to the bitter cold of the upper atmosphere and under the attacks of frost and snow and ice the rocks cracked and crumbled away. The rains beat with greater violence upon the slopes of the hills and carried away the substance of the mountains in torrential streams. There was still no plant covering to modify and resist the power of the rains.

And in the sea, life continued to evolve. The earliest forms have left no fossils by which we can identify them. Probably they were soft-bodied, with no hard parts that could be preserved. Then, too, the rock layers formed in those early days have since been so altered by enormous heat and pressure, under the foldings of the earth's crust, that any fossils they might have contained would have been destroyed.

For the past 500 million years, however, the rocks have preserved the fossil record. By the dawn of the Cambrian period, when the history of living things was first inscribed on rock pages, life in the sea had progressed so far that all the main groups of back-boneless or invertebrate animals had been developed. But there were no animals with backbones, no insects or spiders, and still no plant or animal had been evolved that was capable of venturing onto the forbidding land. So for more than three-fourths of geologic time the continents were desolate and uninhabited, while the sea prepared the life that was later to invade them and make them habitable. Meanwhile, with violent tremblings of the earth and with the fire and smoke of roaring volcanoes, mountains rose and wore away, glaciers moved to and fro over the earth, and the sea crept over the continents and again receded.

It was not until Silurian time, some 350 million years ago, that the first pioneer of land life crept out on the shore. It was an arthropod, one of the great tribe that later produced crabs and lobsters and insects. It must have been something like a modern scorpion, but, unlike some of its descendants, it never wholly severed the ties that united it to the sea. It lived a strange life, half-terrestrial, half-aquatic, something like that of the ghost crabs that speed along the beaches today, now and then dashing into the surf to moisten their gills.

Fish, tapered of body and stream-moulded by the press of running waters, were evolving in Silurian rivers. In times of drought, in the drying pools and lagoons, the shortage of oxygen forced them to develop swim bladders for the storage of air. One form that possessed an air-breathing lung was able to survive the dry periods by burying itself in mud, leaving a passage to the surface through which it breathed.

It is very doubtful that the animals alone would have succeeded in colonizing the land, for only the plants had the power to bring about the first amelioration of its harsh conditions. They helped make soil of the crumbling rocks, they held back the soil from the rains that would have swept it away, and little by little they softened and subdued the bare rock, the lifeless desert. We know very little about the first land plants, but they must have been closely related to some of the larger seaweeds that had learned to live in the coastal shallows, developing strengthened stems and grasping, rootlike holdfasts to resist the drag and pull of the waves. Perhaps it was in some coastal lowlands, periodically drained and flooded, that some such plants found it possible to survive, though separated from the sea. This also seems to have taken place in the Silurian period.

The mountains that had been thrown up by the Laurentian revolution gradually wore away, and as the sediments were washed from their summits and deposited on the lowlands, great areas of the continents sank under the land. The seas crept out of their basins and spread over the lands. Life fared well and was exceedingly abundant in those shallow, sunlit seas. But with the later retreat of the ocean water into the deeper basins, many creatures must have been left stranded in shallow, landlocked bays. Some of these animals found means to survive on land. The lakes, the shores of the rivers, and the coastal swamps of those days were the testing grounds in which plants and animals either became adapted to the new conditions or perished.

As the lands rose and the seas receded, a strange fishlike creature emerged on the land, and over the thousands of years its fins became legs, and instead of gills it developed lungs. In the Devonian sandstone this first amphibian left its footprint.

On land and sea the stream of life poured on. New forms evolved; some old ones declined and disappeared. On land the mosses and the ferns and the seed plants developed. The reptiles for a time dominated the earth, gigantic, grotesque and terrifying. Birds learned to live and move in the ocean of air. The first small mammals lurked inconspicuously in hidden crannies of the earth as though in fear of the reptiles.

When they went ashore the animals that took up a land life carried with them a part of the sea in their bodies, a heritage which they passed on to their children and which even today links each land animal with its origin in the ancient sea. Fish, amphibian and reptile, warm-blooded bird and mammal – each of us carries in our veins a salty stream in which the elements sodium, potassium and calcium are combined in almost the same proportions as in sea water. This is our inheritance from the day, untold millions of years ago, when a remote ancestor, having progressed from the one-celled to the many-celled stage, first developed a circulatory system in which the fluid was merely the water of the sea. In the same way, our lime-hardened skeletons are a heritage from the calcium-rich ocean of Cambrian time. Even the protoplasm that streams within each cell of our bodies has the chemical structure impressed upon all living matter when the first simple creatures were brought forth in the ancient sea. And as life itself began in the sea, so each of us begins his individual life in a miniature ocean within his mother's womb, and in the stages of his embryonic development repeats the steps by which his race evolved, from gill-breathing inhabitants of a water world to creatures able to live on land.

Some of the land animals later returned to the ocean. After perhaps 50 million years of land life, a number of reptiles entered the sea about 170 million years ago, in the Triassic period. They were huge and formidable creatures. Some had oarlike limbs by which they rowed through the water; some were web-footed, with long, serpentine necks. These grotesque monsters disappeared millions of years ago, but we remember them when we come upon a large sea turtle swimming many miles at sea, its barnacle-encrusted shell eloquent of its marine life. Much later, perhaps no more than 50 million years ago, some of the mammals, too, abandoned a land life for the ocean. Their descendants are the sea lions, seals, sea elephants, and whales of today.

Among the land mammals there was a race of creatures that took to an arboreal existence. Their hands underwent remarkable development, becoming skilled in manipulating and examining objects, and along with this skill came a superior brain power that compensated for what these comparatively small mammals lacked in strength. At last, perhaps somewhere in the vast interior of Asia, they descended from the trees and became again terrestrial. The past million years have seen their transformation into beings with body and brain and spirit of man.

Eventually man, too, found his way back to the sea. Standing on its shores, he must have looked out upon it with wonder and curiosity, compounded with an unconscious recognition of his lineage. He could not physically re-enter the ocean as the seals and whales had done. But over the centuries, with all the skill and ingenuity and reasoning powers of his mind, he has sought to explore and investigate even its most remote parts, so that he might re-enter it mentally and imaginatively.

He built boats to venture out on its surface. Later he found ways to descend to the shallow parts of its floor, carrying with him the air that, as a land mammal long unaccustomed to aquatic life, he needed to breathe. Moving in fascination over the deep sea he could not enter, he found ways to probe its depths, he let down nets to capture its life, he invented mechanical eyes and ears that could re-create for his senses a world long lost, but a world that, in the deepest part of his subconscious mind, he had never wholly forgotten.

And yet he has returned to his mother sea only on her own terms. He cannot control or change the ocean as, in his brief tenancy of earth, he has subdued and plundered the continents. In the artificial world of his cities and towns, he often forgets the true nature of his planet and the long vistas of its history, in which the existence of the race of men has occupied a mere moment of time. The sense of all these things comes to him most clearly in the course of a long ocean voyage, when he watches day after day the receding rim of the horizon, ridged and furrowed by waves; when at night he becomes aware of the earth's rotation as the stars pass overhead; or when, alone in this world of water and sky, he feels the loneliness of his earth in space. And then, as never on land, he knows the truth that his world is a water world, a planet dominated by its covering mantle of ocean, in which the continents are but transient intrusions of land above the surface of the all-encircling sea.

DEREK WALCOTT

The Sea is History

Where are your monuments, your battles, martyrs?
Where is your tribal memory? Sirs,
in that gray vault. The sea. The sea
has locked them up. The sea is History.

First, there was the heaving oil,
heavy as chaos;
then, like a light at the end of a tunnel,

the lantern of a caravel,
and that was Genesis.
Then there were the packed cries,
the shit, the moaning:

Exodus.
Bone soldered by coral to bone,
mosaics
mantled by the benediction of the shark's shadow,

that was the Ark of the Covenant.
Then came from the plucked wires
of sunlight on the sea floor

the plangent harp of the Babylonian bondage,
as the white cowries clustered like manacles
on the drowned women,

and those were the ivory bracelets
of the Song of Solomon,
but the ocean kept turning blank pages

looking for History.
Then came the men with eyes heavy as anchors
who sank without tombs,

brigands who barbecued cattle,
leaving their charred ribs like palm leaves on the shore,
then the foaming, rabid maw

of the tidal wave swallowing Port Royal,
and that was Jonah,
but where is your Renaissance?

Sir, it is locked in them sea sands
out there past the reef's moiling shelf,
where the men-o'-war floated down;

strop on these goggles, I'll guide you there myself.
It's all subtle and submarine,
through colonnades of coral,

past the gothic windows of sea fans
to where the crusty grouper, onyx-eyed,
blinks, weighted by its jewels, like a bald queen;

and these groined caves with barnacles
pitted like stone
are our cathedrals,

and the furnace before the hurricanes:
Gomorrah. Bones ground by windmills
into marl and cornmeal,

and that was Lamentations –
that was just Lamentations,
it was not History;

then came, like scum on the river's drying lip,
the brown reeds of villages
mantling and congealing into towns,

and at evening, the midges' choirs,
and above them, the spires
lancing the side of God

as His son set, and that was the New Testament.

Then came the white sisters clapping
to the waves' progress,
and that was Emancipation –

jubilation, O jubilation –
vanishing swiftly
as the sea's lace dries in the sun,

but that was not History,
that was only faith,
and then each rock broke into its own nation;

then came the synod of flies,
then came the secretarial heron,
then came the bullfrog bellowing for a vote,

fireflies with bright ideas
and bats like jetting ambassadors
and the mantis, like khaki police,

and the furred caterpillars of judges
examining each case closely,
and then in the dark ears of ferns

and in the salt chuckle of rocks
with their sea pools, there was the sound
like a rumour without any echo

of History, really beginning.

RICHARD HAKLUYT

The Last Voyage of Thomas Cavendish

The tenth of October being by the accompt of our Captaine and Master very neere the shore, the weather darke, the storme furious, and most of our men having given over to travell, we yeelded our selves to death, without further hope of succour. Our captaine sitting in the gallery very pensive, I came and brought him some Rosa solis to comfort him; for he was so cold, that hee was scarce able to moove a joint. After he had drunke, and was comforted in heart, hee began for the ease of his conscience to make a large repetition of his forepassed time, and with many grievous sighs he concluded in these words: Oh most glorious God, with whose power the mightiest things among men are matters of no moment, I most humbly beseech thee, that the intollerable burthen of my sinnes may through the blood of Jesus Christ be taken from me: and end our daies with speede, or shew us some mercifull signe of thy love and our preservation. Having thus ended, he desired me not to make knowen to any of the company his intollerable griefe and anguish of minde, because they should not thereby be dismayed. And so suddenly, before I went from him the Sunne shined cleere; so that he and the Master both observed the true elevation of the Pole, whereby they knew by what course to recover the Streights. Wherewithall our captaine and Master were so revived, & gave such comfortable speeches to the company, that every man rejoiced, as though we had received a present deliverance. The next day being the 11 of October, we saw

Cabo Deseado being the cape on the South shore (the North shore is nothing but a company of dangerous rocks, Isles, & sholds). This cape being within two leages to leeward off us, our master greatly doubted, that we could not double the same: whereupon the captain told him: You see there is no remedy, either we must double it, or before noon we must die: therefore loose your sails, and let us put it to Gods mercy. The master being a man of good spirit resolutely made quicke dispatch & set sails. Our sailes had not bene halfe an houre aboord, but the footrope of our foresaile brake, so that nothing held but the oylet holes. The seas continually brake over the ships poope, and flew into the sailes with such violence, that we still expected the tearing of our sayles, or oversetting of the ship, and withall to our utter discomfort, wee perceived that wee fell still more and more to leeward, so that wee could not double the cape: wee were nowe come within halfe a mile of the cape, and so neere the shore, that the counter-suffe of the sea would rebound against the shippes side, so that wee were much dismayed with the horror of our present ende. Beeing thus at the very pinch of death, the winde and Seas raging beyond measure, our Master veared some of the maine sheate; and whether it was by that occasion, or by some current, or by the wonderfull power of God, as wee verily thinke it was, the ship quickened her way, and shot past that rocke, where wee thought shee would have shored. Then betweene the cape and the poynt there was a little bay; so that wee were somewhat farther from the shoare: and when we were come so farre as the cape, wee yeelded to death: yet our good God the Father of all mercies delivered us, and wee doubled the cape about the length of our shippe, or very little more. Being shot past the cape, we presently tooke in our sayles, which onely God had preserved unto us: and when we were shot in betweene the high lands, the wind blowing trade, without any inch of sayle, we spooned before the sea, three men being not able to guide the helme, and in sixe houres wee were put five and twenty leagues within the Streights, where wee found a sea answerable to the Ocean.

In this time we freed our ship from water, and after wee had rested a little, our men were not able to moove; their sinewes were stiffe, and their flesh dead, and many of them (which is most lamentable to bee reported) were so eaten with lice, as that in their flesh did lie clusters of lice as big as peason, yea and some as big as beanes. Being in this miserie we were constrained to put into a coove for the refreshing our men. Our Master knowing the shore and every coove very perfectly, put in with the shore, and mored to the trees, as beforetime we had done, laying our ankor to the seaward. Here we continued until the twentieth of October; but not being able any longer to stay through extremitie of famine, the one and twentieth we put off into the chanell, the weather being reasonable calme: but before night it blew most extreamely at Westnorthwest. The storme growing outrageous, our men could scarcely stand by their labour; and the Streights being full of turning reaches we were constrained by discretion of the Captaine and Master in their accounts to guide the ship in the hell-darke night, when we could not see any shore, the chanell being in some places scarse three miles broad. But our captaine, as wee first passed through the Streights drew such an exquisite plat of the same, as I am assured it cannot in any sort be bettered: which plat hee and the Master so often perused, and so carefully regarded, as that in memorie they had every turning and creeke, and in the deepe darke night without any doubting they conveyed the ship through that crooked chanell: so that I conclude, the world hath not any so skilfull pilots for that place, as they are: for otherwise wee could never have passed in such sort as we did.

The 25 wee came to an Island in the Streights named Penguin-isle, whither wee sent our boate to seeke reliefe, for there were great abundance of birds, and the weather was very calme; so wee came to an ankor by the Island in seven fadomes. While our boate was at shore, and we had great store of Penguins, there arose a sudden storme, so that our ship did drive over a breach and our boate sanke at the shore. Captaine Cotton and the Lieutenant being on shore leapt in the boate, and freed the same, and threw away all the birdes, and with great difficultie recovered the ship: my selfe also was in the boate the same time, where for my life I laboured to the best of my power. The ship all this while driving upon the lee-shore, when wee came aboord, we helped to set sayle, and weighed the ankor; for before our comming they could scarse hoise up their yardes, yet with much adoe they set their fore-coarse. Thus in a mighty fret of weather the seven and twentieth day of October wee were free of the Streights, and the thirtieth of October we came to Penguin-isle being three leagues from Port Desire, the place which wee purposed to seeke for our reliefe.

When wee were come to this Isle wee sent our boate on shore, which returned laden with birdes and egges; and our men sayd that the Penguins were so thicke upon the Isle, that shippes might be laden with them; for they could not goe without treading upon the birds, whereat we greatly rejoiced. Then the captaine appointed Charles Parker and Edward Smith, with twenty others to go on shore, and to stay upon the Isle, for the killing and drying of those Penguins, and promised after the ship was in harborough to send the rest, not onely for expedition, but also to save the small store of victuals in the shippe. But Parker, Smith, and the rest of their faction suspected, that this was a devise of the Captaine to leave his men on shore, that by these meanes there might bee victuals for the rest to recover their countrey: and when they remembered, that this was the place where they would have slaine their Captaine and Master, surely (thought they) for revenge hereof will they leave us on shore. Which when our Captaine understood, hee used these speeches unto them: I understand that you are doubtfull of your security through the perversenesse of your owne guilty consciences: it is an extreame griefe unto me, that you should judge mee blood-thirstie, in whome you have seene nothing but kinde conversation: if you have found otherwise, speake boldly, and accuse mee of the wrongs that I have done; if not, why do you then measure me by your owne uncharitable consciences? All the company knoweth indeed, that in this place you practized to the utmost of your powers, to murther me and the master causeles, as God knoweth, which evil in this place we did remit you: & now I may conceive without doing you wrong, that you againe purpose some evill in bringing these matters to repetition: but God has so shortened your confederacie, as that I nothing doubt you: it is for your Masters sake that I have forborne you in your unchristian practizes: and here I protest before God, that for his sake alone I will yet indure this injury, and you shall in no sorte be prejudiced or in any thing be by me commanded: but when we come into England (if God so favour us) your master shall knowe your honesties: in the meane space be voide of these suspicions, for, God I call to witnes, revenge is no part of my thought. They gave him thanks, desiring to go into the harborough with the ship, which he granted. So there were ten left upon the Isle, and the last of October we entred the harborough. Our Master at our last being here having taken carefull notice of every creeke in the river, in a very convenient place, upon sandy oaze, ran the ship on ground, laying our ankor to seaward, and with our running ropes mored her to stakes upon the shore, which hee had fastened for that purpose; where the ship remained till our departure.

The third of November our boat with water, wood, and as many as shee could carry, went for the Isle of Penguins: but being deepe, she durst not proceed, but returned againe the same night. Then Parker, Smith, Townesend, Purpet, with five others, desired that they might goe by land, and that the boate might fetch them when they were against the Isle, it being scarce a mile from the shore. The captaine bade them doe what they thought best, advising them to take weapons with them: for (sayd he) although we have not at any time seene people in this place, yet in the countrey there may be Savages. They answered, that here were great store of Deere, and Ostriches; but if there were Salvages, they would devoure them: notwithstanding the captaine caused them to cary weapons, calievers, swordes, and targets: so the sixt of November they departed by land, and the bote by sea; but from that day to this day wee never heard of our men. The 11 while most of our men were at the Isle, onely the Captaine and Master with sixe others being left in the ship, there came a great multitude of Salvages to the ship, throwing dust in the ayre, leaping and running like brute beasts, having vizards on their faces like dogs faces, or else their faces are dogs faces indeed. We greatly feared least they would set our ship on fire, for they would suddenly make fire, whereat we much marvelled: they came to windward of our ship, and set the bushes on fire, so that we were in a very stinking smoke: but as soone as they came within our shot, we shot at them, & striking one of them in the thigh they all presently fled, so that we never heard nor saw more of them. Hereby we judged that these Canibals had slaine our 9 men. When we considered what they were that thus were slaine, and found that they were the principall men that would have murthered our Captaine and Master, with the rest of their friends, we saw the just judgement of God, and made supplication to his divine Majesty to be mercifull unto us. While we were in this harborough, our Captaine and Master went with the boat to discover how farre this river did run, that if neede should enforce us to leave our ship, we might know how farre we might go by water. So they found, that farther than 20 miles they could not go with the boat. At their returne they sent the boate to the Isle of Penguins; whereby wee understood that the Penguins dryed to our hearts content, and that the multitude of them was infinite. This Penguin hath the shape of a bird, but hath no wings, only two stumps in the place of wings, by which he swimmeth under water with as great swiftnes as any fish. They live upon smelts, whereof there is great abundance upon this coast: in eating they be neither fish nor flesh: they lay great egs, and the bird is of a reasonable bignes, very neere twise so big as a ducke. All the time that wee were in this place, we fared passing well with egs, Penguins, yong Seales, young Gulles, besides other birds, such as I know not: of all which we had great abundance. In this place we found an herbe called Scurvygrasse, which wee fried with egs, using traine oyle in stead of butter. This herbe did so purge ye blood, that it tooke away all kind of swellings, of which many died, & restored us to perfect health of body, so that we were in as good case as when we came first out of England. We stayed in this harbour until the 22 of December, in which time we had dried 20000 Penguins; & the Captaine, the Master, and my selfe had made some salt, by laying salt water upon the rocks in holes, which in 6 dáies would be kerned. Thus God did feed us even as it were with Manna from heaven.

The 22 of December we departed with our ship for the Isle, where with great difficulty, by the skilful industry of our Master we got 14000 of our birds, and had almost lost our captaine in labouring to bring the birds aboord: & had not our Master bene very expert in the set of those wicked tides, which run after many fashions, we had also lost our ship in the same place: but God of his goodnes hath in all our extremities bene our protector. So the 22 at night we departed with 14000 dried Penguins, not being able to fetch the rest, and shaped our course for Brasil. Nowe our captaine rated our victuals, and brought us to such allowance, as that our victuals might last sixe moneths; for our hope was, that within sixe moneths we might recover our countrey, though our sailes were very bad. So the allowance was two ounces & a halfe of meale for a man a day, and to have so twise a weeke, so that 5 ounces did serve for a weeke. Three daies a weeke we had oile, three spoonfuls for a man a day; and 2 dayes in a weeke peason, a pint betweene 4 men a day, and every day 5 Penguins for 4 men, and 6 quartes of water for 4 men a day. This was our allowance; wherewith (we praise God) we lived, though weakly, and very feeble. The 30 of January we arrived at the Ile of Placencia in Brasill, the first place that outward bound we were at: and having made the sholde, our ship lying off at sea, the Captaine with 24 of the company went with the boat on shore, being a whole night before they could recover it. The last of January at sun-rising they suddenly landed, hoping to take the Portugales in their houses, & by that meanes to recover some Casavimeale, or other victuals for our reliefe: but when they came to the houses, they were all razed, and burnt to the ground, so that we thought no man had remained on the Iland. Then the captaine went to the gardens, & brought from thence fruits & roots for the company, and came aboord the ship, and brought her into a fine creeke which he had found out, where we might more her by the trees, and where there was water, and hoopes to trim our caske. Our case being very desperate, we presently laboured for dispatch away; some cut hoopes, which the coopers made, others laboured upon the sailes and ship, every man travelling for his life, & still a guard was kept on shore to defend those that laboured, every man having his weapon like wise by him. The 3 of February our men with 23 shot went againe to the gardens, being 3 miles from us upon the North shore, and fetched Cazavi-roots out of the ground, to relieve our company instead of bread; for we spent not of our meale while we staied here. The 5 of February being munday, our captaine and master hasted the company to their labour; so some went with the Coopers to gather hoopes, and the rest laboured aboord. This night many of our men in the ship dreamed of murther & slaughter: In the morning they reported their dreames, one saying to another; this night I dreamt, that thou wert slaine; another answered, and I dreamed, that thou wert slaine: and this was general through the ship. The captaine hearing this, who like wise had dreamed very strangely himselfe, gave very streight charge, that those which went on shore should take weapons with them, and saw them himselfe delivered into the boat, & sent some of purpose to guard the labourers. All the forenoone they laboured in quietnesse, & when it was ten of the clocke, the heat being extreme, they came to a rocke neere the woods side (for al this countrey is nothing but thick woods) and there they boyled Cazavi-roots, & dined: after dinner some slept, some washed themselves in the sea, all being stripped to their shirts, & no man keeping watch, no match lighted, not a piece charged. Suddenly as they were thus sleeping & sporting, having gotten themselves into a corner out of sight of the ship, there came a multitude of Indians & Portugales upon them, and slew them sleeping: onely two escaped, one very sore hurt, the other not touched, by whom we understood of this miserable massacre: with all speed we manned our boat, & landed to succour our men; but we found them slaine, & laied naked on a ranke one by another, with their faces upward, and a crosse set by them: and withall we saw two very great pinnesses come from the river of Jenero very ful of men; whom we mistrusted came from thence to take us: because there came from Jenero souldiers to Santos, when the Generall had taken the towne and was strong in it. Of 76 persons which departed in our ship out of England, we were now left but 27, having lost 13 in this place, with their chiefe furniture, as muskets, calivers, powder, & shot. Our caske was all in decay, so that we could not take in more water than was in our ship, for want of caske, and that which we had was marvellous ill conditioned: and being there mored by trees for want of cables and ankers, we still expected the cutting of our morings, to be beaten from our decks with our owne furniture, & to be assayled by them of Jenero: what distresse we were now driven into,

I am not able to expresse. To depart with 8 tunnes of water in such bad caske was to sterve at sea, & in staying our case was ruinous. These were hard choises; but being thus perplexed, we made choice rather to fall into the hands of the Lord, then into the hands of men: for his exceeding mercies we had tasted, & of the others cruelty we were not ignorant. So concluding to depart, the 6 of February we were off in the chanell, with our ordinance & small shot in a readines, for any assalt that should come, & having a small gale of winde, we recovered the sea in most deepe distresse. Then bemoning our estate one to another, and recounting over all our extremities, nothing grieved us more, then the losse of our men twise, first by the slaughter of the Canibals at Port Desire, and at this Ile of Placencia by the Indians and Portugals. And considering what they were that were lost, we found that al those that conspired the murthering of our captaine & master were now slain by salvages, the gunner only excepted. Being thus at sea, when we came to cape Frio, the winde was contrary; so that 3 weekes we were grievously vexed with crosse windes, & our water consuming, our hope of life was very small. Some desired to go to Baya, & to submit themselves to the Portugales, rather then to die for thirst: but the captaine with faire perswasions altered their purpose of yeelding to the Portugales. In this distresse it pleased God to send us raine in such plenty, as that we were wel watered, & in good comfort to returne. But after we came neere unto the sun, our dried Penguins began to corrupt, and there bred in them a most lothsome & ugly worme of an inch long. This worme did so mightily increase, and devoure our victuals, that there was in reason no hope how we should avoide famine, but be devoured of these wicked creatures: there was nothing that they did not devour, only yron excepted: our clothes, boots, shooes, hats, shirts, stockings: and for the ship they did so eat the timbers, as that we greatly feared they would undoe us, by gnawing through the ships side. Great was the care and diligence of our captaine, master, and company to consume these vermine, but the more we laboured to kill them, the more they increased; so that at the last we could not sleepe for them, but they would eate our flesh, and bite like Mosquitos. In this wofull case, after we had passed the Equinoctiall toward the North, our men began to fall sick of such a monstrous disease, as I thinke the like was never heard of: for in their ankles it began to swell; from thence in two daies it would be in their breasts, so that they could not draw their breath, and then fell into their cods; and their cods and yardes did swell most grievously, and most dreadfully to behold, so that they could neither stand, lie, nor goe. Whereupon our men grew mad with griefe. Our captain with extreme anguish of his soule, was in such wofull case, that he desired only a speedie end, and though he were scarce able to speake for sorrow, yet he perswaded them to patience, and to give God thankes, & like dutifull children to accept of his chastisement. For all this divers grew raging mad, & some died in most lothsome & furious paine. It were incredible to write our misery as it was: there was no man in perfect health, but the captaine & one boy. The master being a man of good spirit with extreme labour bore out his griefe, so that it grew not upon him. To be short, all our men died except 16, of which there were but 5 able to moove. The captaine was in good health, the master indifferent, captaine Cotton and my selfe swolne and short winded, yet better then the rest that were sicke, and one boy in health: upon us 5 only the labour of the ship did stand. The captaine and master, as occasion served, would take in, and heave out the topsailes, the master onely attended on the sprit-saile, and all of us at the capsten without sheats and tacks. In fine our miserie and weaknesse was so great, that we could not take in, nor heave out a saile: so our top-saile & sprit-sailes were torne all in pieces by the weather. The master and captaine taking their turnes at the helme, were mightily distressed and monstrously grieved with the most wofull lamentation of our sick men. Thus as lost wanderers upon the sea, the 11 of June 1593, it pleased God that we arrived at Bear-haven in Ireland, and there ran the ship on shore: where the Irish men helped us to take in our sailes, and to more our ship for flooting: which slender paines of theirs cost the captaine some ten pounds before he could have the ship in safetie. Thus without victuals, sailes, men, or any furniture God onely guided us into Ireland, where the captaine left the master and three or foure of the company to keepe the ship; and within 5 dayes after he and certaine others had passage in an English fisher-boat to Padstow in Cornewall. In this manner our small remnant by Gods onely mercie were preserved, and restored to our countrey, to whom be all honour and glory world without end.

LUKE DAVIES

Diving the Coolidge

I

The more you dive the more dreams come.
On the island of Espiritu Santo
the ghosts of Spaniards drift in the heavy heat
of a copra breeze, four hundred years old
those ghosts. The *S.S. President Coolidge*
is a living force, a great dream factory
under the water: suddenly you submerge
and a world opens out slow and cool
and implacable. You glide through jeeps,
cranes, ballrooms, a barbershop. You are falling
down the promenade deck, which once
lay level with the gold horizon. Once, in 1933,
a couple kissed here, maybe 1934. Once it was
a luxury liner, then a troop carrier, now
it looms a palace of shadow and curve distorted
by the wordless press of atmospheres. An angel fish
hovers on the edge of vision always there
motionless and every direction is up
and light a distant thing like memory.
A seahorse nibbles on a chandelier.
The ocean dreams, you seem to dream in it.

II

Between dives you read about Daniel Boone.
How once in Kentucky he camped by a creek
in the middle of nowhere, and dreamed.
Deep dreams, strong dreams, unusual for Boone,
who slept with an ear to the smallest noise.
Water unfurling the geometry of fragmentation,
which is consciousness, or the way a stick snaps
and you are not wholly dissolved.
Boone wakes and says, 'I call this Dreaming Creek.'

The sparkle of water clacking on pebbles.
Sunlight like a secret in Kentucky,
flooding the glade and the tiny rapids.

Between and during dives it becomes apparent
that if land is an enclosure of order and light
surrounded by the formlessness of ocean
then Dreaming Creek in Kentucky
– the specific named after the nebulous –
is a small slit in the world, a pathway
from daylight to water, from Daniel Boone
to Espiritu Santo, in the great ocean,
where the more you dive the more dreams come.
The ocean dreams, you seem to dream in it.

Boone sleeping and the water gurgling.
At forty metres your dreams meet his.
Dream sunlight like a secret in Kentucky.

III

You are diving a giant dream called the *Coolidge*
and the more you dive the more dreams come.
You feel you could fall forever, towards
the gathering dark. The thick blue of descent.
In the smoking room, above the fireplace,
you come across a lady and a unicorn
and run your hands along them
where the seaweed sways on their porcelain flesh.
The dark sea all around you filled with flakes,
the statue large through goggles in that gloom.

That night you dream the lady alive:
she is Anastasia, pale and waiflike junkie
you barely knew, Anastasia with nothing,
not even hope, who drank arsenic
seven years ago, who you never think of.
She comes to you in her Elizabethan finery
on a wide windy beach with her unicorn,
and the *Coolidge* refloats and you wear
the Captain's hat. The way dreams work.

You on the bridge gleaming with pride.
Anastasia dead these years and arsenic
a kind of drowning. Seaweed swaying
on her hard white skin. I remember
everything. I remember I felt
nothing. In dream begins compassion.

EDMUND BURKE

On the Sublime and Beautiful

I

Of the passion caused by the sublime.

The passion caused by the great and sublime in *nature*, when those causes operate most powerfully, is astonishment: and astonishment is that state of the soul in which all its motions are suspended, with some degree of horror. In this case the mind is so entirely filled with its object, that it cannot entertain any other, nor by consequence reason on that object which employs it. Hence arises the great power of the sublime, that, far from being produced by them, it anticipates our reasonings, and hurries us on by an irresistible force. Astonishment, as I have said, is the effect of the sublime in its highest degree; the inferior effects are admiration, reverence, and respect.

II

Terror.

No passion so effectually robs the mind of all its powers of acting and reasoning as *fear*. For fear being an apprehension of pain or death, it operates in a manner that resembles actual pain. Whatever therefore is terrible, with regard to sight, is sublime too, whether this cause of terror be endued with greatness of dimensions or not; for it is impossible to look on anything as trifling, or contemptible, that may be dangerous. There are many animals, who, though far from being large, are yet capable of raising ideas of the sublime, because they are considered as objects of terror. As serpents and poisonous animals of almost all kinds. And to things of great dimensions, if we annex an adventitious idea of terror, they become without comparison greater. A level plain of a vast extent on land, is certainly no mean idea; the prospect of such a plain may be as extensive as a prospect of the ocean; but can it ever fill the mind with anything so great as the ocean itself? This is owing to several causes; but it is owing to none more than this, that the ocean is an object of no small terror. Indeed terror is in all cases whatsoever, either more openly or latently, the ruling principle of the sublime. Several languages bear a strong testimony to the affinity of these ideas. They frequently use the same word to signify indifferently the modes of astonishment or admiration and those of terror. *Θάμβος* is in Greek either fear or wonder; *δεινός* is terrible or respectable; *αἰδέω*, to reverence or to fear. *Vereor* in Latin is what *αἰδέω* is in Greek. The Romans used the verb *stupeo*, a term which strongly marks the state of an astonished mind, to express the effect either of simple fear, or of astonishment; the word *attonitus* (thunderstruck) is equally expressive of the alliance of these ideas; and do not the French *étonnement*, and the English *astonishment* and *amazement*, point out as clearly the kindred emotions which attend fear and wonder? They who have a more general knowledge of languages, could produce, I make no doubt, many other and equally striking examples.

WALLACE STEVENS

The Idea of Order at Key West

She sang beyond the genius of the sea.
The water never formed to mind or voice,
Like a body wholly body, fluttering
Its empty sleeves; and yet its mimic motion
Made constant cry, caused constantly a cry,
That was not ours although we understood,
Inhuman, of the veritable ocean.

The sea was not a mask. No more was she.
The song and water were not medleyed sound
Even if what she sang was what she heard,
Since what she sang was uttered word by word.
It may be that in all her phrases stirred
The grinding water and the gasping wind;
But it was she and not the sea we heard.

For she was the maker of the song she sang.
The ever-hooded, tragic-gestured sea
Was merely a place by which she walked to sing.
Whose spirit is this? we said, because we knew
It was the spirit that we sought and knew
That we should ask this often as she sang.
If it was only the dark voice of the sea

That rose, or even coloured by many waves;
If it was only the outer voice of sky
And cloud, of the sunken coral water-walled,
However clear, it would have been deep air,
The heaving speech of air, a summer sound
Repeated in a summer without end
And sound alone. But it was more than that,
More even than her voice, and ours, among
The meaningless plungings of water and the wind,
Theatrical distances, bronze shadows heaped
On high horizons, mountainous atmospheres
Of sky and sea.
    It was her voice that made
The sky acutest at its vanishing.
She measured to the hour its solitude.
She was the single artificer of the world
In which she sang. And when she sang, the sea,
Whatever self it had, became the self
That was her song, for she was the maker. Then we,
As we beheld her striding there alone,
Knew that there never was a world for her
Except the one she sang and, singing, made.

Ramon Fernandez, tell me, if you know,
Why, when the singing ended and we turned
Toward the town, tell why the glassy lights,
The lights in the fishing boats at anchor there,
As the night descended, tilting in the air,
Mastered the night and portioned out the sea,
Fixing emblazoned zones and fiery poles,
Arranging, deepening, enchanting night.

Oh! Blessed rage for order, pale Ramon,
The maker's rage to order words of the sea,
Words of the fragrant portals, dimly-starred,
And of ourselves and of our origins,
In ghostlier demarcations, keener sounds.

SAMUEL TAYLOR COLERIDGE

The Rime of the Ancient Mariner

In seven parts

Facile credo, plures esse Naturas invisibiles quam visibiles in rerum universitate. Sed horum omnium familiam quis nobis enarrabit, et gradus et cognationes et discrimina et singulorum munera? Quid agunt? Quae loca habitant? Harum rerum notitiam semper ambivit ingenium humanum, nunquam attigit. Juvat, interea, non diffiteor, quandoque in animo, tanquam in tabulâ, majoris et melioris mundi imaginem contemplari: ne mens assuefacta hodiernæ vitæ minutiis se contrahat nimis, et tota subsidat in pusillas cogitationes. Sed veritati interea invigilandum est, modusque servandus, ut certa ab incertis, diem a nocte, distinguamus.

T. Burnet, *Archæol. Phil.*, p. 68

I

It is an ancient Mariner,
And he stoppeth one of three.
'By thy long grey beard and glittering eye,
Now wherefore stopp'st thou me?

An ancient Mariner meeteth three gallants bidden to a wedding-feast, and detaineth one.

'The Bridegroom's doors are opened wide,
And I am next of kin;
The guests are met, the feast is set:
May'st hear the merry din.'

He holds him with his skinny hand,
'There was a ship,' quoth he.
'Hold off! unhand me, grey-beard loon!'
Eftsoons his hand dropt he.

He holds him with his glittering eye –
The Wedding-Guest stood still,
And listens like a three years' child:
The Mariner hath his will.

The wedding-guest is spell-bound by the eye of the old sea-faring man, and constrained to hear his tale.

The Wedding-Guest sat on a stone:
He cannot choose but hear;
And thus spake on that ancient man,
The bright-eyed Mariner.

The ship was cheered, the harbour cleared,
Merrily did we drop
Below the kirk, below the hill,
Below the light-house top.

The sun came up upon the left,
Out of the sea came he!
And he shone bright, and on the right
Went down into the sea.

The Mariner tells how the ship sailed southward with a good wind and fair weather, till it reached the line.

Higher and higher every day,
Till over the mast at noon –
The Wedding-Guest here beat his breast,
For he heard the loud bassoon.

The bride hath paced into the hall,
Red as a rose is she;
Nodding their heads before her goes
The merry minstrelsy.

The wedding-guest heareth the bridal music; but the Mariner continueth his tale.

The Wedding-Guest he beat his breast,
Yet he cannot choose but hear;
And thus spake on that ancient man,
The bright-eyed Mariner.

And now the storm-blast came, and he
Was tyrannous and strong:
He struck with his o'ertaking wings,
And chased us south along.

The ship drawn by a storm toward the south pole.

With sloping masts and dipping prow,
As who pursued with yell and blow
Still treads the shadow of his foe,
And forward bends his head,
The ship drove fast, loud roared the blast,
And southward aye we fled.

And now there came both mist and snow,
And it grew wondrous cold:
And ice, mast-high, came floating by,
As green as emerald.

And through the drifts the snowy clifts
Did send a dismal sheen:
Nor shapes of men nor beasts we ken –
The ice was all between.

The land of ice, and of fearful sounds where no living thing was to be seen.

The ice was here, the ice was there,
The ice was all around:
It cracked and growled, and roared and howled,
Like noises in a swound!

At length did cross an Albatross,
Thorough the fog it came;
As if it had been a Christian soul,
We hailed it in God's name.

Till a great sea-bird, called the Albatross, came through the snow-fog, and was received with great joy and hospitality.

It ate the food it ne'er had eat,
And round and round it flew.
The ice did split with a thunder-fit;
The helmsman steered us through!

And a good south wind sprung up behind;
The Albatross did follow,
And ever day, for food or play,
Came to the mariner's hollo!

And lo! the Albatross proveth a bird of good omen, and followeth the ship as it returned northward through fog and floating ice.

In mist or cloud, on mast or shroud,
It perched for vespers nine;
Whiles all the night, through fog-smoke white,
Glimmered the white moon-shine.

'God save thee, ancient Mariner!
From the fiends, that plague thee thus! –
Why look'st thou so?' – With my cross-bow
I shot the Albatross.

The ancient Mariner inhospitably killeth the pious bird of good omen.

II

The Sun now rose upon the right:
Out of the sea came he,
Still hid in mist, and on the left
Went down into the sea.

And the good south wind still blew behind,
But no sweet bird did follow,
Nor any day for food or play
Came to the mariners' hollo!

And I had done a hellish thing,
And it would work 'em woe:
For all averred, I had killed the bird
That made the breeze to blow.
Ah wretch! said they, the bird to slay,
That made the breeze to blow!

His ship-mates cry out against the ancient Mariner, for killing the bird of good luck.

Nor dim nor red, like God's own head,
The glorious Sun uprist:
Then all averred, I had killed the bird
That brought the fog and mist.
'Twas right, said they, such birds to slay,
That bring the fog and mist.

But when the fog cleared off, they justify the same, and thus make themselves accomplices in the crime.

The fair breeze blew, the white foam flew,
The furrow followed free;
We were the first that ever burst
Into that silent sea.

The fair breeze continues; the ship enters the Pacific Ocean, and sails northward, even till it reaches the Line.

Down dropt the breeze, the sails dropt down,
'Twas sad as sad could be;
And we did speak only to break
The silence of the sea!

The ship hath been suddenly becalmed.

All in a hot and copper sky,
The bloody Sun, at noon,
Right up above the mast did stand,
No bigger than the Moon.

Day after day, day after day,
We stuck, nor breath nor motion;
As idle as a painted ship
Upon a painted ocean.

Water, water, every where,
And all the boards did shrink;
Water, water, every where,
Nor any drop to drink.

And the Albatross begins to be avenged.

The very deep did rot: O Christ!
That ever this should be!
Yea, slimy things did crawl with legs
Upon the slimy sea.

About, about, in reel and rout
The death-fires danced at night;
The water, like a witch's oils,
Burnt green, and blue and white.

And some in dreams assurèd were
Of the spirit that plagued us so;
Nine fathom deep he had followed us
From the land of mist and snow.

A spirit had followed them; one of the invisible inhabitants of this planet, neither departed souls nor angels; concerning whom the learned Jew, Josephus, and the Platonic Constantinopolitan, Michael Psellus, may be consulted. They are very numerous, and there is no climate or element without one or more.

And every tongue, through utter drought,
Was withered at the root;
We could not speak, no more than if
We had been choked with soot.

Ah! well a-day! what evil looks
Had I from old and young!
Instead of the cross, the Albatross
About my neck was hung.

The ship-mates, in their sore distress, would fain throw the whole guilt on the ancient Mariner: in sign whereof they hang the dead sea-bird round his neck.

III

There passed a weary time. Each throat
Was parched, and glazed each eye.
A weary time! a weary time!
How glazed each weary eye,
When looking westward, I beheld
A something in the sky.

The ancient Mariner beholdeth a sign in the element afar off.

At first it seemed a little speck,
And then it seemed a mist;
It moved and moved, and took at last
A certain shape, I wist.

A speck, a mist, a shape, I wist!
And still it neared and neared:
As if it dodged a water-sprite,
It plunged and tacked and veered.

With throats unslaked, with black lips baked,
We could nor laugh nor wail;
Through utter drought all dumb we stood!
I bit my arm, I sucked the blood,
And cried, A sail! a sail!

At its nearer approach, it seemeth him to be a ship; and at a dear ransom he freeth his speech from the bonds of thirst.

With throats unslaked, with black lips baked,
Agape they heard me call:
Gramercy! they for joy did grin,
And all at once their breath drew in,
As they were drinking all.

A flash of joy;

See! see! (I cried) she tacks no more!
Hither to work us weal;
Without a breeze, without a tide,
She steadies with upright keel!

And horror follows. For can it be a ship that comes onward without wind or tide?

The western wave was all a-flame.
The day was well nigh done!
Almost upon the western wave
Rested the broad bright Sun;
When that strange shape drove suddenly
Betwixt us and the Sun.

And straight the Sun was flecked with bars,
(Heaven's Mother send us grace!)
As if through a dungeon-grate he peered
With broad and burning face.

It seemeth him but the skeleton of a ship.

Alas! (thought I, and my heart beat loud)
How fast she nears and nears!
Are those her sails that glance in the Sun,
Like restless gossameres?

Are those her ribs through which the Sun
Did peer, as through a grate?
And is that Woman all her crew?
Is that a Death? and are there two?
Is Death that woman's mate?

And its ribs are seen as bars on the face of the setting Sun. The spectre-woman and her death-mate, and no other on board the skeleton-ship.

Her lips were red, her looks were free,
Her locks were yellow as gold:
Her skin was as white as leprosy,
The Night-mare Life-in-Death was she,
Who thicks man's blood with cold.

Like vessel, like crew!

The naked hulk alongside came,
And the twain were casting dice;
'The game is done! I've, I've won!'
Quoth she, and whistles thrice.

Death and Life-in-death have diced for the ship's crew, and she (the latter) winneth the ancient Mariner.

The Sun's rim dips; the stars rush out:
At one stride comes the dark;
With far-heard whisper, o'er the sea,
Off shot the spectre-bark.

No twilight within the courts of the Sun.

We listened and looked sideways up!
Fear at my heart, as at a cup,
My life-blood seemed to sip!
The stars were dim, and thick the night,
The steersman's face by his lamp gleamed white;
From the sails the dew did drip –
Till clomb above the eastern bar
The hornèd Moon, with one bright star
Within the nether tip.

At the rising of the Moon.

One after one, by the star-dogged Moon,
Too quick for groan or sigh,
Each turned his face with a ghastly pang,
And cursed me with his eye.

One after another,

Four times fifty living men,
(And I heard nor sigh nor groan)
With heavy thump, a lifeless lump,
They dropped down one by one.

His ship-mates drop down dead.

The souls did from their bodies fly, –
They fled to bliss or woe!
And every soul, it passed me by,
Like the whizz of my cross-bow!

But Life-in-Death begins her work on the ancient Mariner.

IV

'I fear thee, ancient Mariner!
I fear thy skinny hand!
And thou art long, and lank, and brown,
As is the ribbed sea-sand.

The wedding-guest feareth that a spirit is talking to him.

I fear thee and thy glittering eye,
And thy skinny hand, so brown.' –
Fear not, fear not, thou Wedding-Guest!
This body dropt not down.

But the ancient Mariner assureth him of his bodily life, and proceedeth to relate his horrible penance.

Alone, alone, all, all alone,
Alone on a wide wide sea!
And never a saint took pity on
My soul in agony.

The many men, so beautiful!
And they all dead did lie:
And a thousand thousand slimy things
Lived on; and so did I.

He despiseth the creatures of the calm.

I looked upon the rotting sea,
And drew my eyes away;
I looked upon the rotting deck,
And there the dead men lay.

And envieth that they should live, and so many lie dead.

I looked to heaven, and tried to pray;
But or ever a prayer had gusht,
A wicked whisper came, and made
My heart as dry as dust.

I closed my lids, and kept them close,
And the balls like pulses beat;
For the sky and the sea, and the sea and the sky
Lay like a load on my weary eye,
And the dead were at my feet.

The cold sweat melted from their limbs,
Nor rot nor reek did they:
The look with which they looked on me
Had never passed away.

But the curse liveth for him in the eye of the dead men.

An orphan's curse would drag to hell
A spirit from on high;
But oh! more horrible than that
Is the curse in a dead man's eye!
Seven days, seven nights, I saw that curse,
And yet I could not die.

The moving Moon went up the sky,
And no where did abide:
Softly she was going up,
And a star or two beside –

In his loneliness and fixedness he yearneth towards the journeying Moon, and the stars that still sojourn, yet still move onward; and every where the blue sky belongs to them, and is their appointed rest, and their native country and their own natural homes, which they enter unannounced, as lords that are certainly expected and yet there is a silent joy at their arrival.

Her beams bemocked the sultry main,
Like April hoar-frost spread;
But where the ship's huge shadow lay,
The charmèd water burnt alway
A still and awful red.

Beyond the shadow of the ship,
I watched the water-snakes:
They moved in tracks of shining white,
And when they reared, the elfish light
Fell off in hoary flakes.

By the light of the Moon he beholdeth God's creatures of the great calm.

Within the shadow of the ship
I watched their rich attire:
Blue, glossy green, and velvet black,
They coiled and swam; and every track
Was a flash of golden fire.

O happy living things! no tongue
Their beauty might declare:
A spring of love gushed from my heart,
And I blessed them unaware:
Sure my kind saint took pity on me,
And I blessed them unaware.

Their beauty and their happiness.

He blesseth them in his heart.

The selfsame moment I could pray;
And from my neck so free
The Albatross fell off, and sank
Like lead into the sea.

The spell begins to break.

V

Oh sleep! it is a gentle thing,
Beloved from pole to pole!
To Mary Queen the praise be given!
She sent the gentle sleep from Heaven,
That slid into my soul

The silly buckets on the deck,
That had so long remained,
I dreamt that they were filled with dew;
And when I awoke, it rained.

By grace of the holy Mother, the ancient Mariner is refreshed with rain.

My lips were wet, my throat was cold,
My garments all were dank;
Sure I had drunken in my dreams,
And still my body drank.

I moved, and could not feel my limbs:
I was so light – almost
I thought that I had died in sleep,
And was a blessed ghost.

And soon I heard a roaring wind:
It did not come anear;
But with its sound it shook the sails,
That were so thin and sere.

He heareth sounds and seeth strange sights and commotions in the sky and the element.

The upper air burst into life!
And a hundred fire-flags sheen,
To and fro they were hurried about!
And to and fro, and in and out,
The wan stars danced between.

And the coming wind did roar more loud,
And the sails did sigh like sedge;
And the rain poured down from one black cloud
The Moon was at its edge.

The thick black cloud was cleft, and still
The Moon was at its side:
Like waters shot from some high crag,
The lightning fell with never a jag,
A river steep and wide.

The loud wind never reached the ship,
Yet now the ship moved on!
Beneath the lightning and the moon
The dead men gave a groan.

The bodies of the ship's crew are inspired, and the ship moves on;

They groaned, they stirred, they all uprose,
Nor spake, nor moved their eyes;
It had been strange, even in a dream,
To have seen those dead men rise.

The helmsman steered, the ship moved on;
Yet never a breeze up blew;
The mariners all 'gan work the ropes,
Where they were wont to do;
They raised their limbs like lifeless tools –
We were a ghastly crew.

The body of my brother's son
Stood by me, knee to knee:
The body and I pulled at one rope,
But he said nought to me.

'I fear thee, ancient Mariner!'
Be calm, thou Wedding-Guest!
'Twas not those souls that fled in pain,
Which to their corses came again,
But a troop of spirits blest:

But not by the souls of the men, nor by demons of earth or middle air, but by a blessed troop of angelic spirits, sent down by the invocation of the guardian saint.

For when it dawned – they dropped their arms,
And clustered round the mast;
Sweet sounds rose slowly through their mouths,
And from their bodies passed.

Around, around, flew each sweet sound
Then darted to the Sun;
Slowly the sounds came back again,
Now mixed, now one by one.

Sometimes a-dropping from the sky
I heard the sky-lark sing;
Sometimes all little birds that are,
How they seemed to fill the sea and air
With their sweet jargoning!

And now 'twas like all instruments,
Now like a lonely flute;
And now it is an angel's song,
That makes the heavens be mute.

It ceased; yet still the sails made on
A pleasant noise till noon,
A noise like of a hidden brook
In the leafy month of June,
That to the sleeping woods all night
Singeth a quiet tune.

Till noon we quietly sailed on,
Yet never a breeze did breathe:
Slowly and smoothly went the ship,
Moved onward from beneath.

Under the keel nine fathom deep,
From the land of mist and snow,
The spirit slid: and it was he
That made the ship to go.
The sails at noon left off their tune,
And the ship stood still also.

The lonesome spirit from the south pole carries on the ship as far as the line, in obedience to the angelic troop, but still requireth vengeance.

The Sun, right up above the mast,
Had fixed her to the ocean:
But in a minute she 'gan stir,
With a short uneasy motion –
Backwards and forwards half her length
With a short uneasy motion.

Then like a pawing horse let go,
She made a sudden bound:
It flung the blood into my head,
And I fell down in a swound.

How long in that same fit I lay,
I have not to declare;
But ere my living life returned,
I heard, and in my soul discerned
Two voices in the air.

The Polar Spirit's fellow demons, the invisible inhabitants of the element, take part in his wrong; and two of them relate, one to the other, that penance long and heavy for the ancient Mariner hath been accorded to the Polar Spirit, who returneth southward.

'Is it he?' quoth one, 'Is this the man?
By him who died on cross,
With his cruel bow he laid full low
The harmless Albatross.

'The spirit who bideth by himself
In the land of mist and snow,
He loved the bird that loved the man
Who shot him with his bow.'

The other was a softer voice,
As soft as honey-dew:
Quoth he, 'The man hath penance done,
And penance more will do.'

VI

First voice
But tell me, tell me! speak again,
Thy soft response renewing –
What makes that ship drive on so fast?
What is the ocean doing?

Second voice
Still as a slave before his lord,
The ocean hath no blast;
His great bright eye most silently
Up to the Moon is cast –

If he may know which way to go;
For she guides him smooth or grim.
See, brother, see! how graciously
She looketh down on him.

First voice
But why drives on that ship so fast,
Without or wave or wind?

The Mariner hath been cast into a trance; for the angelic power causeth the vessel to drive northward faster than human life could endure.

Second voice
The air is cut away before,
And closes from behind.

Fly, brother, fly! more high, more high!
Or we shall be belated:
For slow and slow that ship will go,
When the Mariner's trance is abated.

I woke, and we were sailing on
As in a gentle weather:
'Twas night, calm night, the moon was high;
The dead men stood together.

The supernatural motion is retarded; the Mariner awakes, and his penance begins anew.

All stood together on the deck,
For a charnel-dungeon fitter:
All fixed on me their stony eyes,
That in the Moon did glitter.

The pang, the curse, with which they died,
Had never passed away:
I could not draw my eyes from theirs,
Nor turn them up to pray.

And now this spell was snapt: once more
I viewed the ocean green,
And looked far forth, yet little saw
Of what had else been seen –

The curse is finally expiated.

Like one, that on a lonesome road
Doth walk in fear and dread,
And having once turned round walks on,
And turns no more his head;
Because he knows, a frightful fiend
Doth close behind him tread.

But soon there breathed a wind on me,
Nor sound nor motion made:
Its path was not upon the sea,
In ripple or in shade.

It raised my hair, it fanned my cheek
Like a meadow-gale of spring –
It mingled strangely with my fears,
Yet it felt like a welcoming.

Swiftly, swiftly flew the ship,
Yet she sailed softly too:
Sweetly, sweetly blew the breeze –
On me alone it blew.

Oh! dream of joy! is this indeed
The light-house top I see?
Is this the hill? is this the kirk?
Is this mine own countree?

And the ancient Mariner beholdeth his native country.

We drifted o'er the harbour-bar,
And I with sobs did pray –
O let me be awake, my God!
Or let me sleep alway.

The harbour-bay was clear as glass,
So smoothly it was strewn!
And on the bay the moonlight lay,
And the shadow of the moon.

The rock shone bright, the kirk no less,
That stands above the rock:
The moonlight steeped in silentness
The steady weathercock.

And the bay was white with silent light
Till rising from the same,
Full many shapes, that shadows were,
In crimson colours came.

The angelic spirits leave the dead bodies,

A little distance from the prow
Those crimson shadows were:
I turned my eyes upon the deck –
Oh, Christ! what saw I there!

And appear in their own forms of light.

Each corse lay flat, lifeless and flat,
And, by the holy rood!
A man all light, a seraph-man,
On every corse there stood.

This seraph-band, each waved his hand:
It was a heavenly sight!
They stood as signals to the land,
Each one a lovely light;

This seraph-band, each waved his hand,
No voice did they impart –
No voice; but oh! the silence sank
Like music on my heart.

But soon I heard the dash of oars,
I heard the Pilot's cheer;
My head was turned perforce away,
And I saw a boat appear.

The Pilot and the Pilot's boy,
I heard them coming fast:
Dear Lord in Heaven! it was a joy
The dead men could not blast.

I saw a third – I heard his voice:
It is the Hermit good!
He singeth loud his godly hymns
That he makes in the wood.
He'll shrieve my soul, he'll wash away
The Albatross's blood.

VII

This Hermit good lives in that wood
Which slopes down to the sea.
How loudly his sweet voice he rears!
He loves to talk with marineres
That come from a far countree.

The Hermit of the wood,

He kneels at morn, and noon, and eve –
He hath a cushion plump:
It is the moss that wholly hides
The rotted old oak-stump.

The skiff-boat neared: I heard them talk,
'Why, this is strange, I trow!
Where are those lights so many and fair,
That signal made but now?'

'Strange, by my faith!' the Hermit said –
'And they answered not our cheer!
The planks looked warped! and see those sails,
How thin they are and sere!
I never saw aught like to them,
Unless perchance it were

Approacheth the ship with wonder.

Brown skeletons of leaves that lag
My forest-brook along;
When the ivy-tod is heavy with snow,
And the owlet whoops to the wolf below,
That eats the she-wolf's young.'

'Dear Lord! it hath a fiendish look –
(The Pilot made reply)
I am a-feared' – 'Push on, push on!'
Said the Hermit cheerily.

The boat came closer to the ship,
But I nor spake nor stirred;
The boat came close beneath the ship,
And straight a sound was heard.

Under the water it rumbled on,
Still louder and more dread:
It reached the ship, it split the bay;
The ship went down like lead.

The ship suddenly sinketh.

Stunned by that loud and dreadful sound,
Which sky and ocean smote,
Like one that hath been seven days drowned
My body lay afloat;
But swift as dreams, myself I found
Within the Pilot's boat.

The ancient Mariner is saved in the Pilot's boat.

Upon the whirl, where sank the ship,
The boat spun round and round;
And all was still, save that the hill
Was telling of the sound.

I moved my lips – the Pilot shrieked
And fell down in a fit;
The holy Hermit raised his eyes,
And prayed where he did sit.

I took the oars: the Pilot's boy,
Who now doth crazy go,
Laughed loud and long, and all the while
His eyes went to and fro.
'Ha! ha!' quoth he, 'full plain I see,
The Devil knows how to row.'

And now, all in my own countree,
I stood on the firm land!
The Hermit stepped forth from the boat,
And scarcely he could stand.

'O shrieve me, shrieve me, holy man!'
The Hermit crossed his brow.
'Say quick,' quoth he, 'I bid thee say –
What manner of man art thou?'

The ancient Mariner earnestly entreateth the Hermit to shrieve him; and the penance of life falls on him.

Forthwith this frame of mine was wrenched
With a woful agony,
Which forced me to begin my tale;
And then it left me free.

Since then, at an uncertain hour,
That agony returns:
And till my ghastly tale is told,
This heart within me burns.

And ever and anon throughout his future life an agony constraineth him to travel from land to land.

I pass, like night, from land to land;
I have strange power of speech;
That moment that his face I see,
I know the man that must hear me:
To him my tale I teach.

What loud uproar bursts from that door!
The wedding-guests are there:
But in the garden-bower the bride
And bride-maids singing are:
And hark the little vesper bell,
Which biddeth me to prayer!

O Wedding-Guest! this soul hath been
Alone on a wide wide sea:
So lonely 'twas, that God himself
Scarce seemed there to be.

O sweeter than the marriage-feast,
'Tis sweeter far to me,
To walk together to the kirk
With a goodly company! –

To walk together to the kirk,
And all together pray,
While each to his great Father bends,
Old men, and babes, and loving friends,
And youths and maidens gay!

Farewell, farewell! but this I tell
To thee, thou Wedding-Guest!
He prayeth well, who loveth well
Both man and bird and beast.

And to teach, by his own example, love and reverence to all things that God made and loveth.

He prayeth best, who loveth best
All things both great and small;
For the dear God who loveth us,
He made and loveth all.'

The Mariner, whose eye is bright,
Whose beard with age is hoar,
Is gone: and now the Wedding-Guest
Turned from the bridegroom's door.

He went like one that hath been stunned,
And is of sense forlorn:
A sadder and a wiser man,
He rose the morrow morn.

MATTHEW FONTAINE MAURY

The Physical Geography of the Sea and its Meteorology

The Gulf Stream

70. There is a river in the ocean : in the severest droughts it never fails, and in the mightiest floods it never overflows; its banks and its bottom are of cold water, while its current is of warm; the Gulf of Mexico is its fountain, and its mouth is in the Arctic Seas. It is the Gulf Stream. There is in the world no other such majestic flow of waters. Its current is more rapid than the Mississippi or the Amazon, and its volume more than a thousand times greater. Its waters, as far out from the Gulf as the Carolina coasts, are of an indigo blue. They are so distinctly marked that their line of junction with the common sea-water may be traced by the eye. Often one half of the vessel may be perceived floating in Gulf Stream water, while the other half is in common water of the sea – so sharp is the line, and such the want of affinity between those waters, and such, too, the reluctance, so to speak, on the part of those of the Gulf Stream to mingle with the littoral waters of the sea.

Its colour.

71. At the salt-works in France, and along the shores of the Adriatic, where the '*salines*' are carried on by the process of solar evaporation, there is a series of vats or pools through which the water is passed as it comes from the sea, and is reduced

How caused.

to the briny state. The longer it is exposed to evaporation, the salter it grows, and the deeper is the hue of its blue, until crystallization is about to commence, when the now deep blue water puts on a reddish tint. Now the waters of the Gulf Stream are salter than the littoral waters of the sea through which they flow, and hence we can account for the deep indigo blue which all navigators observe off the Carolina coasts. The salt-makers are in the habit of judging of the richness of the sea-water in salt by its color – the greener the hue, the fresher the water. We have in this, perhaps, an explanation of the contrasts which the waters of the Gulf Stream present with those of the Atlantic, as well as of the light green of the North Sea and other Polar waters; also of the dark blue of the trade-wind regions, and especially of the Indian Ocean, which poets have described as the 'black waters'.

Speculations concerning the Gulf Stream.

72. What is the cause of the Gulf Stream has always puzzled philosophers. Many are the theories and numerous the speculations that have been advanced with regard to it. Modern investigations and examinations are beginning to throw some light upon the subject, though all is not yet entirely clear. But they seem to encourage the opinion that this stream, as well as all the *constant* currents of the sea, is due *mainly* to the *constant* difference produced by temperature and saltness in the specific gravity of water in certain parts of the ocean. Such difference of specific gravity is inconsistent with aqueous equilibrium, and to maintain this equilibrium these great currents are set in motion. The agents which derange equilibrium in the waters of the sea, by altering specific gravity, reach from the equator to the poles, and in their operations they are as ceaseless as heat and cold, consequently they call for a system of perpetual currents to undo their perpetual work.

Agencies concerned.

73. These agents, however, are not the *sole* cause of currents. The winds *help* to make currents by pressing upon the waves

and drifting before them the water of the sea; so do the rains, by raising its level here and there; and so does the atmosphere, by pressing with more or less superincumbent force upon different parts of the ocean at the same moment, and as indicated by the changes of the barometric column. But when the winds and the rains cease, and the barometer is stationary, the currents that were the consequence also cease. The currents thus created are therefore *ephemeral.* But the changes of temperature and of saltness, and the work of other agents which affect the specific gravity of sea-water and derange its equilibrium, are as ceaseless in their operations as the sun in his course, and in their effects they are as endless. Philosophy points to them as the *chief* cause of the Gulf Stream and of all the *constant* currents of the sea.

Early writers.

74. Early writers, however, maintained that the Mississippi River was the father of the Gulf Stream. Its floods, they said, produce it; for the velocity of this river in the sea might, it was held, be computed by the rate of the current of the river on the land.

Objection to the fresh-water theory.

75. Captain Livingston overturned this hypothesis by showing that the volume of water which the Mississippi River empties into the Gulf of Mexico is not equal to the three thousandth part of that which escapes from it through the Gulf Stream. Moreover, the water of the Gulf Stream is salt – that of the Mississippi, fresh; and the advocates of this freshwater theory forgot that just as much salt as escapes from the Gulf of Mexico through this stream, must enter the Gulf through some other channel from the main ocean; for, if it did not, the Gulf of Mexico, in process of time, unless it had a salt bed at the bottom, or was fed with salt springs from below – neither of which is probable – would become a fresh-water basin.

76. The above quoted argument of Captain Livingston, however,

Livingston's hypothesis.

was held to be conclusive; and upon the remains of the hypothesis which he had so completely overturned, he set up another, which, in turn, has also been upset. In it he ascribed the velocity of the Gulf Stream as depending 'on the motion of the sun in the ecliptic, and the influence he has on the waters of the Atlantic'.

Franklin's theory.

77. But the opinion that came to be most generally received and deep-rooted in the mind of seafaring people was the one repeated by Dr. Franklin, and which held that the Gulf Stream is the escaping of the waters that have been *forced* into the Caribbean Sea by the trade-winds, and that it is the pressure of those winds upon the water which drives up into that sea a head, as it were, for this stream.

Objections to it.

78. We know of instances in which waters have been accumulated on one side of a lake, or in one end of a canal, at the expense of the other. The pressure of the trade-winds may *assist* to give the Gulf Stream its initial velocity, but are they of themselves adequate to such an effect? Examination shows that they are not. With the view of ascertaining the average number of days during the year that the N.E. trade-winds of the Atlantic operate upon the currents between 25° N. and the equator, log-books containing no less than 380,284* observations on the force and direction of the wind in that ocean were examined. The data thus afforded were carefully compared and discussed. The results show that within those latitudes, and on the average, the wind from the N.E. quadrant is in excess of the winds from the S.W. only 111 days out of the 365. During the rest of the year the S.W. counteract the effect of the N.E. winds upon the currents. Now can the N.E. trades, by blowing for less than one third of the time, cause the Gulf Stream to run all the time, and without varying its velocity either to their force or their prevalence?

*Nautical Monographs, Washington Observatory, No. 1.

79. Sir John Herschel maintains* that they can; that the trade-winds are the *sole cause*† of the Gulf Stream; not, indeed, by causing 'a head of water' in the West Indian seas, but by rolling particles of water before them somewhat as billiard balls are rolled over the table. He denies to evaporation, temperature, salts, and sea-shells, any effective influence whatever upon the circulation of the waters in the ocean. According to him, the winds are the supreme current-producing power in the sea.‡

Herschel's explanation.

80. This theory would require all the currents of the sea to set with the winds, or, when deflected, to be deflected from the shore, as billiard balls are from the cushions of the table, making the littoral angles of incidence and reflection equal. Now, so far from this being the case, *not* ONE of the *constant* currents of the sea either makes such a rebound or sets with the winds. The Gulf Stream sets, as it comes out of the Gulf of Mexico, and for hundreds of miles after it enters the Atlantic, against the trade-winds; for a part of the way it runs right in the 'wind's eye'. The Japan current, 'the Gulf Stream of the Pacific', does the same. The Mozambique current runs to the south, against the S.E. trade-winds, and it changes not with the monsoons. The ice-bearing currents of the north oppose the winds in their course. Humboldt's current has its genesis in the ex-tropical regions of the south, where the 'brave west winds' blow with almost if not with quite the regularity of the trades, but with double their force. And this current, instead of setting to the S.E. before these winds, flows north in spite of them. These are the main and constant currents of the sea – the great arteries and jugulars through which its

Objections to it.

*Article 'Physical Geography', 8th edition Encyclopædia Britannica.

†'The dynamics of the Gulf Stream have of late, in the work of Lieut. Maury, already mentioned, been made the subject of much (we can not but think misplaced) wonder, as if there could be any possible ground for doubting that it owes its origin *entirely* to the trade-winds.' – Art. 57, Phys. Geography, 8th edition Encyc. Brit.

‡Art. 65, Phys. Geography, Encyc. Brit.

circulation is conducted. In every instance, and regardless of winds, those currents that are warm flow toward the poles, those that are cold set toward the equator. And this they do, not by the force of the winds, but in spite of them, and by the force of those very agencies that make the winds to blow. They flow thus by virtue of those efforts which the sea is continually making to restore that equilibrium to its waters which heat and cold, the forces of evaporation, and the secretion of its inhabitants are everlastingly destroying.

EDGAR ALLAN POE

A Descent into the Maelström

The ways of God in Nature, as in Providence, are not as *our* ways; nor are the models that we frame any way commensurate to the vastness, profundity, and unsearchableness of His works, *which have a depth in them greater than the well of Democritus.*

Joseph Glanville

We had now reached the summit of the loftiest crag. For some minutes the old man seemed too much exhausted to speak. 'Not long ago,' said he at length, 'and I could have guided you on this route as well as the youngest of my sons; but, about three years past, there happened to me an event such as never happened before to mortal man – or at least such as no man ever survived to tell of – and the six hours of deadly terror which I then endured have broken me up body and soul. You suppose me a *very* old man – but I am not. It took less than a single day to change these hairs from a jetty black to white, to weaken my limbs, and to unstring my nerves, so that I tremble at the least exertion, and am frightened at a shadow. Do you know I can scarcely look over this little cliff without getting giddy?'

The 'little cliff,' upon whose edge he had so carelessly thrown himself down to rest that the weightier portion of his body hung over it, while he was only kept from falling by the tenure of his elbow on its extreme and slippery edge – this 'little cliff' arose, a sheer unobstructed precipice of black shining rock, some fifteen or sixteen hundred feet from the world of crags beneath us. Nothing would have tempted me to within half-a-dozen yards of its brink. In truth, so deeply was I excited by the perilous position of my companion, that I fell at full length upon the ground, clung to the shrubs around me, and dared not even glance upward at the sky – while I struggled in vain to divest myself of the idea that the very foundations of the mountain were in danger from the fury of the winds. It was long before I could reason myself into sufficient courage to sit up and look out into the distance.

'You must get over these fancies,' said the guide, 'for I have brought you here that you might have the best possible view of the scene of that event I mentioned – and to tell you the whole story with the spot just under your eye.

'We are now,' he continued in that particularising manner which distinguished him – 'we are now close upon the Norwegian coast – in the sixty-eighth degree of latitude – in the great province of Nordland – and in the dreary district of Lofoden. The mountain upon whose top we sit is Helseggen, the Cloudy. Now raise yourself up a little higher – hold on to the grass if you feel giddy – so – and look out, beyond the belt of vapour beneath us, into the sea.'

I looked dizzily, and beheld a wide expanse of ocean, whose waters wore so inky a hue as to bring at once to my mind the Nubian geographer's account of the *Mare Tenebrarum*. A panorama more deplorably desolate no human imagination can conceive. To the right and left, as far as the eye could reach, there lay outstretched, like ramparts of the world, lines of horridly black and beetling cliff, whose character of gloom was but the more forcibly illustrated by the surf which reared high up against its white and ghastly crest, howling and shrieking for ever. Just opposite the promontory upon whose apex we were placed, and at a distance of some five or six miles out at sea, there was visible a small, bleak-looking island; or, more properly, its position was discernible through the wilderness of surge in which it was enveloped. About two miles nearer the land arose another of smaller size, hideously craggy and barren, and encompassed at various intervals by a cluster of dark rocks.

The appearance of the ocean, in the space between the more distant island and the shore, had something very unusual about it. Although at the time so strong a gale was blowing landward that a brig in the remote offing lay to under a double-reefed try-sail, and constantly plunged her whole hull out of sight, still there was here nothing like a regular swell, but only a short, quick, angry cross dashing of water in every direction – as well in the teeth of the wind as otherwise. Of foam there was little except in the immediate vicinity of the rocks.

'The island in the distance,' resumed the old man, 'is called by the Norwegians Vurrgh. The one midway is Moskoe. That a mile to the northward is Ambaaren. Yonder are Islesen, Hotholm, Keildhelm, Suarven, and Buckholm. Farther off – between Moskoe and Vurrgh – are Otterholm, Flimen, Sandflesen, and Stockholm. These are the true names of the places – but why it has been thought necessary to name them at all, is more than either you or I can understand. Do you hear anything? Do you see any change in the water?'

We had now been about ten minutes upon the top of Helseggen, to which we had ascended from the interior of Lofoden, so that we had caught no glimpse of the sea until it had burst upon us from the summit. As the old man spoke, I became aware of a loud and gradually increasing sound, like the moaning of a vast herd of buffaloes upon an American prairie; and at the same moment I perceived that what seamen term the *chopping* character of the ocean beneath us, was rapidly changing into a current which set to the eastward. Even while I gazed this current acquired a monstrous velocity. Each moment added to its speed – to its headlong impetuosity. In five minutes the whole sea as far as Vurrgh was lashed into ungovernable fury: but it was between Moskoe and the coast that the main uproar held its sway. Here the vast bed of the waters, seamed and scarred into a thousand conflicting channels, burst suddenly into frenzied convulsion – heaving, boiling, hissing, – gyrating in gigantic and innumerable vortices, and all whirling and plunging on to the eastward with a rapidity which water never elsewhere assumes except in precipitous descents.

In a few minutes more, there came over the scene another radical alteration. The general surface grew somewhat more smooth, and the whirlpools one by one disappeared, while prodigious streaks of foam became apparent where none had been seen before. These streaks, at length, spreading out to a great distance, and entering into combination, took unto themselves the gyratory motion of the subsided vortices, and seemed to form the germ of another more vast. Suddenly – very suddenly – this assumed a distinct and definite existence in a circle of more than a mile in diameter. The edge of the whirl was represented by a broad belt of gleaming spray; but no particle of this slipped into the mouth of the terrific funnel, whose interior, as far as the eye could fathom it, was a smooth, shining, and jet-black wall of water, inclined to the horizon at an angle of some forty-five degrees, speeding dizzily round and round with a swaying and sweltering motion, and sending forth to the winds an appalling voice, half-shriek, half-roar, such as not even the mighty cataract of Niagara ever lifts up in its agony to Heaven.

The mountain trembled to its very base, and the rock rocked. I threw myself upon my face, and clung to the scant herbage in an excess of nervous agitation.

'This,' said I at length, to the old man – 'this *can* be nothing else than the great whirlpool of the Maelström.'

'So it is sometimes termed,' said he. 'We Norwegians call it the Moskoe-ström, from the island of Moskoe in the midway.'

The ordinary accounts of this vortex had by no means prepared me for what I saw. That of Jonas Ramus, which is perhaps the most circumstantial of any, cannot impart the faintest conception either of the magnificence, or of the horror of the scene – or of the wild bewildering sense of *the novel* which confounds the beholder. I am not sure from what point of view the writer in question surveyed it, nor at what time; but it could neither have been from the summit of Helseggen, nor during a storm. There are some passages of his description, nevertheless, which may be quoted for their details, although their effect is exceedingly feeble in conveying an impression of the spectacle.

'Between Lofoden and Moskoe,' he says, 'the depth of the water is between thirty-five and forty fathoms; but on the other side, towards Ver (Vurrgh) this depth decreases so as not to afford a convenient passage for a vessel, without the risk of splitting on the rocks, which happens even in the calmest weather. When it is in flood, the stream runs up the country between Lofoden and Moskoe with a boisterous rapidity, but the roar of its impetuous ebb to the sea is scarce equalled by the loudest and most dreadful cataracts – the noise being heard several leagues off, and the vortices or pits are of such an extent and depth, that if a ship comes within its attraction it is inevitably absorbed and carried down to the bottom and there beaten to pieces against the rocks, and when the water relaxes the fragments thereof are thrown up again. But these intervals of tranquillity are only at the turn of the ebb and flood, and in calm weather, and last but a quarter of an hour, its violence gradually returning. When the stream is most boisterous, and its fury heightened by a storm, it is dangerous to come within a Norway mile of it. Boats, yachts and ships have been carried away by not guarding against it before they were within its reach. It likewise happens frequently that whales come too near the stream, and are overpowered by its violence, and then it is impossible to describe their howlings and bellowings in their fruitless struggles to disengage themselves. A bear once, attempting to swim from Lofoden to Moskoe, was caught by the stream and borne down, while he roared terribly, so as to be heard on shore. Large stocks of firs and pine trees, after being absorbed by the current, rise again broken and torn to such a degree as if bristles grew upon them. This plainly shows the bottom to consist of craggy rocks, among which they are whirled to and fro. This stream is regulated by the flux and reflux of the sea – it being constantly high and low water every six hours. In the year 1645, early in the morning of Sexagesima Sunday, it raged with such noise and impetuosity that the very stones of the houses on the coast fell to the ground.'

In regard to the depth of the water, I could not see how this could have been ascertained at all in the immediate vicinity of the vortex. The 'forty fathoms' must have reference only to portions of the channel close upon the shore either of Moskoe or Lofoden. The depth in the centre of the Moskoe-ström must be immeasurably greater; and no better proof of this fact is necessary than can be obtained from even the sidelong glance into the abyss of the whirl which may be had from the highest crag of Helseggen. Looking down from this pinnacle upon the howling Phlegethon below, I could not help smiling at the simplicity with which the honest Jonas Ramus records, as a matter difficult of belief, the anecdotes of the whales and the bears; for it appeared to me, in fact, a self-evident thing that the largest ship of the line in existence coming within the influence of that deadly attraction could resist it as little as a feather the hurricane, and must disappear bodily and at once.

The attempts to account for the phenomenon – some of which I remember seemed to me sufficiently plausible in perusal, now wore a very different and unsatisfactory aspect. The idea generally received is that this, as well as three smaller vortices among the Ferroe Islands, 'have no other cause than the collision of waves rising and falling at flux and reflux against a ridge of rocks and shelves, which confines the water so that it precipitates itself like a cataract; and thus the higher the flood rises the deeper must the fall be, and the natural result of all is as a whirlpool or vortex, the prodigious suction of which is sufficiently known by lesser experiments'. These are the words of the Encyclopædia Britannica. – Kircher and others imagine that in the centre of the channel of the Maelström is an abyss penetrating the globe, and issuing in some very remote part – the Gulf of Bothnia being somewhat decidedly named in one instance. This opinion, idle in itself, was the one to which, as I gazed, my imagination most readily assented; and, mentioning it to the guide, I was rather surprised to hear him say that, although it was the view almost universally entertained of the subject by the Norwegians, it nevertheless was not his own. As to the former notion he confessed his inability to comprehend it; and here I agreed with him – for, however conclusive on paper, it becomes altogether unintelligible, and even absurd, amid the thunder of the abyss.

'You have had a good look at the whirl now,' said the old man, 'and if you will creep round this crag so as to get in its lee, and deaden the roar of the water, I will tell you a story that will convince you I ought to know something of the Moskoe-ström.'

I placed myself as desired, and he proceeded.

'Myself and my two brothers once owned a schooner-rigged smack of about seventy tons burden, with which we were in the habit of fishing among the islands, beyond Moskoe, nearly to Vurrgh. In all violent eddies at sea there is good fishing at proper opportunities if one has only the courage to attempt it, but among the whole of the Lofoden coastmen, we three were the only ones who made a regular business of going out to the islands, as I tell you. The usual grounds are a great way lower down to the southward. There fish can be got at all hours, without much risk, and therefore these places are preferred. The choice spots over here among the rocks, however, not only yield the finest variety, but in far greater abundance, so that we often got in a single day what the more timid of the craft could not scrape together in a week. In fact, we made it a matter of desperate speculation – the risk of life standing instead of labour, and courage answering for capital.

'We kept the smack in a cove about five miles higher up the coast than this; and it was our practice, in fine weather, to take advantage of the fifteen minutes' slack to push across the main channel of the Moskoe-ström, far above the pool, and then drop down upon anchorage somewhere near Otterholm, or Sandflesen, where the eddies are not so violent as elsewhere. Here we used to remain until nearly time for slack-water again when we weighed and made for home. We never set out upon this expedition without a steady side wind for going and coming – one that we felt sure would not fail us before our return – and we seldom made a miscalculation upon this point. Twice during six years we were forced to stay all night at anchor on account of a dead calm, which is a rare thing indeed just about here; and once we had to remain on the grounds nearly a week, starving to death, owing to a gale which blew up shortly after our arrival, and made the channel too boisterous to be thought of. Upon this occasion we should have been driven out to sea in spite of everything (for the whirlpools threw us round and round so violently that at length we fouled our anchor and dragged it) if it had not been that we drifted into one of the innumerable cross currents – here today and gone tomorrow – which drove us under the lee of Flimen, where, by good luck, we brought up.

'I could not tell you the twentieth part of the difficulties we encountered "on the grounds" – it is a bad spot to be in, even in good weather – but we made shift always to run the gauntlet of the

Moskoe-ström itself without accident; although at times my heart has been in my mouth when we happened to be a minute or so behind or before the slack. The wind sometimes was not as strong as we thought it at starting, and then we made rather less way than we could wish, while the current rendered the smack unmanageable. My eldest brother had a son eighteen years old, and I had two stout boys of my own. These would have been of great assistance at such times in using the sweeps, as well as afterward in fishing, but somehow, although we ran the risk ourselves, we had not the heart to let the young ones get into the danger – for, after all is said and done, it *was* a horrible danger, and that is the truth.

'It is now within a few days of three years since what I am going to tell you occurred. It was on the tenth day of July 18—, a day which the people of this part of the world will never forget – for it was one in which blew the most terrible hurricane that ever came out of the heavens; and yet all the morning, and indeed until late in the afternoon, there was a gentle and steady breeze from the south-west, while the sun shone brightly, so that the oldest seaman among us could not have foreseen what was to follow.

'The three of us – my two brothers and myself – had crossed over to the islands about 2 o'clock P.M., and had soon nearly loaded the smack with fine fish, which, we all remarked, were more plentiful that day than we had ever known them. It was just seven *by my watch* when we weighed and started for home, so as to make the worst of the Ström at slack water, which we knew would be at eight.

'We set out with a fresh wind on our starboard quarter, and for some time spanked along at a great rate, never dreaming of danger, for indeed we saw not the slightest reason to apprehend it. All at once we were taken aback by a breeze from over Helseggen. This was most unusual – something that had never happened to us before – and I began to feel a little uneasy without exactly knowing why. We put the boat on the wind, but could make no headway at all for the eddies, and I was just upon the point of proposing to return to the anchorage, when, looking astern, we saw the whole horizon covered with a singular copper-coloured cloud that rose with the most amazing velocity.

'In the meantime the breeze that had headed us off fell away, and we were dead becalmed, drifting about in every direction. This state of things however, did not last long enough to give us time to think about it. In less than a minute the storm was upon us – in less than two the sky was entirely overcast – and what with this and the driving spray it became suddenly so dark that we could not see each other in the smack.

'Such a hurricane as then blew it is folly to attempt describing. The oldest seaman in Norway never experienced anything like it. We had let our sails go by the run before it cleverly took us; but, at the first puff, both our masts went by the board as if they had been sawed off – the mainmast taking with it my youngest brother, who had lashed himself to it for safety.

'Our boat was the lightest feather of a thing that ever sat upon water. It had a complete flush deck, with only a small hatch near the bow, and this hatch it had always been our custom to batten down when about to cross the Ström by way of precaution against the chopping seas. But for this circumstance we should have foundered at once – for we lay entirely buried for some moments. How my elder brother escaped destruction I cannot say, for I never had an opportunity of ascertaining. For my part, as soon as I had let the foresail run, I threw myself flat on deck, with my feet against the narrow gunwale of the bow, and with my hands grasping a ring-bolt near the foot of the fore-mast. It was mere instinct that prompted me to do this – which was undoubtedly the very best thing I could have done – for I was too much flurried to think.

'For some moments we were completely deluged, as I say, and all this time I held my breath, and clung to the bolt. When I could stand it no longer I raised myself upon my knees, still keeping hold with my hands, and thus got my head clear. Presently our little boat gave herself a shake, just as a dog does in coming out of the water, and thus rid herself in some measure of the seas. I was now trying to get the better of the stupor that had come over me, and to collect my senses so as to see what was to be done, when I felt somebody grasp my arm. It was my elder brother, and my heart leaped for joy, for I had made sure that he was overboard – but the next moment all this joy was turned into horror – for he put his mouth close to my ear, and screamed out the word "*Moskoe-ström!*"

'No one ever will know what my feelings were at that moment. I shook from head to foot, as if I had had the most violent fit of the ague. I knew what he meant by that one word well enough – I knew what he wished to make me understand. With the wind that now drove us on, we were bound for the whirl of the Ström, and nothing could save us!

'You perceive that in crossing the Ström *channel*, we always went a long way up above the whirl, even in the calmest weather, and then had to wait and watch carefully for the slack – but now we were driving right upon the pool itself, and in such a hurricane as this! "To be sure," I thought, "we shall get there just about the slack – there is some little hope in that" – but in the next moment I cursed myself for being so great a fool as to dream of hope at all. I knew very well that we were doomed had we been ten times a ninety-gun ship.

'By this time the first fury of the tempest had spent itself, or perhaps we did not feel it so much as we scudded before it, but at all events the seas, which at first had been kept down by the wind and lay flat and frothing now got up into absolute mountains. A singular change, too, had come over the heavens. Around in every direction it was still as black as pitch, but nearly overhead there burst out, all at once, a circular rift of clear sky – as clear as I ever saw, and of a deep bright blue – and through it there blazed forth the full moon with a lustre that I never before knew her to wear. She lit up everything about us with the greatest distinctness – but, O God, what a scene it was to light up!

'I now made one or two attempts to speak to my brother – but, in some manner which I could not understand, the din had so increased that I could not make him hear a single word, although I screamed at the top of my voice in his ear. Presently he shook his head, looking as pale as death, and held up one of his fingers, as if to say "*listen!*"

'At first I could not make out what he meant, but soon a hideous thought flashed upon me. I dragged my watch from its fob. It was not going. I glanced at its face by the moonlight, and then burst into tears as I flung it far away into the ocean. *It had run down at seven o'clock! We were behind the time of the slack, and the whirl of the Ström was in full fury!*

'When a boat is well built, properly trimmed, and not deep laden, the waves in a strong gale, when she is going large, seem always to slip from beneath her – which appears very strange to a landsman – and this is what is called *riding*, in sea-phrase. Well, so far we had ridden the swells very cleverly, but presently a gigantic sea happened to take us right under the counter, and bore us with it as it rose – up – up – as if into the sky. I would not have believed that any wave could rise so high. And then down we came with a sweep, a slide, and a plunge, that made me feel sick and dizzy, as if I was falling from some lofty mountain-top in a dream. But while we were up I had thrown a quick glance around – and that one glance was all sufficient. I saw our exact position in an instant. The Moskoe-ström whirlpool was about a quarter of a mile dead ahead, but no more like the everyday Moskoe-ström than the whirl as you now see it is like a mill-race. If I had not known where we were, and what we had to expect, I should not have recognised the place at all. As it was, I involuntarily closed my eyes in horror. The lids clenched themselves together as if in a spasm.

'It could not have been more than two minutes afterwards until we suddenly felt the waves subside, and were enveloped in foam. The boat made a sharp half turn to larboard, and then shot off in its new direction like a thunderbolt. At the same moment the roaring noise of the water was completely drowned in a kind of shrill shriek – such a sound as you might imagine given out by the wastepipes of many thousand steam-vessels letting off their steam all together. We were now in the belt of surf that always surrounds the whirl; and I thought of course that another moment would plunge us into the abyss – down which we could only see indistinctly on account of the amazing velocity with which we were borne along. The boat did not seem to sink into the water at all, but to skim like an air-bubble upon the surface of the surge. Her starboard side was next the whirl, and on the larboard arose the world of ocean we had left. It stood like a huge writhing wall between us and the horizon.

'It may appear strange, but now, when we were in the very jaws of the gulf, I felt more composed than when we were only approaching it. Having made up my mind to hope no more, I got rid of a great deal of that terror which unmanned me at first. I suppose it was despair that strung my nerves.

'It may look like boasting – but what I tell you is truth – I began to reflect how magnificent a thing it was to die in such a manner, and how foolish it was in me to think of so paltry a consideration as my own individual life in view of so wonderful a manifestation of God's power. I do believe that I blushed with shame when this idea crossed my mind. After a little while I became possessed with the keenest curiosity about the whirl itself. I positively felt a *wish* to explore its depths, even at the sacrifice I was going to make; and my principal grief was that I should never be able to tell my old companions on shore about the mysteries I should see. These no doubt were singular fancies to occupy a man's mind in such extremity, and I have often thought since that the revolution of the boat around the pool might have rendered me a little light-headed.

'There was another circumstance which tended to restore my self-possession, and this was the cessation of the wind, which could not reach us in our present situation – for, as you saw yourself, the belt of surf is considerably lower than the general bed of the ocean, and this latter now towered above us, a high, black, mountainous ridge. If you have never been at sea in a heavy gale you can form no idea of the confusion of mind occasioned by the wind and spray together. They blind, deafen, and strangle you, and take away all power of action or reflection. But we were now, in a great measure, rid of these annoyances – just as death-condemned felons in prison are allowed petty indulgences, forbidden them while their doom is yet uncertain.

'How often we made the circuit of the belt it is impossible to say. We careered round and round for perhaps an hour, flying rather than floating, getting gradually more and more into the middle of the surge, and then nearer and nearer to its horrible inner edge. All this time I had never let go of the ring-bolt. My brother was at the stern, holding on to a small empty water-cask which had been securely lashed under the coop of the counter, and was the only thing on deck that had not been swept overboard when the gale first took us. As we approached the brink of the pit he let go his hold upon this, and made for the ring, from which, in the agony of his terror, he endeavoured to force my hands, as it was not large enough to afford us both a secure grasp. I never felt deeper grief than when I saw him attempt this act – although I knew he was a madman when he did it – a raving maniac through sheer fright. I did not care, however, to contest the point with him. I knew it could make no difference whether either of us held on at all, so I let him have the bolt, and went astern to the cask. This there was no great difficulty in doing, for the smack flew round steadily enough, and upon an even keel, only swaying to and fro with the immense sweeps and swelters of the whirl. Scarcely had I secured myself in my new position when we gave a wild lurch to starboard, and rushed headlong into the abyss. I muttered a hurried prayer to God, and thought all was over.

'As I felt the sickening sweep of the descent I had instinctively tightened my hold upon the barrel, and closed my eyes. For some seconds I dared not open them, while I expected instant destruction, and wondered that I was not already in my death-struggles with the water. But moment after moment elapsed. I still lived. The sense of falling had ceased; and the motion of the vessel seemed much as it had been before while in the belt of foam, with the exception that she now lay more along. I took courage, and looked once again upon the scene.

'Never shall I forget the sensations of awe, horror, and admiration with which I gazed about me. The boat appeared to be hanging, as if by magic, midway down, upon the interior surface of a funnel vast in circumference, prodigious in depth, and whose perfectly smooth sides might have been mistaken for ebony but for the bewildering rapidity with which they spun around, and for the gleaming and ghastly radiance they shot forth, as the rays of the full moon, from that circular rift amid the clouds which I have already described, streamed in a flood of golden glory along the black walls, and far away down into the inmost recesses of the abyss.

'At first I was too much confused to observe anything accurately. The general burst of terrific grandeur was all that I beheld. When I recovered myself a little, however, my gaze fell instinctively downward. In this direction I was able to obtain an unobstructed view from the manner in which the smack hung on the inclined surface of the pool. She was quite upon an even keel – that is to say, her deck lay in a plane parallel with that of the water – but this latter sloped at an angle of more than forty-five degrees, so that we seemed to be lying upon our beam-ends. I could not help observing, nevertheless, that I had scarcely more difficulty in maintaining my hold and footing in this situation than if we had been upon a dead level, and this, I suppose, was owing to the speed at which we revolved.

'The rays of the moon seemed to search the very bottom of the profound gulf; but still I could make out nothing distinctly, on account of a thick mist in which everything there was enveloped, and over which there hung a magnificent rainbow, like that narrow and tottering bridge which Mussulmen say is the only pathway between Time and Eternity. This mist or spray was no doubt occasioned by the clashing of the great walls of the funnel as they all met together at the bottom, but the yell that went up to the Heavens from out of that mist I dare not attempt to describe.

'Our first slide into the abyss itself, from the belt of foam above, had carried us a great distance down the slope, but our farther descent was by no means proportionate. Round and round we swept – not with any uniform movement – but in dizzying swings and jerks, that sent us sometimes only a few hundred yards – sometimes nearly the complete circuit of the whirl. Our progress downward at each revolution was slow but very perceptible.

'Looking about me upon the wide waste of liquid ebony on which we were thus borne, I perceived that our boat was not the only object in the embrace of the whirl. Both above and below us were visible fragments of vessels, large masses of building timber and trunks of trees, with many smaller articles, such as pieces of house furniture, broken boxes, barrels, and staves. I have already described the unnatural curiosity which had taken the place of my original terrors. It appeared to grow upon me as I drew nearer and nearer to my dreadful doom. I now began to watch, with a strange interest, the numerous things that floated in our company. I *must* have been delirious, for I even sought *amusement* in speculating upon the relative velocities of their several descents towards the foam below. "This fir-tree," I found myself at one time saying, "will certainly be the next thing that takes the awful plunge and disappears," – and then I was disappointed to find that the wreck of a Dutch merchant ship overtook it and went down before. At length, after making several guesses of this nature, and being deceived in all, this fact – the fact of my invariable miscalculation – set me upon a train of reflection that made my limbs again tremble, and my heart beat heavily once more.

'It was not a new terror that thus affected me, but the dawn of a more exciting *hope*. This hope arose partly from memory and partly from present observation. I called to mind the great variety of buoyant matter that strewed the coast of Lofoden, having been absorbed and then thrown forth by the Moskoe-ström. By far the greater number of the articles were shattered in the most extraordinary way – so chafed and roughened as to have the appearance of being stuck full of splinters – but then I distinctly recollected that there were *some* of them which were not disfigured at all. Now I could not account for this difference except by supposing that the roughened fragments were the only ones which had been *completely absorbed* – that the others had entered the whirl at so late a period of the tide, or, for some reason, had descended so slowly after entering, that they did not reach the bottom before the turn of the flood came, or of the ebb, as the case might be. I conceived it possible, in either instance, that they might thus be whirled up again to the level of the ocean, without undergoing the fate of those which had been drawn in more early, or absorbed more rapidly. I made also three important observations. The first was that, as a general rule, the larger the bodies were the more rapid their descent; the second, that, between two masses of equal extent, the one spherical and the other *of any other shape*, the superiority in speed of descent was with the sphere; the third, that, between two masses of equal size, the one cylindrical and the other of any other shape, the cylinder was absorbed the more slowly. Since my escape I have had several conversations on this subject with an old schoolmaster of the district, and it was from him that I learned the use of the words "cylinder" and "sphere." He explained to me – although I have forgotten the explanation – how what I observed was in fact the natural consequence of the forms of the floating fragments, and showed me how it happened that a cylinder swimming in a vortex offered more resistance to its suction, and was drawn in with greater difficulty than an equally bulky body of any form whatever.*

'There was one startling circumstance which went a great way in enforcing these observations and rendering me anxious to turn them to account, and this was that at every revolution we passed something like a barrel, or else the yard or the mast of a vessel, while many of these things which had been on our level when I first opened my eyes upon the wonders of the whirlpool were now high up above us, and seemed to have moved but little from their original station.

'I no longer hesitated what to do. I resolved to lash myself securely to the water-cask upon which I now held, to cut it loose from the counter, and to throw myself with it into the water. I attracted my brother's attention by signs, pointed to the floating barrels that came near us, and did everything in my power to make him understand what I was about to do. I thought at length that he comprehended my design, but, whether this was the case or not, he shook his head despairingly, and refused to move from his station by the ring-bolt. It was impossible to reach him, the emergency admitted of no delay, and so, with a bitter struggle, I resigned him to his fate, fastened

*See Archimedes '*De Incidentibus in Fluido.*' – lib. 2.

myself to the cask by means of the lashings which secured it to the counter, and precipitated myself with it into the sea without another moment's hesitation.

'The result was precisely what I had hoped it might be. As it is myself who now tell you this tale – as you see that I *did* escape – and as you are already in possession of the mode in which this escape was effected, and must therefore anticipate all that I have further to say, I will bring my story quickly to conclusion. It might have been an hour or thereabout after my quitting the smack, when, having descended to a vast distance beneath me; it made three or four wild gyrations in rapid succession and, bearing my loved brother with it, plunged headlong at once and for ever into the chaos of foam below. The barrel to which I was attached sunk very little farther than half the distance between the bottom of the gulf and the spot at which I leaped overboard, before a great change took place in the character of the whirlpool. The slope of the sides of the vast funnel became momently less and less steep. The gyrations of the whirl grew gradually less and less violent. By degrees the froth and the rainbow disappeared, and the bottom of the gulf seemed slowly to uprise. The sky was clear, the winds had gone down, and the full moon was setting radiantly in the west, when I found myself on the surface of the ocean, in full view of the shores of Lofoden, and above the spot where the pool of the Moskoe-ström *had been*. It was the hour of the slack – but the sea still heaved in mountainous waves from the effects of the hurricane. I was borne violently into the channel of the Ström, and in a few minutes was hurried down the coast into the "grounds" of the fishermen. A boat picked me up, exhausted from fatigue and (now that the danger was removed) speechless from the memory of its horror. Those who drew me on board were my old mates and daily companions, but they knew me no more than they would have known a traveller from the spirit-land. My hair which had been raven black the day before, was as white as you see it now. They say, too, that the whole expression of my countenance had changed. I told them my story – they did not believe it. I now tell it to *you*, and I can scarcely expect you to put more faith in it than did the merry fishermen of Lofoden.'

ELIZABETH BISHOP

At the Fishhouses

Although it is a cold evening,
down by one of the fishhouses
an old man sits netting,
his net, in the gloaming almost invisible,
a dark purple-brown,
and his shuttle worn and polished.
The air smells so strong of codfish
it makes one's nose run and one's eyes water.
The five fishhouses have steeply peaked roofs
and narrow, cleated gangplanks slant up
to storerooms in the gables
for the wheelbarrows to be pushed up and down on.
All is silver: the heavy surface of the sea,
swelling slowly as if considering spilling over,
is opaque, but the silver of the benches,
the lobster pots, and masts, scattered
among the wild jagged rocks,
is of an apparent translucence
like the small old buildings with an emerald moss
growing on their shoreward walls.
The big fish tubs are completely lined
with layers of beautiful herring scales
and the wheelbarrows are similarly plastered

with creamy iridescent coats of mail,
with small iridescent flies crawling on them.
Up on the little slope behind the houses,
set in the sparse bright sprinkle of grass,
is an ancient wooden capstan,
cracked, with two long bleached handles
and some melancholy stains, like dried blood,
where the ironwork has rusted.
The old man accepts a Lucky Strike.
He was a friend of my grandfather.
We talk of the decline in the population
and of codfish and herring
while he waits for a herring boat to come in.
There are sequins on his vest and on his thumb.
He has scraped the scales, the principal beauty,
from unnumbered fish with that black old knife,
the blade of which is almost worn away.

Down at the water's edge, at the place
where they haul up the boats, up the long ramp
descending into the water, thin silver
tree trunks are laid horizontally
across the grey stones, down and down
at intervals of four or five feet.

Cold dark deep and absolutely clear,
element bearable to no mortal,
to fish and to seals . . . One seal particularly
I have seen here evening after evening.
He was curious about me. He was interested in music;
like me a believer in total immersion,

so I used to sing him Baptist hymns.
I also sang 'A Mighty Fortress Is Our God.'
He stood up in the water and regarded me
steadily, moving his head a little.
Then he would disappear, then suddenly emerge
almost in the same spot, with a sort of shrug
as if it were against his better judgment.
Cold dark deep and absolutely clear,
the clear grey icy water . . . Back, behind us,
the dignified tall firs begin.
Bluish, associating with their shadows,
a million Christmas trees stand
waiting for Christmas. The water seems suspended
above the rounded grey and blue-grey stones.
I have seen it over and over, the same sea, the same,
slightly, indifferently swinging above the stones,
icily free above the stones,
above the stones and then the world.
If you should dip your hand in,
your wrist would ache immediately,
your bones would begin to ache and your hand would burn
as if the water were a transmutation of fire
that feeds on stones and burns with a dark grey flame.
If you tasted it, it would first taste bitter,
then briny, then surely burn your tongue.
It is like what we imagine knowledge to be:
dark, salt, clear, moving, utterly free,
drawn from the cold hard mouth
of the world, derived from the rocky breasts
forever, flowing and drawn, and since
our knowledge is historical, flowing, and flown.

JOSEPH CONRAD

Lord Jim

A marvellous stillness pervaded the world, and the stars, together with the serenity of their rays, seemed to shed upon the earth the assurance of everlasting security. The young moon recurved, and shining low in the west, was like a slender shaving thrown up from a bar of gold, and the Arabian Sea, smooth and cool to the eye like a sheet of ice, extended its perfect level to the perfect circle of a dark horizon. The propeller turned without a check, as though its beat had been part of the scheme of a safe universe; and on each side of the *Patna* two deep folds of water, permanent and sombre on the unwrinkled shimmer, enclosed within their straight and diverging ridges a few white swirls of foam bursting in a low hiss, a few wavelets, a few ripples, a few undulations that, left behind, agitated the surface of the sea for an instant after the passage of the ship, subsided splashing gently, calmed down at last into the circular stillness of water and sky with the black speck of the moving hull remaining everlastingly in its centre.

Jim on the bridge was penetrated by the great certitude of unbounded safety and peace that could be read on the silent aspect of nature like the certitude of fostering love upon the placid tenderness of a mother's face. Below the roof of awnings, surrendered to the wisdom of white men and to their courage, trusting the power of their unbelief and the iron shell of their fire-ship, the pilgrims of an exacting faith slept on mats, on blankets, on bare planks, on every deck, in all the dark corners, wrapped in dyed cloths, muffled in soiled rags, with their heads resting on small bundles, with their faces pressed to bent forearms: the men, the women, the children; the old with the young, the decrepit with the lusty – all equal before sleep, death's brother.

A draught of air, fanned from forward by the speed of the ship, passed steadily through the long gloom between the high bulwarks, swept over the rows of prone bodies; a few dim flames in globe-lamps were hung short here and there under the ridge-poles, and in the blurred circles of light thrown down and trembling slightly to the unceasing vibration of the ship appeared a chin upturned, two closed eyelids, a dark hand with silver rings, a meagre limb draped in a torn covering, a head bent back, a naked foot, a throat bared and stretched as if offering itself to the knife. The well-to-do had made for their families shelters with heavy boxes and dusty mats; the poor reposed side by side with all they had on earth tied up in a rag under their heads; the lone old men slept, with drawn-up legs, upon their prayer-carpets, with their hands over their ears and one elbow on each side of the face: a father, his shoulders up and his knees under his forehead, dozed dejectedly by a boy who slept on his back with tousled hair and one arm commandingly extended; a woman covered from head to foot, like a corpse, with a piece of white sheeting, had a naked child in the hollow of each arm; the Arab's belongings, piled right aft, made a heavy mound of broken outlines, with a cargo-lamp swung above, and a great confusion of vague forms behind: gleams of paunchy brass pots, the foot-rest of a deck-chair, blades of spears, the straight scabbard of an old sword leaning against a heap of pillows, the spout of a tin coffee-pot. The patent log on the taffrail periodically rang a single tinkling stroke for every mile traversed on an errand of faith. Above the mass of sleepers a faint and patient sigh at times floated, the exhalation of a troubled dream; and short metallic clangs bursting out suddenly in the depths of the ship, the harsh scrape of a shovel, the violent slam of a furnace-door, exploded brutally, as if the men handling the mysterious things below had their breasts full of fierce anger: while the slim high hull of the steamer went on evenly ahead, without a sway of her bare masts, cleaving continuously the great calm of the waters under the inaccessible serenity of the sky.

Jim paced athwart, and his footsteps in the vast silence were loud to his own ears, as if echoed by the watchful stars: his eyes roaming about the line of the horizon, seemed to gaze hungrily into the unattainable, and did not see the shadow of the coming event. The only shadow on the sea was the shadow of the black smoke pouring heavily from the funnel its immense streamer, whose end was constantly dissolving in the air. Two Malays, silent and almost motionless, steered, one on each side of the wheel, whose brass rim shone fragmentarily in the oval of light thrown out by the binnacle. Now and then a hand, with black fingers alternately letting go and catching hold of revolving spokes, appeared in the illumined part; the links of wheel-chains ground heavily in the grooves of the barrel. Jim would glance at the compass, would glance around the unattainable horizon, would stretch himself till his joints cracked, with a leisurely twist of the body, in the very excess of wellbeing; and, as if made audacious by the invincible aspect of the peace, he felt he cared for nothing that could happen to him to the end of his days. From time to time he glanced idly at a chart pegged out with four drawing-pins on a low three-legged table abaft the steering-gear case. The sheet of paper portraying the depths of the sea presented a shiny surface under the light of a bull's-eye lamp lashed to a stanchion, a surface as level and smooth as the glimmering surface of the waters. Parallel rulers with a pair of dividers reposed on it; the ship's position at last noon was marked with a small black cross, and the straight pencil-line drawn firmly as far as Perim figured the course of the ship – the path of souls towards the holy place, the promise of salvation, the reward of eternal life – while the pencil with its sharp end touching the Somali coast lay round and still like a naked ship's spar floating in the pool of a sheltered dock. 'How steady she goes,' thought Jim with wonder, with something like gratitude for this high peace of sea and sky. At such times his thoughts would be full of valorous deeds: he loved these dreams and the success of his imaginary achievements. They were the best parts of life, its secret truth, its hidden reality. They had a gorgeous virility, the charm of vagueness, they passed before him with a heroic tread; they carried his soul away with them and made it drunk with the divine philtre of an unbounded confidence in itself. There was nothing he could not face. He was so pleased with the idea that he smiled, keeping perfunctorily his eyes ahead; and when he happened to glance back he saw the white streak of the wake drawn as straight by the ship's keel upon the sea as the black line drawn by the pencil upon the chart.

The ash-buckets racketed, clanking up and down the stokehold ventilators, and this tin-pot clatter warned him the end of his watch was near. He sighed with content, with regret as well at having to part from that serenity which fostered the adventurous freedom of his thoughts. He was a little sleepy too, and felt a pleasurable languor running through every limb as though all the blood in his body had turned to warm milk. His skipper had come up noiselessly, in pyjamas and with his sleeping-jacket flung wide open. Red of face, only half awake, the left eye partly closed, the right staring stupid and glassy, he hung his big head over the chart and scratched his ribs sleepily.

There was something obscene in the sight of his naked flesh. His bared breast glistened soft and greasy as though he had sweated out his fat in his sleep. He pronounced a professional remark in a voice harsh and dead, resembling the rasping sound of a wood-file on the edge of a plank; the fold of his double chin hung like a bag triced up close under the hinge of his jaw. Jim started, and his answer was full of deference; but the odious and fleshy figure, as though seen for the first time in a revealing moment, fixed itself in his memory for ever as the incarnation of everything vile and base that lurks in the world we love: in our own hearts we trust for our salvation, in the men that surround us, in the sights that fill our eyes, in the sounds that fill our ears, and in the air that fills our lungs.

The thin gold shaving of the moon floating slowly downwards had lost itself on the darkened surface of the waters, and the eternity beyond the sky seemed to come down nearer to the earth, with the augmented glitter of the stars, with the more profound sombreness in the lustre of the half-transparent dome covering the flat disc of an opaque sea. The ship moved so smoothly that her onward motion was imperceptible to the senses of men, as though she had been a crowded planet speeding through the dark spaces of ether behind the swarm of suns, in the appalling and calm solitudes awaiting the breath of future creations. 'Hot is no name for it down below,' said a voice.

Jim smiled without looking round. The skipper presented an unmoved breadth of back: it was the renegade's trick to appear pointedly unaware of your existence unless it suited his purpose to turn at you with a devouring glare before he let loose a torrent of foamy, abusive jargon that came like a gush from a sewer. Now he emitted only a sulky grunt; the second engineer at the head of the bridge-ladder, kneading with damp palms a dirty sweat-rag, unabashed, continued the tale of his complaints. The sailors had a good time of it up here, and what was the use of them in the world he would be blowed if he could see. The poor devils of engineers had to get the ship along anyhow, and they could very well do the rest too; by gosh they – 'Shut up!' growled the German stolidly. 'Oh yes! Shut up – and when anything goes wrong you fly to us, don't you?' went on the other. He was more than half cooked, he expected; but anyway, now, he did not mind how much he sinned, because these last three days he had passed through a fine course of training for the place where the bad boys go when they die – b'gosh, he had – besides being made jolly well deaf by the blasted racket below. The durned, compound, surface-condensing, rotten scrap-heap rattled and banged down there like an old deck-winch, only more so; and what made him risk his life every night and day that God made amongst the refuse of a breaking-up yard flying round at fifty-seven revolutions, was more than *he* could tell. He must have been born reckless, b'gosh. He . . . 'Where did you get drink?' inquired the German, very savage, but motionless in the light of the binnacle, like a clumsy effigy of a man cut out of a block of fat. Jim went on smiling at the retreating horizon; his heart was full of generous impulses, and his thought was contemplating his own superiority. 'Drink!' repeated the engineer with amiable scorn: he was hanging on with both hands to the rail, a shadowy figure with flexible legs. 'Not from you, Captain. You're far too mean, b'gosh. You would let a good man die sooner than give him a drop of Schnapps. That's what you Germans call economy. Penny wise, pound foolish.' He became sentimental. The chief had given him a four-finger nip about ten o'clock – 'only one, s'elp me!' – good old chief; but as to getting the old fraud out of his bunk – a five-ton crane couldn't do it. Not it. Not tonight anyhow. He was sleeping sweetly like a little child, with a bottle of prime brandy under his pillow. From the thick throat of the commander of the *Patna* came a low rumble, on which the sound of the word *Schwein* fluttered high and low like a capricious feather in a faint stir of air. He and the chief engineer had been cronies for a good few years – serving the same jovial, crafty, old Chinaman, with horn-rimmed goggles and strings of red silk plaited into the venerable grey hairs of his pigtail. The quay-side opinion in the *Patna's* home-port was that these two in the way of brazen peculation 'had done together pretty well everything you can think of.' Outwardly they were badly matched: one dull-eyed, malevolent, and of soft fleshy curves; the other lean, all hollows, with a head long and bony like the head of an old horse, with sunken cheeks, with sunken temples, with an indifferent glazed glance of sunken eyes. He had been stranded out East somewhere – in Canton, in Shanghai, or perhaps in Yokohama; he probably did not care to remember himself the exact locality, nor yet the cause of his shipwreck. He had been, in mercy to his youth, kicked quietly out of his ship twenty years ago or more, and it might have been so much worse for him that the memory of the episode had in it hardly a trace of misfortune. Then, steam navigation expanding in these seas and men of his craft being scarce at first, he had 'got on' after a sort. He was eager to let strangers know in a dismal mumble that he was 'an old stager out here.' When he moved a skeleton seemed to sway loose in his clothes; his walk was mere wandering, and he was given to wander thus around the engine-room skylight, smoking, without relish, doctored tobacco in a brass bowl at the end of a cherrywood stem four feet long, with the imbecile gravity of a thinker evolving a system of philosophy from the hazy glimpse of a truth. He was usually anything but free with his private store of liquor; but on that night he had departed from his principles, so that his second, a weak-headed child of Wapping, what with the unexpectedness of the treat and the strength of the stuff, had become very happy, cheeky and talkative. The fury of the New South Wales German was extreme; he puffed like an exhaust-pipe, and Jim, faintly amused by the scene, was impatient for the time when he could get below: the last ten minutes of the watch were irritating like a gun that hangs fire; those men did not belong to the world of heroic adventure; they weren't bad chaps though. Even the skipper himself . . . His gorge rose at the mass of panting flesh from which issued gurgling mutters, a cloudy trickle of filthy expressions; but he was too pleasurably languid to dislike actively this or any other thing. The quality of these men did not matter; he rubbed shoulders with them, but they could not touch him; he shared the air they breathed, but he was different . . . Would the skipper go for the engineer? . . . The life was easy and he was too sure of himself – too sure of himself to . . . The line dividing his meditation from a surreptitious doze on his feet was thinner than a thread in a spider's web.

The second engineer was coming by easy transitions to the consideration of his finances and of his courage.

'Who's drunk? I? No, no, Captain! That won't do. You ought to know by this time the chief ain't free-hearted enough to make a sparrow drunk, b'gosh. I've never been the worse for liquor in my life; the stuff ain't made yet that would make *me* drunk. I could drink liquid fire against your whisky peg for peg, b'gosh, and keep as cool as a cucumber. If I thought I was drunk I would jump overboard – do away with myself, b'gosh. I would! Straight! And I won't go off the bridge. Where do you expect me to take the air on a night like this, eh? On deck amongst that vermin down there? Likely – ain't it! And I am not afraid of anything you can do.'

The German lifted two heavy fists to heaven and shook them a little without a word.

'I don't know what fear is,' pursued the engineer, with the enthusiasm of sincere conviction. 'I am not afraid of doing all the bloomin' work in this rotten hooker, b'gosh! And a jolly good thing for you that there are some of us about the world that aren't afraid of their lives, or where would you be – you and this old thing here with her plates like brown paper – brown paper, s'elp me? It's all very fine for you – you get a power of pieces out of her one way and another; but what about me – what do I get? A measly hundred and fifty dollars a-month and find yourself. I wish to ask you respectfully – respectfully, mind – who wouldn't chuck a dratted job like this? 'Tain't safe, s'elp me, it ain't! Only I am one of them fearless fellows . . .'

He let go the rail and made ample gestures as if demonstrating in the air the shape and extent of his valour; his thin voice darted in prolonged squeaks upon the sea, he tiptoed back and forth for the better emphasis of utterance, and suddenly pitched down head-first as though he had been clubbed from behind. He said 'Damn!' as he tumbled; an instant of silence followed upon his screeching: Jim and the skipper staggered forward by common accord, and catching themselves up, stood very stiff and still gazing, amazed, at the undisturbed level of the sea. Then they looked upwards at the stars.

What had happened? The wheezy thump of the engines went on. Had the earth been checked in her course? They could not understand; and suddenly the calm sea, the sky without a cloud, appeared formidably insecure in their immobility, as if poised on the brow of yawning destruction. The engineer rebounded vertically full length and collapsed again into a vague heap. This heap said 'What's that?' in the muffled accents of profound grief. A faint noise as of thunder, of thunder infinitely remote, less than a sound, hardly more than a vibration, passed slowly, and the ship quivered in response, as if the thunder had growled deep down in the water. The eyes of the two Malays at the wheel glittered towards the white men, but their dark hands remained closed on the spokes. The sharp hull driving on its way seemed to rise a few inches in succession through its whole length, as though it had become pliable, and settled down again rigidly to its work of cleaving the smooth surface of the sea. Its quivering stopped, and the faint noise of thunder ceased all at once, as though the ship had steamed across a narrow belt of vibrating water and of humming air.

JOSHUA SLOCUM

Sailing Alone Around the World

On the evening of July 5 the *Spray*, after having steered all day over a lumpy sea, took it into her head to go without the helmsman's aid. I had been steering southeast by south, but the wind hauling forward a bit, she dropped into a smooth lane, heading southeast, and making about eight knots, her very best work. I crowded on sail to cross the track of the liners without loss of time, and to reach as soon as possible the friendly Gulf Stream. The fog lifting before night, I was afforded a look at the sun just as it was touching the sea. I watched it go down and out of sight. Then I turned my face eastward, and there, apparently at the very end of the bowsprit, was the smiling full moon rising out of the sea. Neptune himself coming over the bows could not have startled me more. 'Good evening, sir,' I cried; 'I'm glad to see you.' Many a long talk since then I have had with the man in the moon; he had my confidence on the voyage.

About midnight the fog shut down again denser than ever before. One could almost 'stand on it.' It continued so for a number of days, the wind increasing to a gale. The waves rose high, but I had a good ship. Still, in the dismal fog I felt myself drifting into loneliness, an insect on a straw in the midst of the elements. I lashed the helm, and my vessel held her course, and while she sailed I slept.

During these days a feeling of awe crept over me. My memory worked with startling power. The ominous, the insignificant, the great, the small, the wonderful, the commonplace – all appeared before my mental vision in magical succession. Pages of my history were recalled which had been so long forgotten that they seemed to belong to a previous existence. I heard all the voices of the past laughing, crying, telling what I had heard them tell in many corners of the earth.

The loneliness of my state wore off when the gale was high and I found much work to do. When fine weather returned, then came the sense of solitude, which I could not shake off. I used my voice often, at first giving some order about the affairs of a ship, for I had been told that from disuse I should lose my speech. At the meridian altitude of the sun I called aloud, 'Eight bells,' after the custom on a ship at sea. Again from my cabin I cried to an imaginary man at the helm, 'How does she head, there?' and again, 'Is she on her course?' But getting no reply, I was reminded the more palpably of my condition. My voice sounded hollow on the empty air, and I dropped the practice. However, it was not long before the thought came to me that when I was a lad I used to sing; why not try that now, where it would disturb no one? My musical talent had never bred envy in others, but out on the Atlantic, to realize what it meant, you should have heard me sing. You should have seen the porpoises leap when I pitched my voice for the waves and the sea and all that was in it. Old turtles, with large eyes, poked their heads up out of the sea as I sang 'Johnny Boker', and 'We'll Pay Darby Doyl for his Boots', and the like. But the porpoises were, on the whole, vastly more appreciative than the turtles; they jumped a deal higher. One day when I was humming a favourite chant, I think it was 'Babylon's a-Fallin'', a porpoise jumped higher than the bowsprit. Had the *Spray* been going a little faster she would have scooped him in. The sea-birds sailed around rather shy.

July 10, eight days at sea, the *Spray* was twelve hundred miles east of Cape Sable. One hundred and fifty miles a day for so small a vessel must be considered good sailing. It was the greatest run the *Spray* ever made before or since in so few days. On the evening of July 14, in better humor than ever before, all hands cried, 'Sail ho!' The sail was a barkantine, three points on the weather bow, hull down. Then came the night. My ship was sailing along now without attention to the helm. The wind was south; she was heading east. Her sails were trimmed like the sails of the nautilus. They drew steadily all night. I went frequently on deck, but found all well. A merry breeze kept on from the south. Early in the morning of the 15th the *Spray* was close aboard the stranger, which proved to be *La Vaguisa* of Vigo, twenty-three days from Philadelphia, bound for Vigo. A lookout from his masthead had spied the *Spray* the evening before. The captain, when I came near enough, threw a line to me and sent a bottle of wine across slung by the neck, and very good wine it was. He also sent his card, which bore the name of Juan Gantes. I think he was a good man, as Spaniards go. But when I asked him to report me 'all well' (the *Spray* passing him in a lively manner), he hauled his shoulders much above his head; and when his mate, who knew of my expedition, told him that I was alone, he crossed himself and made for his cabin. I did not see him again. By sundown he was as far astern as he had been ahead the evening before.

There was now less and less monotony. On July 16 the wind was northwest and clear, the sea smooth, and a large bark, hull down, came in sight on the lee bow, and at 2:30 P.M. I spoke the stranger. She was the bark *Java* of Glasgow, from Peru for Queenstown for orders. Her old captain was bearish, but I met a bear once in Alaska that looked pleasanter. At least, the bear seemed pleased to meet me, but this grizzly old man! Well, I suppose my hail disturbed his siesta, and my little sloop passing his great ship had somewhat the effect on him that a red rag has upon a bull. I had the advantage over heavy ships, by long odds, in the light winds of this and the two previous days. The wind was light; his ship was heavy and foul, making poor headway, while the *Spray*, with a great mainsail bellying even to light winds, was just skipping along as nimbly as one could wish. 'How long has it been calm about here?' roared the captain of the *Java*, as I came within hail of him. 'Dunno, cap'n,' I shouted back as loud as I could bawl. 'I haven't been here long.' At this the mate on the forecastle wore a broad grin. 'I left Cape Sable fourteen days ago,' I added. (I was now well across toward the Azores.) 'Mate,' he roared to his chief officer – 'mate, come here and listen to the Yankee's yarn. Haul down the flag, mate, haul down the flag!' In the best of humour, after all, the *Java* surrendered to the *Spray*.

The acute pain of solitude experienced at first never returned. I had penetrated a mystery, and, by the way, I had sailed through a fog. I had met Neptune in his wrath, but he found that I had not treated him with contempt, and so he suffered me to go on and explore.

HART CRANE

At Melville's Tomb

Often beneath the wave, wide from this ledge
The dice of drowned men's bones he saw bequeath
An embassy. Their numbers as he watched,
Beat on the dusty shore and were obscured.

And wrecks passed without sound of bells,
The calyx of death's bounty giving back
A scattered chapter, livid hieroglyph,
The portent wound in corridors of shells.

Then in the circuit calm of one vast coil,
Its lashings charmed and malice reconciled,
Frosted eyes there were that lifted altars;
And silent answers crept across the stars.

Compass, quadrant and sextant contrive
No farther tides . . . High in the azure steeps
Monody shall not wake the mariner.
This fabulous shadow only the sea keeps.

HERMAN MELVILLE

Moby-Dick

Days, weeks passed, and under easy sail, the ivory Pequod had slowly swept across four several cruising-grounds; that off the Azores; off the Cape de Verdes; on the Plate (so called), being off the mouth of the Rio de la Plata; and the Carrol Ground, an unstaked, watery locality, southerly from St. Helena.

It was while gliding through these latter waters that one serene and moonlight night, when all the waves rolled by like scrolls of silver; and, by their soft, suffusing seethings, made what seemed a silvery silence, not a solitude: on such a silent night a silvery jet was seen far in advance of the white bubbles at the bow. Lit up by the moon, it looked celestial; seemed some plumed and glittering god uprising from the sea. Fedallah first descried this jet. For of these moonlight nights, it was his wont to mount to the main-mast head, and stand a look-out there, with the same precision as if it had been day. And yet, though herds of whales were seen by night, not one whaleman in a hundred would venture a lowering for them. You may think with what emotions, then, the seamen beheld this old Oriental perched aloft at such unusual hours; his turban and the moon, companions in one sky. But when, after spending his uniform interval there for several successive nights without uttering a single sound; when, after all this silence, his unearthly voice was heard announcing that silvery, moon-lit jet, every reclining mariner started to his feet as if some winged spirit had lighted in the rigging, and hailed the mortal crew.

'There she blows!' Had the trump of judgment blown, they could not have quivered more; yet still they felt no terror; rather pleasure. For though it was a most unwonted hour, yet so impressive was the cry, and so deliriously exciting, that almost every soul on board instinctively desired a lowering.

Walking the deck with quick, side-lunging strides, Ahab commanded the t'gallant sails and royals to be set, and every stunsail spread. The best man in the ship must take the helm. Then, with every mast-head manned, the piled-up craft rolled down before the wind. The strange, upheaving, lifting tendency of the taffrail breeze filling the hollows of so many sails, made the buoyant, hovering deck to feel like air beneath the feet; while still she rushed along, as if two antagonistic influences were struggling in her – one to mount direct to heaven, the other to drive yawingly to some horizontal goal. And had you watched Ahab's face that night, you would have thought that in him also two different things were warring. While his one live leg made lively echoes along the deck, every stroke of his dead limb sounded like a coffin-tap. On life and death this old man walked. But though the ship so swiftly sped, and though from every eye, like arrows, the eager glances shot, yet the silvery jet was no more seen that night. Every sailor swore he saw it once, but not a second time.

This midnight-spout had almost grown a forgotten thing, when, some days after, lo! at the same silent hour, it was again announced: again it was descried by all; but upon making sail to overtake it, once more it disappeared as if it had never been. And so it served us night after night, till no one heeded it but to wonder at it. Mysteriously jetted into the clear moonlight, or starlight, as the case might be; disappearing again for one whole day, or two days, or three; and somehow seeming at every distinct repetition to be advancing still further and further in our van, this solitary jet seemed for ever alluring us on.

Nor with the immemorial superstition of their race, and in accordance with the preternaturalness, as it seemed, which in many things invested the Pequod, were there wanting some of the seamen who swore that whenever and wherever descried; at however remote times, or in however far apart latitudes and longitudes, that unnearable spout was cast by one self-same whale; and that whale, Moby-Dick. For a time, there reigned, too, a sense of peculiar dread at this flitting apparition, as if it were treacherously beckoning us on and on, in order that the monster might turn round upon us, and rend us at last in the remotest and most savage seas.

These temporary apprehensions, so vague but so awful, derived a wondrous potency from the contrasting serenity of the weather, in which, beneath all its blue blandness, some thought there lurked a devilish charm, as for days and days we voyaged along, through seas so wearily, lonesomely mild, that all space, in repugnance to our vengeful errand, seemed vacating itself of life before our urn-like prow.

But, at last, when turning to the eastward, the Cape winds began howling around us, and we rose and fell upon the long, troubled seas that are there; when the ivory-tusked Pequod sharply bowed to the blast, and gored the dark waves in her madness, till, like showers of silver chips, the foam-flakes flew over her bulwarks; then all this desolate vacuity of life went away, but gave place to sights more dismal than before.

Close to our bows, strange forms in the water darted hither and thither before us; while thick in our rear flew the inscrutable sea-ravens. And every morning, perched on our stays, rows of these birds were seen; and spite of our hootings, for a long time obstinately clung to the hemp, as though they deemed our ship some drifting, uninhabited craft; a thing appointed to desolation, and therefore fit roosting-place for their homeless selves. And heaved and heaved, still unrestingly heaved the black sea, as if its vast tides were a conscience;

and the great mundane soul were in anguish and remorse for the long sin and suffering it had bred.

Cape of Good Hope, do they call ye? Rather Cape Tormentoso, as called of yore; for long allured by the perfidious silences that before had attended us, we found ourselves launched into this tormented sea, where guilty beings transformed into those fowls and these fish, seemed condemned to swim on everlastingly without any haven in store, or beat that black air without any horizon. But calm, snow-white, and unvarying; still directing its fountain of feathers to the sky; still beckoning us on from before, the solitary jet would at times be descried.

During all this blackness of the elements, Ahab, though assuming for the time the almost continual command of the drenched and dangerous deck, manifested the gloomiest reserve; and more seldom than ever addressed his mates. In tempestuous times like these, after everything above and aloft has been secured, nothing more can be done but passively to await the issue of the gale. Then Captain and crew become practical fatalists. So, with his ivory leg inserted into its accustomed hole, and with one hand firmly grasping a shroud, Ahab for hours and hours would stand gazing dead to windward, while an occasional squall of sleet or snow would all but congeal his very eyelashes together. Meantime, the crew driven from the forward part of the ship by the perilous seas that burstingly broke over its bows, stood in a line along the bulwarks in the waist; and the better to guard against the leaping waves, each man had slipped himself into a sort of bowline secured to the rail, in which he swung as in a loosened belt. Few or no words were spoken; and the silent ship, as if manned by painted sailors in wax, day after day tore on through all the swift madness and gladness of the demoniac waves. By night the same muteness of humanity before the shrieks of the ocean prevailed; still in silence the men swung in the bowlines; still wordless Ahab stood up to the blast. Even when wearied nature seemed demanding repose he would not seek that repose in his hammock. Never could Starbuck forget the old man's aspect, when one night going down into the cabin to mark how the barometer stood, he saw him with closed eyes sitting straight in his floor-screwed chair; the rain and half-melted sleet of the storm from which he had some time before emerged, still slowly dripping from the unremoved hat and coat. On the table beside him lay unrolled one of those charts of tides and currents which have previously been spoken of. His lantern swung from his tightly clenched hand. Though the body was erect, the head was thrown back so that the closed eyes were pointed towards the needle of the tell-tale that swung from a beam in the ceiling.*

Terrible old man! thought Starbuck with a shudder, sleeping in this gale, still thou steadfastly eyest thy purpose.

* The cabin-compass is called the tell-tale, because without going to the compass at the helm, the Captain, while below, can inform himself of the course of the ship.

OWEN CHASE

Narrative of the Most Extraordinary and Distressing Shipwreck of the Whale-Ship Essex

I have not been able to recur to the scenes which are now to become the subject of description, although a considerable time has elapsed, without feeling a mingled emotion of horror and astonishment at the almost incredible destiny that has preserved me and my surviving companions from a terrible death. Frequently, in my reflections on the subject, even after this lapse of time, I find myself shedding tears of gratitude for our deliverance, and blessing God, by whose divine aid and protection we were conducted through a series of unparalleled suffering and distress, and restored to the bosoms of our families and friends. There is no knowing what a stretch of pain and misery the human mind is capable of contemplating, when it is wrought upon by the anxieties of preservation; nor what pangs and weaknesses the body is able to endure, until they are visited upon it; and when at last deliverance comes, when the dream of hope is realized, unspeakable gratitude takes possession of the soul, and tears of joy choke the utterance. We require to be taught in the school of some signal suffering, privation, and despair, the great lessons of constant dependence upon an almighty forbearance and mercy. In the midst of the wide ocean, at night, when the sight of the heavens was shut out, and the dark tempest came upon us; then it was, that we felt ourselves ready to exclaim, 'Heaven have mercy upon us, for nought but that can save us now.' But I proceed to the recital. – On the 20th of November, (cruising in latitude 0° 40' S. longitude 119° 0' W.) a shoal of whales was discovered off the lee-bow. The weather at this time was extremely fine and clear, and it was about 8 o'clock in the morning, that the man at the mast-head gave the usual cry of, 'there she blows.' The ship was immediately put away, and we ran down in the direction for them. When we had got within half a mile of the place where they were observed, all our boats were lowered down, manned, and we started in pursuit of them. The ship, in the mean time, was brought to the wind, and the main-top-sail hove aback, to wait for us. I had the harpoon in the second boat; the captain preceded me in the first. When I arrived at the spot where we calculated they were, nothing was at first to be seen. We lay on our oars in anxious expectation of discovering them come up somewhere near us. Presently one rose, and spouted a short distance ahead of my boat; I made all speed towards it, came up with, and struck it; feeling the harpoon in him, he threw himself, in an agony, over towards the boat, (which at that time was up alongside of him,) and giving a severe blow with his tail, struck the boat near the edge of the water, amidships, and stove a hole in her. I immediately took up the boat hatchet, and cut the line, to disengage the boat from the whale, which by this time was running off with great velocity. I succeeded in getting clear of him, with the loss of the harpoon and line; and finding the water to pour fast in the boat, I hastily stuffed three or four of our jackets in the hole, ordered one man to keep constantly bailing, and the rest to pull immediately for the ship; we succeeded in keeping the boat free, and shortly gained the ship. The captain and the second mate, in the other two boats, kept up the pursuit, and soon struck another whale. They being at this time a considerable distance to leeward, I went forward, braced around the main-yard, and put the ship off in a direction for them; the boat which had been stove was immediately hoisted in, and after examining the hole, I found that I could, by nailing a piece of canvass over it, get her ready to join in a fresh pursuit, sooner than by lowering down the other remaining boat which belonged to the ship. I accordingly turned her over upon the quarter, and was in the act of nailing on the canvass, when I observed a very large spermaceti whale, as well as I could judge, about eighty-five feet in length; he broke water about twenty rods off our weather-bow, and was lying quietly, with his head in a direction for the ship. He spouted two or three times, and then disappeared. In less than two or three seconds he came up again, about the length of the ship off, and made directly for us, at the rate of about three knots. The ship was then going with about the same velocity. His appearance and attitude gave us at first no alarm; but while I stood watching his movements, and observing him but a ship's length off, coming down for us with great celerity, I involuntarily ordered the boy at the helm to put it hard up; intending to sheer off and avoid him. The words were scarcely out of my mouth, before he came down upon us with full speed, and struck the ship with his head, just forward of the fore-chains; he gave us such an appalling and tremendous jar, as nearly threw us all on our faces. The ship brought up as suddenly and violently as if she had struck a rock, and trembled for a few seconds like a leaf. We looked at each other with perfect amazement, deprived almost of the power of speech. Many minutes elapsed before we were able to realize the dreadful accident; during which time he passed under the ship, grazing her keel as he went along, came up alongside of her to leeward, and lay on the top of the water, (apparently stunned with the violence of the blow,) for the space of a minute; he then suddenly started off, in a direction to leeward. After a few moments' reflection, and recovering, in some measure, from the sudden consternation that had seized us, I of course concluded that he had stove a hole in the ship, and that it would be necessary to set the pumps going. Accordingly they were rigged, but had not been in operation more than one minute, before I perceived the head of the ship to be gradually settling down in the water; I then ordered the signal to be set for the other boats, which, scarcely had I despatched, before I again discovered the whale, apparently in convulsions, on the top of the water, about one hundred rods to leeward. He was enveloped in the foam of the sea, that his continual and violent thrashing about in the water had created around him, and I could distinctly see him smite his jaws together, as if distracted with rage and fury. He remained a short time in this situation, and then started off with great velocity, across the bows of the ship, to windward. By this time the ship had settled down a considerable distance in the water, and I gave her up as lost. I however, ordered the pumps to be kept constantly going, and endeavoured to collect my thoughts for the occasion. I turned to the boats, two of which we then had with the ship, with an intention of clearing them away, and getting all things ready to embark in them, if there should be no other resource left; and while my attention was thus engaged for a moment, I was aroused with the cry of a man at the hatch-way, 'here he is – he is making for us again.' I turned around, and saw him about one hundred rods directly ahead of us, coming down apparently with twice his ordinary speed, and to me at that moment, it appeared with tenfold fury and vengeance in his aspect. The surf flew in all directions about him, and his course towards us was marked by a white foam of a rod in width, which he made with the continual violent thrashing of his tail; his head was about half out of water, and in that way he came upon, and again struck the ship. I was in hopes when I descried him making for us, that by a dexterous movement of putting the ship away immediately, I should be able to cross the line of his approach, before he could get up to us, and thus avoid, what I knew, if he should strike us again, would prove our inevitable destruction, I bawled out to the helmsman, 'hard up!' but she had not fallen off more than a point, before we took the second shock. I should judge the speed of the ship to have been at this time about three knots, and that of the whale about six. He struck her to windward, directly under the cat-head, and completely stove in her bows. He passed under the ship again, went off to leeward, and we saw no more of him. Our situation at this juncture can be more readily imagined than described. The shock to our feelings was such, as I am sure none can have an adequate conception of, that were not there: the misfortune befel us at a moment when we least dreamt of any accident; and from the pleasing anticipations we had formed, of realizing the certain profits of our labour, we were dejected by a sudden, most mysterious, and overwhelming calamity. Not a moment, however, was to be lost in endeavouring to provide for the extremity to which it was now certain we were reduced. We were more than a thousand miles from the nearest land, and with nothing but a light open boat, as the resource of safety for myself and companions. I ordered the men to cease pumping, and every one to provide for himself; seizing a hatchet at the same time, I cut away the lashings of the spare boat, which lay bottom up, across two spars directly over the quarter deck, and cried out to those near me, to take her as she came down. They did so accordingly, and bore her on their shoulders as far as the waist of the ship. The steward had in the mean time gone down into the cabin twice, and saved two quadrants, two practical navigators, and the captain's trunk and mine; all which were hastily thrown into the boat, as she lay on the deck, with the two compasses which I snatched from the binnacle. He attempted to descend again; but the water by this time had rushed in, and he returned without being able to effect his purpose. By the time we had got the boat to the waist, the ship had filled with water, and was going down on her beam-ends: we shoved our boat as quickly as possible from the plank-shear into the water, all hands jumping in her at the same time, and launched off clear of the ship. We were scarcely two boat's lengths distant from her, when she fell over to windward, and settled down in the water.

Amazement and despair now wholly took possession of us. We contemplated the frightful situation the ship lay in, and thought with horror upon the sudden and dreadful calamity that had overtaken us. We looked upon each other, as if to gather some consolatory sensation from an interchange of sentiments, but every countenance was marked with the paleness of despair. Not a word was spoken for several minutes by any of us; all appeared to be bound in a spell of stupid consternation; and from the time we were first attacked by the whale, to the period of the fall of the ship, and of our leaving her in the boat, more than ten minutes could not certainly have elapsed! God only knows in what way, or by what means, we were enabled to accomplish in that short time what we did; the cutting away and transporting the boat from where she was deposited would of itself, in ordinary circumstances, have consumed as much time as that, if the whole ship's crew had been employed in it. My companions had not saved a single article but what they had on their backs; but to me it was a source of infinite satisfaction, if any such could be gathered from the horrors of our gloomy situation, that we had been fortunate enough to have preserved our compasses, navigators, and quadrants. After the first shock of my feelings was over, I enthusiastically contemplated them as the probable instruments of our salvation; without them all would have been dark and hopeless. Gracious God! what a picture of distress and suffering now presented itself to my imagination. The crew of the ship were saved, consisting of twenty human souls. All that remained to conduct these twenty beings through the stormy terrors of the ocean, perhaps many thousand miles, were three open light boats. The prospect of obtaining any provisions or water from the ship, to subsist upon during the time, was at least now doubtful. How many long and watchful nights, thought I, are to be passed? How many tedious days of partial starvation are to be endured, before the least relief or mitigation of our sufferings can be reasonably anticipated. We lay at this time in our boat, about two ship's lengths off from the wreck, in perfect silence, calmly contemplating her situation, and absorbed in our own melancholy reflections, when the other boats were discovered rowing up to us. They had but shortly before discovered that some accident had befallen us, but of the nature of which they were entirely ignorant. The sudden and mysterious disappearance of the ship was first discovered by the boat-steerer in the captain's boat, and with a horror-struck countenance and voice, he suddenly exclaimed, 'Oh, my God! where is the ship?' Their operations upon this were instantly suspended, and a general cry of horror and despair burst from the lips of every man, as their looks were directed for her, in vain, over every part of the ocean. They immediately made all haste towards us. The captain's boat was the first that reached us. He stopped about a boat's length off, but had no power to utter a single syllable: he was so completely overpowered with the spectacle before him, that he sat down in his boat, pale and speechless. I could scarcely recognise his countenance, he appeared to be so much altered, awed, and overcome, with the oppression of his feelings, and the dreadful reality that lay before him. He was in a short time however enabled to address the inquiry to me, 'My God, Mr. Chase, what is the matter?' I answered, 'We have been stove by a whale.' I then briefly told him the story. After a few moment's reflection he observed, that we must cut away her masts, and endeavour to get something out of her to eat. Our thoughts were now all accordingly bent on endeavours to save from the wreck whatever we might possibly want, and for this purpose we rowed up and got on to her. Search was made for every means of gaining access to her hold; and for this purpose the lanyards were cut loose, and with our hatchets we commenced to cut away the masts, that she might right up again, and enable us to scuttle her decks. In doing which we were occupied about three quarters of an hour, owing to our having no axes, nor indeed any other instruments, but the small hatchets belonging to the boats. After her masts were gone she came up about two-thirds of the way upon an even keel. While we were employed about the masts the captain took his quadrant, shoved off from the ship, and got an observation. We found ourselves in latitude 0° 40' S. longitude 119° W. We now commenced to cut a hole through the planks, directly above two large casks of bread, which most fortunately were between decks, in the waist of the ship, and which being in the upper side, when she upset, we had strong hopes was not wet. It turned out according to our wishes, and from these casks we obtained six hundred pounds of hard bread. Other parts of the deck were then scuttled, and we got without difficulty as much fresh water as we dared to take in the boats, so that each was supplied with about sixty-five gallons; we got also from one of the lockers a musket, a small canister of powder, a couple of files, two rasps, about two pounds of boat nails, and a few turtle. In the afternoon the wind came on to blow a strong breeze; and having obtained every thing that occurred to us could then be got out, we began to make arrangements for our safety during the night. A boat's line was made fast to the ship, and to the other end of it one of the boats was moored, at about fifty fathoms to leeward; another boat was then attached to the first one, about eight fathoms astern; and the third boat, the like distance astern of her. Night came on just as we had finished our operations; and such a night as it was to us! so full of feverish and distracting inquietude, that we were deprived entirely of rest. The wreck was constantly before my eyes. I could not, by any effort, chase away the horrors of the preceding day from my mind: they haunted me the live-long night. My companions – some of them were like sick women; they had no idea of the extent of their deplorable situation. One or two slept unconcernedly, while others wasted the night in unavailing murmurs. I now had full leisure to examine, with some degree of coolness, the dreadful circumstances of our disaster. The scenes of yesterday passed in such quick succession in my mind that it was not until after many hours of severe reflection that I was able to discard the idea of the catastrophe as a dream. Alas! it was one from which there was no awaking; it was too certainly true, that but yesterday we had existed as it were, and in one short moment had been cut off from all the hopes and prospects of the living! I have no language to paint out the horrors of our situation. To shed tears was indeed altogether unavailing, and withal unmanly; yet I was not able to deny myself the relief they served to afford me. After several hours of idle sorrow and repining I began to reflect upon the accident, and endeavoured to realize by what unaccountable destiny or design, (which I could not at first determine,) this sudden and most deadly attack had been made upon us: by an animal, too, never before suspected of premeditated violence, and proverbial for its insensibility and inoffensiveness. Every fact seemed to warrant me in concluding that it was any thing but chance which directed his operations; he made two several attacks upon the ship, at a short interval between them, both of which, according to their direction, were calculated to do us the most injury, by being made ahead, and thereby combining the speed of the two objects for the shock; to effect which, the exact manoeuvres which he made were necessary. His aspect was most horrible, and such as indicated resentment and fury. He came directly from the shoal which we had just before entered, and in which we had struck three of his companions, as if fired with revenge for their sufferings. But to this it may be observed, that the mode of fighting which they always adopt is either with repeated strokes of their tails, or snapping of their jaws together; and that a case, precisely similar to this one, has never been heard of amongst the oldest and most experienced whalers. To this I would answer, that the structure and strength of the whale's head is admirably designed for this mode of attack; the most prominent part of which is almost as hard and as tough as iron; indeed, I can compare it to nothing else but the inside of a horse's hoof, upon which a lance or harpoon would not make the slightest impression. The eyes and ears are removed nearly one-third the length of the whole fish, from the front part of the head, and are not in the least degree endangered in this mode of attack. At all events, the whole circumstances taken together, all happening before my own eyes, and producing, at the time, impressions in my mind of decided, calculating mischief, on the part of the whale (many of which impressions I cannot now recall,) induce me to be satisfied that I am correct in my opinion. It is certainly, in all its bearings, a hitherto unheard of circumstance, and constitutes, perhaps, the most extraordinary one in the annals of the fishery.

HENRY DAVID THOREAU

Cape Cod

The light-house lamps were still burning, though now with a silvery lustre, when I rose to see the sun come out of the Ocean; for he still rose eastward of us; but I was convinced that he must have come out of a dry bed beyond that stream, though he seemed to come out of the water.

> 'The sun once more touched the fields,
> Mounting to heaven from the fair flowing
> Deep-running Ocean.'

Now we saw countless sails of mackerel fishers abroad on the deep, one fleet in the north just pouring round the Cape, another standing down toward Chatham, and our host's son went off to join some lagging member of the first which had not yet left the Bay.

Before we left the light-house we were obliged to anoint our shoes faithfully with tallow, for walking on the beach, in the salt water and the sand, had turned them red and crisp. To counterbalance this, I have remarked that the seashore, even where muddy, as it is not here, is singularly clean; for, notwithstanding the spattering of the water and mud and squirting of the clams, while walking to and from the boat, your best black pants retain no stain nor dirt, such as they would acquire from walking in the country.

We have heard that a few days after this, when the Provincetown

Bank was robbed, speedy emissaries from Provincetown made particular inquiries concerning us at this light-house. Indeed, they traced us all the way down the Cape, and concluded that we came by this unusual route down the back side and on foot, in order that we might discover a way to get off with our booty when we had committed the robbery. The Cape is so long and narrow, and so bare withal, that it is well-nigh impossible for a stranger to visit it without the knowledge of its inhabitants generally, unless he is wrecked on to it in the night. So, when this robbery occurred, all their suspicions seem to have at once centred on us two travellers who had just passed down it. If we had not chanced to leave the Cape so soon, we should probably have been arrested. The real robbers were two young men from Worcester County who travelled with a centre-bit, and are said to have done their work very neatly. But the only bank that we pried into was the great Cape Cod sand-bank, and we robbed it only of an old French crown piece, some shells and pebbles, and the materials of this story.

Again we took to the beach for another day (October 13), walking along the shore of the resounding sea, determined to get it into us. We wished to associate with the Ocean until it lost the pond-like look which it wears to a countryman. We still thought that we could see the other side. Its surface was still more sparkling than the day before, and we beheld 'the countless smilings of the ocean waves'; though some of them were pretty broad grins, for still the wind blew and the billows broke in foam along the beach. The nearest beach to us on the other side, whither we looked, due east, was on the coast of Galicia, in Spain, whose capital is Santiago, though by old poets' reckoning it should have been Atlantis or the Hesperides; but heaven is found to be farther west now. At first we were abreast of that part of Portugal *entre Douro e Mino*, and then Galicia and the port of Pontevedra opened to us as we walked along; but we did not enter, the breakers ran so high. The bold headland of Cape Finisterre, a little north of east, jutted toward us next, with its vain brag, for we flung back, – 'Here is Cape Cod, – Cape Land's-Beginning.' A little indentation toward the north – for the land loomed to our imaginations by a common mirage, – we knew was the Bay of Biscay, and we sang:

> 'There we lay, till next day,
> In the Bay of Biscay O!'

A little south of east was Palos, where Columbus weighed anchor, and farther yet the pillars which Hercules set up; concerning which when we inquired at the top of our voices what was written on them, – for we had the morning sun in our faces, and could not see distinctly, – the inhabitants shouted *Ne plus ultra* (no more beyond), but the wind bore to us the truth only, *plus ultra* (more beyond), and over the Bay westward was echoed *ultra* (beyond). We spoke to them through the surf about the Far West, the true Hesperia, ἕω πέρας or end of the day, the This Side Sundown, where the sun was extinguished in the *Pacific*, and we advised them to pull up stakes and plant those pillars of theirs on the shore of California, whither all our folks were gone, – the only *ne* plus ultra now. Whereat they looked crestfallen on their cliffs, for we had taken the wind out of all their sails.

We could not perceive that any of their leavings washed up here, though we picked up a child's toy, a small dismantled boat, which may have been lost at Pontevedra.

The Cape became narrower and narrower as we approached its wrist between Truro and Provincetown and the shore inclined more decidedly to the west. At the head of East Harbor Creek, the Atlantic is separated but by half a dozen rods of sand from the tide-waters of the Bay. From the Clay Pounds the bank flatted off for the last ten miles to the extremity at Race Point, though the highest parts, which are called 'islands' from their appearance at a distance on the sea, were still seventy or eighty feet above the Atlantic, and afforded a good view of the latter, as well as a constant view of the Bay, there being no trees nor a hill sufficient to interrupt it. Also the sands began to invade the land more and more, until finally they had entire possession from sea to sea, at the narrowest part. For three or four miles between Truro and Provincetown there were no inhabitants from shore to shore, and there were but three or four houses for twice that distance.

As we plodded along, either by the edge of the ocean, where the sand was rapidly drinking up the last wave that wet it, or over the sandhills of the bank, the mackerel fleet continued to pour round the Cape north of us, ten or fifteen miles distant, in countless numbers, schooner after schooner, till they made a city on the water. They were so thick that many appeared to be afoul of one another; now all standing on this tack, now on that. We saw how well the New-Englanders had followed up Captain John Smith's suggestions with regard to the fisheries, made in 1616, – to what a pitch they had carried 'this contemptible trade of fish,' as he significantly styles it, and were now equal to the Hollanders whose example he holds up for the English to emulate; notwithstanding that 'in this faculty,' as he says, 'the former are so naturalized, and of their vents so certainly acquainted, as there is no likelihood they will ever be paralleled, having two or three thousand busses, flat-bottoms, sword-pinks, todes and such like, that breeds them sailors, mariners, soldiers, and merchants, never to be wrought out of that trade and fit for any other.' We thought that it would take all these names and more to describe the numerous craft which we saw. Even then, some years before our 'renowned sires' with their 'peerless dames' stepped on Plymouth Rock, he wrote, 'Newfoundland doth yearly freight neir eight hundred sail of ships with a silly, lean, skinny, poor-john, and cor fish', though all their supplies must be annually transported from Europe. Why not plant a colony here then, and raise those supplies on the spot? 'Of all the four parts of the world,' says he, 'that I have yet seen, not inhabited, could I have but means to transport a colony, I would rather live here than anywhere. And if it did not maintain itself, were we but once indifferently well fitted, let us starve.' Then 'fishing before your doors', you 'may every night sleep quietly ashore, with good cheer and what fires you will, or, when you please, with your wives and family.' Already he anticipates 'the new towns in New England in memory of their old,' – and who knows what may be discovered in the 'heart and entrails' of the land, 'seeing even the very edges', etc., etc.

All this has been accomplished, and more, and where is Holland now? Verily the Dutch have taken it. There was no long interval between the suggestion of Smith and the eulogy of Burke.

Still one after another the mackerel schooners hove in sight round the head of the Cape, 'whitening all the sea road', and we watched each one for a moment with an undivided interest. It seemed a pretty sport. Here in the country it is only a few idle boys or loafers that go a-fishing on a rainy day; but there it appeared as if every able-bodied man and helpful boy in the Bay had gone out on a pleasure excursion in their yachts, and all would at last land and have a chowder on the Cape. The gazetteer tells you gravely how many of the men and boys of these towns are engaged in the whale, cod and mackerel fishery, how many go to the banks of Newfoundland, or the coast of Labrador, the Straits of Belle Isle or the Bay of Chaleurs (Shalore, the sailors call it); as if I were to reckon up the number of boys in Concord who are engaged during the summer in the perch, pickerel, bream, horn-pout and shiner fishery, of which no one keeps the statistics, – though I think that it is pursued with as much profit to the moral and intellectual man (or boy), and certainly with less danger to the physical one.

One of my playmates, who was apprenticed to a printer, and was somewhat of a wag, asked his master one afternoon if he might go a-fishing, and his master consented. He was gone three months. When he came back, he said that he had been to the Grand Banks, and went to setting type again as if only an afternoon had intervened.

I confess I was surprised to find that so many men spent their whole day, ay, their whole lives almost, a-fishing. It is remarkable what a serious business men make of getting their dinners, and how universally shiftlessness and a grovelling taste take refuge in a merely ant-like industry. Better go without your dinner, I thought, than be thus everlastingly fishing for it like a cormorant. Of course, *viewed from the shore*, our pursuits in the country appear not a whit less frivolous.

I once sailed three miles on a mackerel cruise myself. It was a Sunday evening after a very warm day in which there had been frequent thunder-showers, and I had walked along the shore from Cohasset to Duxbury. I wished to get over from the last place to Clark's Island, but no boat could stir, they said, at that stage of the tide, they being left high on the mud. At length I learned that the tavern-keeper, Winsor, was going out mackereling with seven men that evening, and would take me. When there had been due delay, we one after another straggled down to the shore in a leisurely manner, as if waiting for the tide still, and in India-rubber boots, or carrying our shoes in our hands, waded to the boats, each of the crew bearing an armful of wood, and one a bucket of new potatoes besides. Then they resolved that each should bring one more armful of wood, and that would be enough. They had already got a barrel of water, and had some more in the schooner. We shoved the boats a dozen rods over the mud and water till they floated, then rowing half a mile to the vessel climbed aboard, and there we were in a mackerel schooner, a fine stout vessel of forty-three tons, whose name I forget. The baits were not dry on the hooks. There was the mill in which they ground the mackerel, and the trough to hold it, and the long-handled dipper to cast it overboard with; and already in the harbor we saw the surface rippled with schools of small mackerel, the real *Scomber vernalis.* The crew proceeded leisurely to weigh anchor and raise their two sails, there being a fair but very slight wind; – and the sun now setting clear and shining on the vessel after the thunder-showers, I thought that I could not have commenced the voyage under more favourable auspices. They had four dories and commonly fished in them, else they fished on the starboard side aft where their lines hung ready, two to a man. The boom swung round once or twice, and Winsor cast overboard the foul juice of mackerel mixed with rain-water which remained in his trough, and then we gathered about the helmsman and told stories. I remember that the compass was affected by iron in its neighborhood and varied a few degrees. There was one among us just returned from California, who was now going as passenger for his health and amusement. They expected to be gone about a week, to begin fishing the next morning, and to carry their fish fresh to Boston. They landed me at Clark's Island, where the Pilgrims landed, for my companions wished to get some milk for the voyage. But I had seen the whole of it. The rest was only going to sea and catching the mackerel. Moreover, it was as well that I did not remain with them, considering the small quantity of supplies they had taken.

Now I saw the mackerel fleet *on its fishing-ground,* though I was not at first aware of it. So my experience was complete.

It was even more cold and windy today than before, and we were frequently glad to take shelter behind a sand-hill. None of the elements were resting. On the beach there is a ceaseless activity, always something going on, in storm and in calm, winter and summer, night and day. Even the sedentary man here enjoys a breadth of view which is almost equivalent to motion. In clear weather the laziest may look across the Bay as far as Plymouth at a glance, or over the Atlantic as far as human vision reaches, merely raising his eyelids; or if he is too lazy to look after all, he can hardly help *hearing* the ceaseless dash and roar of the breakers. The restless ocean may at any moment cast up a whale or a wrecked vessel at your feet. All the reporters in the world, the most rapid stenographers, could not report the news it brings. No creature could move slowly where there was so much life around. The few wreckers were either going or coming, and the ships and the sand-pipers, and the screaming gulls overhead; nothing stood still but the shore. The little beach-birds trotted past close to the water's edge, or paused but an instant to swallow their food, keeping time with the elements. I wondered how they ever got used to the sea, that they ventured so near the waves. Such tiny inhabitants the land brought forth! except one fox. And what could a fox do, looking on the Atlantic from that high bank? What is the sea to a fox? Sometimes we met a wrecker with his cart and dog, – and his dog's faint bark at us wayfarers, heard through the roaring of the surf, sounded ridiculously faint. To see a little trembling dainty-footed cur stand on the margin of the ocean, and ineffectually bark at a beach-bird, amid the roar of the Atlantic! Come with design to bark at a whale, perchance! That sound will do for farmyards. All the dogs looked out of place there, naked and as if shuddering at the vastness; and I thought that they would not have been there had it not been for the countenance of their masters. Still less could you think of a cat bending her steps that way, and shaking her wet foot over the Atlantic; yet even this happens sometimes, they tell me. In summer I saw the tender young of the Piping Plover, like chickens just hatched, mere pinches of down on two legs, running in troops, with a faint peep, along the edge of the waves. I used to see packs of half-wild dogs haunting the lonely beach on the south shore of Staten Island, in New York Bay, for the sake of the carrion there cast up; and I remember that once, when for a long time I had heard a furious barking in the tall grass of the marsh, a pack of half a dozen large dogs burst forth on to the beach, pursuing a little one which ran straight to me for protection, and I afforded it with some stones, though at some risk to myself; but the next day the little one was the first to bark at me. Under these circumstances I could not but remember the words of the poet: –

'Blow, blow, thou winter wind
Thou art not so unkind
    As *his* ingratitude;
Thy tooth is not so keen,
Because thou art not seen,
    Although thy breath be rude.

'Freeze, freeze, thou bitter sky,
Thou dost not bite so nigh
    As benefits forgot;
Though thou the waters warp,
Thy sting is not so sharp
    As friend remembered not.'

Sometimes, when I was approaching the carcass of a horse or ox which lay on the beach there, where there was no living creature in sight, a dog would unexpectedly emerge from it and slink away with a mouthful of offal.

The seashore is a sort of neutral ground, a most advantageous point from which to contemplate this world. It is even a trivial place. The waves forever rolling to the land are too far-traveled and untamable to be familiar. Creeping along the endless beach amid the sun-squawl and the foam, it occurs to us that we, too, are the product of sea-slime.

It is a wild, rank place, and there is no flattery in it. Strewn with crabs, horse-shoes, and razor-clams, and whatever the sea casts up, – a vast *morgue*, where famished dogs may range in packs, and crows come daily to glean the pittance which the tide leaves them. The carcasses of men and beasts together lie stately up upon its shelf, rotting and bleaching in the sun and waves, and each tide turns them in their beds, and tucks fresh sand under them. There is naked Nature, – inhumanly sincere, wasting no thought on man, nibbling at the cliffy shore where gulls wheel amid the spray.

We saw this forenoon what, at a distance, looked like a bleached log with a branch still left on it. It proved to be one of the principal bones of a whale, whose carcass, having been stripped of blubber at sea and cut adrift, had been washed up some months before. It chanced that this was the most conclusive evidence which we met with to prove, what the Copenhagen antiquaries assert, that these shores were the *Furdustrandas*, which Thorhall, the companion of Thorfinn during his expedition to Vinland in 1007, sailed past in disgust. It appears that after they had left the Cape and explored the country about Straum-Fiordr (Buzzard's Bay!), Thorhall, who was disappointed at not getting any wine to drink there, determined to sail north again in search of Vinland. Though the antiquaries have given us the original Icelandic, I prefer to quote their translation, since theirs is the only Latin which I know to have been aimed at Cape Cod.

> 'Cum parati erant, sublato
> velo, cecinit Thorhallus:
> Eò redeamus, ubi conterranei
> sunt nostri! faciamus aliter,
> expansi arenosi peritum,
> lata navis explorare curricula:
> dum procellam incitantes gladii
> moræ impatientes, qui terram
> collaudant, Furdustrandas
> inhabitant et coquunt balænas.'

In other words, 'When they were ready and their sail hoisted, Thorhall sang: Let us return thither where our fellow-countrymen are. Let us make a bird[1] skillful to fly through the heaven of sand[2], to explore the broad track of ships; while warriors who impel to the tempest of swords[3], who praise the land, inhabit Wonder Strands, *and cook whales*.' And so he sailed north past Cape Cod, as the antiquaries say, 'and was shipwrecked on to Ireland.'

Though once there were more whales cast up here, I think that it was never more wild than now. We do not associate the idea of antiquity with the ocean, nor wonder how it looked a thousand years ago, as we do of the land, for it was equally wild and unfathomable always. The Indians have left no traces on its surface, but it is the same to the civilized man and the savage. The aspect of the shore only has changed. The ocean is a wilderness reaching round the globe, wilder than a Bengal jungle, and fuller of monsters, washing the very wharves of our cities and the gardens of our seaside residences. Serpents, bears, hyenas, tigers, rapidly vanish as civilization advances, but the most populous and civilized city cannot scare a shark far from its wharves. It is no further advanced than Singapore, with its tigers, in this respect. The Boston papers had never told me that there were seals in the harbor. I had always associated these with the Esquimaux and other outlandish people. Yet from the parlour windows all along the coast you may see families of them sporting on the flats. They were as strange to me as the merman would be. Ladies who never walk in the woods, sail over the sea. To go to sea! Why, it is to have the experience of Noah, – to realize the deluge. Every vessel is an ark.

[1] i.e. a vessel

[2] The sea, which is arched over its sandy bottom like a heaven.

[3] Battle.

WILLIAM FALCONER

An Universal Dictionary of the Marine

CURRENT, (*courans*, Fr. *currens*, Lat.) in navigation, a certain progressive movement of the water of the sea, by which all bodies floating therein are compelled to alter their course, or velocity, or both, and submit to the laws imposed on them by the current.

In the sea, currents are either natural and general, as arising from the diurnal rotation of the earth about its axis; or accidental and particular, caused by the waters being driven against promontories, or into gulfs and streights; where, wanting room to spread, they are driven back, and thus disturb the ordinary flux of the sea.

'Currents are various, and directed towards different parts of the ocean, of which some are constant, and others periodical. The most extraordinary current of the sea is that by which part of the Atlantic or African ocean moves about Guinea from Cape Verd towards the curvature or bay of Africa, which they call Fernando Poo, viz. from west to east, contrary to the general motion. And such is the force of this current, that when ships approach too near the shore, it carries them violently towards that bay, and deceives the mariners in their reckoning.

'There is a great variety of shifting currents, which do not last, but return at certain periods; and these do, most of them, depend upon, and follow the anniversary winds or monsoons, which by blowing in one place may cause a current in another*.' *Varenius*.

In the streights of Gibraltar the currents almost constantly drive to the eastward, and carry ships into the Mediterranean: they are also found to drive the same way into St George's-channel.

The setting, or progressive motion of the current, may be either quite down to the bottom, or to a certain determinate depth.

As the knowledge of the direction and velocity of currents is a very material article in navigation, it is highly necessary to discover both, in order to ascertain the ship's situation and course with as much accuracy as possible. The most successful method which has been hitherto

* 'At Java, in the streights of Sunda, when the monsoons blow from the west, viz. in the month of May, the currents set to the eastward, contrary to the general motion.

'Also between the island of Celebes and Madura, when the western monsoons set in, viz. in December, January, and February, or when the winds blow from the N.W. or between the north and west, the currents set to the S.E. or between the south and east.

'At Ceylon, from the middle of March to October, the currents set to the southward, and in the other parts of the year to the northward; because at this time the southern monsoons blow, and at the other, the northern.

'Between Cochin-China and Malacca, when the western monsoons blow, viz. from April to August, the currents set eastward against the general motion, but the rest of the year set westward; the monsoon conspiring with the general motion. They run so strongly in these seas, that unexperienced sailors mistake them for waves that beat upon the rocks known by the name of breakers.

'So for some months after the fifteenth of February the currents set from the Maldivies towards India on the east, against the general motion of the sea.

'On the shore of China and Cambodia, in the months of October, November, and December, the currents set to the N.W. and from January to the S.W. when they run with such a rapidity of motion about the shoals of Parcel, that it seems swifter than that of an arrow.

'At Pulo Condore, upon the coast of Cambodia, though the monsoons are shifting, yet the currents set strongly towards the east, even when they blow to a contrary point.

'Along the coasts of the bay of Bengal, as far as the cape Romania, at the extreme point of Malacca, the current runs southward in November and December.

'When the Monsoons blow from China to Malacca, the sea runs swiftly from Pulo Cambi to Pulo Condore, on the coast of Cambodia.

'In the bay of Sans Bras, not far from the cape of Good Hope, there is a current particularly remarkable, where the sea runs from east to west to the landward; and this more vehemently as it becomes opposed by the winds from a contrary direction. The cause is undoubtedly owing to some adjacent shore, which is higher than this.' *Varenius.*

These currents constantly follow the winds, and set to the same point with the monsoon, or trade-wind, at sea.

attempted by mariners for this purpose, is as follows. A common iron pot, which may contain four or five gallons, is suspended by a small rope fastened to its ears or handles, so as to hang directly upright, as when placed upon the fire. This rope, which may be from 70 to 100 fathoms in length, being prepared for the experiment, is coiled in the boat, which is hoisted out of the ship at a proper opportunity, when there is little or no wind to ruffle the surface of the sea. The pot being then thrown overboard into the water, and immediately sinking, the line is slackened till about seventy or eighty fathoms run out, after which the line is fastened to the boat's stern, by which she is accordingly restrained, and rides as at anchor. The velocity of the current is then easily tried by the *log* and half-minute glass, the usual method of discovering the rate of a ship's sailing at sea. The course of the stream is next obtained by means of the compass provided for this operation.

Having thus found the setting and drift of the current, it remains to apply this experiment to the purposes of navigation. If the ship sails along the direction of the current, then the motion of ship is increased by as much as is the drift or velocity of the current.

If a current sets directly against the ship's course, then her motion is retarded in proportion to the strength of the current. Hence it is plain, 1. If the velocity of the current be less than that of the ship, then the ship will advance so much as is the difference of these velocities. 2. If the velocity of the current be more than that of the ship, then will the ship fall as much *astern* as is the difference of these velocities. 3. If the velocity of the current be equal to that of the ship, then will the ship stand still, the one velocity destroying the other.

If the current thwarts the course of a ship, it not only diminishes or increases her velocity, but gives her a new direction, compounded of the course she steers, and the setting of the current, as appears by the following.

LEMMA

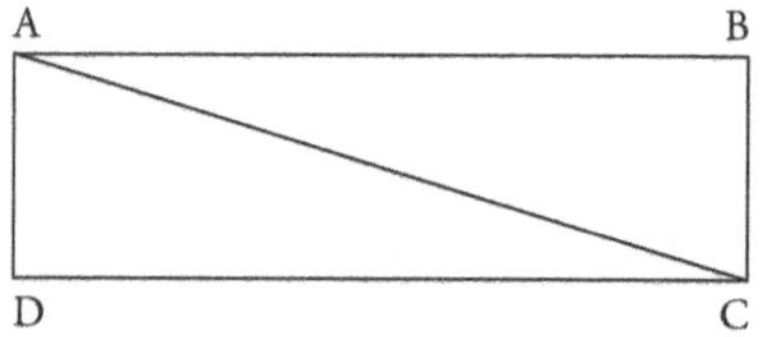

If a body at A be impelled by two forces at the same time, the one in the direction A B, carrying it from A to B in a certain space of time, and the other in the direction A D, pushing it from A to D in the same time; complete the parallelogram A B C D, and draw the diagonal A C: then the body at A, (which let us suppose a ship agitated by the wind and current; A B being the line along which she advances as impressed by the wind, and A D the line upon which she is driven by the current) will move along the diagonal A C, and will be in the point C, at the end of the time in which it would have moved along A D or A B, as impelled by either of those forces, (the wind or current) separately.

NAVIGATION, (*navigation*, Fr.) the art of directing the movements of a ship by the action of the wind upon the sails.

Navigation is then applied, with equal propriety, to the arrangement of the sails, according to the state of the wind; and to the directing and measuring a ship's course by the laws of geometry; or it may comprehend both, being then considered as the theory and practice thereof.

Since every sea-officer is presumed to be furnished with books of navigation, in which that science is copiously described, it would be superfluous to enter into a particular detail of it in this place. As it would also be a fruitless task to those who are entirely ignorant of the roles of trigonometry, and those who are versed in that science generally understand the principles of navigation already, it appears not to come within the limits of our design. It suffices to say, that the course of a ship, and the distance she has run thereon, are measured by the angles and sides of a right-angled plain triangle, in which the hypothenuse is converted into the distance; the perpendicular, into the difference of latitude; the base, into the departure from the meridian, the angle, formed by the perpendicular and hypothenuse, into the course; and the opposite angle, contained between the hypothenuse and base, into its complement of the course.

The course of the ship is determined by the *compass*; and the *log-line*, or a solar observation, ascertains the distance. Hence the hypothenuse and angles are given, to find the base and perpendicular: a problem well known in trigonometry.

That part of navigation, which regards the piloting or conducting a ship along the sea-coast, can only be acquired by a thorough knowledge of that particular coast, after repeated voyages. The most necessary articles thereof are already described in the article COASTING: it is sufficient to observe, that the bearings and distances from various parts of the shore are generally ascertained in the night, either by *light-houses*, or by the different depths of the water, and the various sorts of ground at the bottom; as shells of different sizes and colours, sand, gravel, clay, stones, ooze or shingle. In the day the ship's place is known by the appearance of the land, which is set by the compass, whilst the distance is estimated by the master or pilot.

WIND, (*vent*, Fr.) a stream or current of air which may be felt; and usually blows from one part of the horizon to its opposite part.

The horizon, besides being divided into 360 degrees, like all other circles, is by mariners supposed to be divided into four quadrants, called the north-east, north-west, south-east, and south-west quarters. Each of these quarters they divide into eight equal parts, called points, and each point into four equal parts, called quarter-points. So that the horizon is divided into 32 points, which are called *rhumbs* or *winds*; to each wind is assigned a name, which shows from what point of the horizon the wind blows. The points of north, south, east and west, are called *cardinal points*; and are at the distance of 90 degrees, or eight points from one another.

Winds are either constant or variable, general or particular. Constant winds are such as blow the same way, at least for one or more days; and variable winds are such as frequently shift within a day. A general or *reigning* wind is that which blows the same way, over a large tract of the earth, almost the whole year. A particular wind is what blows, in any place, sometimes one way, and sometimes another, indifferently. If the wind blows gently, it is called a breeze; if it blows harder, it is called a gale, or a stiff gale; and if it blows with violence, it is called a storm or hard gale.*

The following observations on the wind have been made by skilful seamen; and particularly the great Dr. Halley.

1st. Between the limits of 60 degrees, namely, from 30° of north latitude to 30° of south latitude, there is a constant east wind throughout the year, blowing on the Atlantic and Pacific oceans; and this is called the *trade-wind.*

For as the sun, in moving from east to west, heats the air more immediately under him, and thereby expands it; the air to the eastward is constantly rushing towards the west to restore the equilibrium, or natural state of the atmosphere; and this occasions a perpetual east wind in those limits.

2d. The trade-winds near their northern limits blow between the north and east, and near the southern limits they blow between the south and east.

For as the air is expanded by the heat of the sun near the equator; therefore the air from the northward and southward will both tend towards the equator to restore the equilibrium. Now these motions

*The swiftness of the wind in a great storm is not more than 50 or 60 miles in an hour; and a common brisk gale is about 15 miles an hour. *Robertson's Navigation.*

from the north and south, joined with the foregoing easterly motion, will produce the motions observed near the said limits between the north and east, and between the south and west.

3d. These general motions of the wind are disturbed on the continents, and near their coasts.

For the nature of the soil may either cause the air to be heated or cooled; and hence will arise motions that may be contrary to the foregoing general one.

4th. In some parts of the Indian ocean there are periodical winds, which are called Monsoons; that is, such as blow half the year one way, and the other half-year the contrary way.

For air that is cool and dense, will force the warm and rarified air in a continual stream upwards, where it must spread itself to preserve the equilibrium: so that the upper course or current of the air shall be contrary to the under current; for the upper air must move from those parts where the greatest heat is; and so, by a kind of circulation, the N.E. trade-wind below will be attended with a S.W. above; and a S.E. below with a N.W. above: And this is confirmed by the experience of seamen, who, as soon as they get out of the trade-winds, generally find a wind blowing from the opposite quarter.

5th. In the Atlantic ocean, near the coasts of Africa, at about 100 leagues from shore, between the latitude of 28° and 10° north, seamen constantly meet with a fresh gale of wind blowing from the N.E.

6th. Those bound to the Caribbee islands, across the Atlantic ocean, find, as they approach the American side, that the said N.E. wind becomes easterly; or seldom blows more than a point from the east, either to the northward or southward.

Those trade-winds, on the American side, are extended to 30, 31, or even to 32° of N. latitude; which is about 4° further than what they extend to on the African side: Also, to the southward of the equator, the trade-winds extend three or four degrees further towards the coast of Brasil on the American side, than they do near the Cape of Good Hope on the African side.

7th. Between the latitudes of 4° north and 4° south, the wind always blows between the south and east. On the African side the winds are nearest the south; and on the American side nearest the east. In these seas Dr. Halley observed, that when the wind was eastward, the weather was gloomy, dark, and rainy, with hard gales of wind; but when the wind veered to the southward, the weather generally became serene, with gentle breezes next to a calm.

These winds are somewhat changed by the seasons of the year; for when the sun is far northward, the Brasil S.E. wind gets to the south, and the N.E. wind to the east; and when the sun is far south, the S.E. wind gets to the east, and the N.E. winds on this side of the equator veer more to the north.

8th. Along the coast of Guinea, from Sierre Leone to the island of St. Thomas, (under the equator) which is above 500 leagues, the southerly and south-west winds blow perpetually: for the S.E. trade-wind having passed the equator, and approaching the Guinea coast within 80 or 100 leagues, inclines towards the shore, and becomes south, then S.E. and by degrees, as it approaches the land, it veers about to south, S.S.W. and when very near the land it is S.W. and sometimes W.S.W. This tract is troubled with frequent calms, and violent sudden gusts of wind, called tornadoes, blowing from all points of the horizon.

The reason of the wind setting in west on the coast of Guinea is, in all probability, owing to the nature of the coast, which, being greatly heated by the sun, rarifies the air exceedingly, and consequently the cool air from off the sea will keep rushing in to restore the equilibrium.

9th. Between the 4th and 10th degrees of north latitude, and between the longitude of Cape Verd, and the easternmost of the Cape

Verd isles, there is a tract of sea which seems to be condemned to perpetual calms, attended with terrible thunder and lightnings, and such frequent rains, that this part of the sea is called the *rains*. In sailing through these six degrees, ships are said to have been sometimes detained whole months.

The cause of this is apparently, that the westerly winds setting in on this coast, and meeting the general easterly wind in this track, balance each other, and so produce the calms; and the vapours carried thither by each wind, meeting and condensing, occasion the almost constant rains.

The last three observations show the reason of two things which mariners experience in sailing from Europe to India, and in the Guinea trade.

And first. The difficulty which ships in going to the southward, especially in the months of July and August, find in passing between the coast of Guinea and Brazil, notwithstanding the width of this sea is more than 500 leagues. This happens, because the S.E. winds at that time of the year commonly extend some degrees beyond the ordinary limits of 4° N. latitude; and besides coming so much southerly, as to be sometimes south, sometimes a point or two to the west; it then only remains to ply to windward: And if, on the one side, they steer W.S.W. they get a wind more and more easterly; but then there is danger of falling in with the Brasilian coast, or shoals: and if they steer E.S.E. they fall into the neighbourhood of the coast of Guinea, from whence they cannot depart without running easterly as far as the island of St. Thomas; and this is the constant practice of all the Guinea ships.

Secondly. All ships departing from Guinea for Europe, their direct course is northward; but on this course they cannot proceed, because the coast bending nearly east and west, the land is to the northward. Therefore, as the winds on this coast are generally between the S. and W.S.W. they are obliged to steer S.S.E. or south, and with these courses they run off the shore; but in so doing, they always find the winds more and more contrary; so that when near the shore, they can lie south; but at a greater distance they can make no better than S.E. and afterwards E.S.E. with which courses they commonly fetch the island of St. Thomas and Cape Lopez, where finding the winds to the eastward of the south, they sail westerly with it, till coming to the latitude of four degrees south, where they find the S.E. wind blowing perpetually.

On account of these general winds, all those that use the West India trade, and even those bound to Virginia, reckon it their best course to get as soon as they can to the southward, that so they may be certain of a fair and fresh gale to run before it to the westward: And for the same reason those homeward-bound from America endeavour to gain the latitude of 30 degrees, where they first find the winds begin to be variable; though the most ordinary winds in the north Atlantic ocean come from between the south and west.

10th. Between the southern latitudes of 10 and 30 degrees in the Indian ocean, the general trade-wind about the S.E. *by* S. is found to blow all the year long in the same manner as in the like latitudes in the Ethiopic ocean: and during the six months from May to December, these winds reach to within two degrees of the equator; but during the other six months, from November to June, a N.W. wind blows in the tract lying between the 3d and 10th degrees of southern latitude, in the meridian of the north end of Madagascar; and between the 2d and 12th degree of south latitude, near the longitude of Sumatra and Java.

11th. In the tract between Sumatra and the African coast, and from three degrees of south latitude quite northward to the Asiatic coasts, including the Arabian sea and the Gulf of Bengal, the Monsoons blow from September to April on the N.E. and from March to October on the S.W. In the former half-year the wind is more steady and gentle, and the weather clearer, than in the latter six months: and the wind is more strong and steady in the Arabian sea than in the Gulf of Bengal.

12th. Between the island of Madagascar and the coast of Africa, and thence northward as far as the equator, there is a tract, wherein from April to October there is a constant fresh S.S.W. wind; which to the northward changes into the W.S.W. wind, blowing at times in the Arabian sea.

13th. To the eastward of Sumatra and Malacca on the north of the equator, and along the coasts of Cambodia and China, quite through the Philippines, as far as Japan, the Monsoons blow northerly and southerly; the northern one setting in about October or November, and the southern about May: These winds are not quite so certain as those in the Arabian seas.

14th. Between Sumatra and Java to the west, and New Guinea to the east, the same northerly and southerly winds are observed; but the first half year Monsoon inclines to the N.W. and the latter to the S.E. These winds begin a month or six weeks after those in the Chinese sea set in, and are quite as variable.

15th. These contrary winds do not shift from one point to its opposite all at once; and in some places the time of the change is attended with calms, in others by variable winds: and it often happens on the shores of Coromandel and China, towards the end of the Monsoons, that there are more violent storms, greatly resembling the hurricanes in the West Indies; wherein the wind is so excessively strong, that hardly any thing can resist its force.

All navigation in the Indian ocean must necessarily be regulated by these winds; for if the mariners should delay their voyages till the contrary Monsoon begins, they must either sail back, or go into harbour, and wait for the return of the trade-wind.

STEPHEN CRANE

The Open Boat

A Tale intended to be after the fact. Being the experience of four men from the sunk steamer Commodore

I

None of them knew the colour of the sky. Their eyes glanced level, and were fastened upon the waves that swept toward them. These waves were of the hue of slate, save for the tops, which were of foaming white, and all of the men knew the colors of the sea. The horizon narrowed and widened, and dipped and rose, and at all times its edge was jagged with waves that seemed thrust up in points like rocks. Many a man ought to have a bath-tub larger than the boat which here rode upon the sea. These waves were most wrongfully and barbarously abrupt and tall, and each froth-top was a problem in small-boat navigation.

The cook squatted in the bottom and looked with both eyes at the six inches of gunwale which separated him from the ocean. His sleeves were rolled over his fat forearms, and the two flaps of his unbuttoned vest dangled as he bent to bail out the boat. Often he said: 'Gawd! That was a narrow clip.' As he remarked it he invariably gazed eastward over the broken sea.

The oiler, steering with one of the two oars in the boat, sometimes raised himself suddenly to keep clear of water that swirled in over the stern. It was a thin little oar and it seemed often ready to snap.

The correspondent, pulling at the other oar, watched the waves and wondered why he was there.

The injured captain, lying in the bow, was at this time buried in that profound dejection and indifference which comes, temporarily at least, to even the bravest and most enduring when, willy nilly, the firm fails, the army loses, the ship goes down. The mind of the master of a vessel is rooted deep in the timbers of her, though he commanded for a day or a decade, and this captain had on him the stern impression of a scene in the greys of dawn of seven turned faces, and later a stump of a top-mast with a white ball on it that slashed to and fro at the waves, went low and lower, and down. Thereaster there was something strange in his voice. Although steady, it was deep with mourning, and of a quality beyond oration or tears.

'Keep 'er a little more south, Billie,' said he.

'A little more south, sir,' said the oiler in the stern.

A seat in this boat was not unlike a seat upon a bucking broncho, and by the same token, a broncho is not much smaller. The craft pranced and reared, and plunged like an animal. As each wave came, and she rose for it, she seemed like a horse making at a fence outrageously high. The manner of her scramble over these walls of water is a mystic thing, and, moreover, at the top of them were ordinarily these problems in white water, the foam racing down from the summit of each wave, requiring a new leap, and a leap from the air. Then, after scornfully bumping a crest, she would slide, and race, and splash down a long incline, and arrive bobbing and nodding in front of the next menace.

A singular disadvantage of the sea lies in the fact that after successfully surmounting one wave you discover that there is another behind it just as important and just as nervously anxious to do something effective in the way of swamping boats. In a ten-foot dinghy one can get an idea of the resources of the sea in the line of waves that is not probable to the average experience which is never at sea in a dinghy. As each slatey wall of water approached, it shut all else from the view of the men in the boat, and it was not difficult to imagine that this particular wave was the final outburst of the ocean, the last effort of the grim water. There was a terrible grace in the move of the waves, and they came in silence, save for the snarling of the crests.

In the wan light, the faces of the men must have been grey. Their eyes must have glinted in strange ways as they gazed steadily astern. Viewed from a balcony, the whole thing would doubtless have been weirdly picturesque. But the men in the boat had no time to see it, and if they had had leisure there were other things to occupy their minds. The sun swung steadily up the sky, and they knew it was broad day because the colour of the sea changed from slate to emerald-green, streaked with amber lights, and the foam was like tumbling snow. The process of the breaking day was unknown to them. They were aware only of this effect upon the colour of the waves that rolled toward them.

In disjointed sentences the cook and the correspondent argued as to the difference between a life-saving station and a house of refuge. The cook had said: 'There's a house of refuge just north of the Mosquito Inlet Light, and as soon as they see us, they'll come off in their boat and pick us up.'

'As soon as who see us?' said the correspondent.

'The crew,' said the cook.

'Houses of refuge don't have crews,' said the correspondent. 'As I understand them, they are only places where clothes and grub are stored for the benefit of shipwrecked people. They don't carry crews.'

'Oh, yes, they do,' said the cook.

'No, they don't,' said the correspondent.

'Well, we're not there yet, anyhow,' said the oiler, in the stern.

'Well,' said the cook, 'perhaps it's not a house of refuge that I'm thinking of as being near Mosquito Inlet Light. Perhaps it's a life-saving station.'

'We're not there yet,' said the oiler, in the stern.

II

As the boat bounced from the top of each wave, the wind tore through the hair of the hatless men, and as the craft plopped her stern down again the spray splashed past them. The crest of each of these waves was a hill, from the top of which the men surveyed, for a moment, a broad tumultuous expanse, shining and wind-riven. It was probably splendid. It was probably glorious, this play of the free sea, wild with lights of emerald and white and amber.

'Bully good thing it's an on-shore wind,' said the cook; 'If not, where would we be? Wouldn't have a show.'

'That's right,' said the correspondent.

The busy oiler nodded his assent.

Then the captain, in the bow, chuckled in a way that expressed humour, contempt, tragedy, all in one. 'Do you think we've got much of a show now, boys?' said he.

Whereupon the three were silent, save for a trifle of hemming and hawing. To express any particular optimism at this time they felt to be childish and stupid, but they all doubtless possessed this sense of the situation in their mind. A young man thinks doggedly at such times. On the other hand, the ethics of their condition was decidedly against any open suggestion of hopelessness. So they were silent.

'Oh, well,' said the captain, soothing his children, 'We'll get ashore all right.'

But there was that in his tone which made them think, so the oiler quoth: 'Yes! If this wind holds!'

The cook was bailing: 'Yes! If we don't catch hell in the surf.'

Canton flannel gulls flew near and far. Sometimes they sat down on the sea, near patches of brown seaweed that rolled on the waves with a movement like carpets on a line in a gale. The birds sat comfortably in groups, and they were envied by some in the dinghy, for the wrath of the sea was no more to them than it was to a covey of prairie chickens a thousand miles inland. Often they came very close and stared at the men with black bead-like eyes. At these times they were uncanny and sinister in their unblinking scrutiny, and the men hooted angrily at them, telling them to be gone. One came, and evidently decided to alight on the top of the captain's head. The bird flew parallel to the boat and did not circle, but made short sidelong jumps in the air in chicken-fashion. His black eyes were wistfully fixed upon the captain's head. 'Ugly brute,' said the oiler to the bird. 'You look as if you were made with a jack-knife.' The cook and the correspondent swore darkly at the creature. The captain naturally wished to knock it away with the end of the heavy painter; but he did not dare do it, because anything resembling an emphatic gesture would have capsized this freighted boat, and so with his open hand, the captain gently and carefully waved the gull away. After it had been discouraged from the pursuit the captain breathed easier on account of his hair, and others breathed easier because the bird struck their minds at this time as being somehow gruesome and ominous.

In the meantime the oiler and the correspondent rowed. And also they rowed.

They sat together in the same seat, and each rowed an oar. Then the oiler took both oars; then the correspondent took both oars; then the oiler; then the correspondent. They rowed and they rowed. The very ticklish part of the business was when the time came for the reclining one in the stern to take his turn at the oars. By the very last star of truth, it is easier to steal eggs from under a hen than it was to change seats in the dinghy. First the man in the stern slid his hand along the thwart and moved with care, as if he were of Sevres. Then the man in the rowing seat slid his hand along the other thwart. It was all done with most extraordinary care. As the two sidled past each other, the whole party kept watchful eyes on the coming wave, and the captain cried: 'Look out now! Steady there!'

The brown mats of seaweed that appeared from time to time were like islands, bits of earth. They were travelling, apparently, neither one way nor the other. They were, to all intents, stationary. They informed the men in the boat that it was making progress slowly toward the land.

The captain, rearing cautiously in the bow, after the dinghy soared on a great swell, said that he had seen the light-house at Mosquito Inlet. Presently the cook remarked that he had seen it. The correspondent was at the oars then, and for some reason he too wished to look at the lighthouse, but his back was toward the far shore and the waves were important, and for some time he could not seize an opportunity to turn his head. But at last there came a wave more gentle than the others, and when at the crest of it he swiftly scoured the western horizon.

'See it?' said the captain.

'No,' said the correspondent slowly, 'I didn't see anything.'

'Look again,' said the captain. He pointed. 'It's exactly in that direction.'

At the top of another wave, the correspondent did as he was bid, and this time his eyes chanced on a small still thing on the edge of the swaying horizon. It was precisely like the point of a pin. It took an anxious eye to find a light house so tiny.

'Think we'll make it, captain?'

'If this wind holds and the boat don't swamp, we can't do much else,' said the captain.

The little boat, lifted by each towering sea, and splashed viciously by the crests, made progress that in the absence of seaweed was not apparent to those in her. She seemed just a wee thing wallowing, miraculously top-up, at the mercy of five oceans. Occasionally, a great spread of water, like white flames, swarmed into her.

'Bail her, cook,' said the captain serenely.

'All right, captain,' said the cheerful cook.

III

It would be difficult to describe the subtle brotherhood of men that was here established on the seas. No one said that it was so. No one mentioned it. But it dwelt in the boat, and each man felt it warm him. They were a captain, an oiler, a cook, and a correspondent, and they were friends, friends in a more curiously iron-bound degree than may be common. The hurt captain, lying against the water-jar in the bow, spoke always in a low voice and calmly, but he could never command a more ready and swiftly obedient crew than the motley three of the dinghy. It was more than a mere recognition of what was best for the common safety. There was surely in it a quality that was personal and heartfelt. And after this devotion to the commander of the boat there was this comradeship that the correspondent, for instance, who had been taught to be cynical of men, knew even at the time was the best experience of his life. But no one said that it was so. No one mentioned it.

'I wish we had a sail,' remarked the captain. 'We might try my overcoat on the end of an oar and give you two boys a chance to rest.' So the cook and the correspondent held the mast and spread wide the overcoat. The oiler steered, and the little boat made good way with her new rig. Sometimes the oiler had to scull sharply to keep a sea from breaking into the boat, but otherwise sailing was a success.

Meanwhile the lighthouse had been growing slowly larger. It had now almost assumed colour, and appeared like a little grey shadow on the sky. The man at the oars could not be prevented from turning his head rather often to try for a glimpse of this little grey shadow.

At last, from the top of each wave the men in the tossing boat could see land. Even as the lighthouse was an upright shadow on the sky, this land seemed but a long black shadow on the sea. It certainly was thinner than paper. 'We must be about opposite New Smyrna,' said the cook, who had coasted this shore often in schooners. 'Captain, by the way, I believe they abandoned that life-saving station there about a year ago.'

'Did they?' said the captain.

The wind slowly died away. The cook and the correspondent were not now obliged to slave in order to hold high the oar. But the waves continued their old impetuous swooping at the dinghy, and the little craft, no longer under way, struggled woundily over them. The oiler or the correspondent took the oars again.

Shipwrecks are a propos of nothing. If men could only train for them and have them occur when the men had reached pink condition, there would be less drowning at sea. Of the four in the dinghy none had slept any time worth mentioning for two days and two nights previous to embarking in the dinghy, and in the excitement of clambering about the deck of a foundering ship they had also forgotten to eat heartily.

For these reasons, and for others, neither the oiler nor the correspondent was fond of rowing at this time. The correspondent wondered ingenuously how in the name of all that was sane could there be people who thought it amusing to row a boat. It was not an amusement; it was a diabolical punishment, and even a genius of mental aberrations could never conclude that it was anything but a horror to the muscles and a crime against the back. He mentioned to the boat in general how the amusement of rowing struck him, and the weary-faced oiler smiled in full sympathy. Previously to the foundering, by the way, the oiler had worked double-watch in the engine-room of the ship.

'Take her easy, now, boys,' said the captain. 'Don't spend yourselves. If we have to run a surf you'll need all your strength, because we'll sure have to swim for it. Take your time.'

Slowly the land arose from the sea. From a black line it became a line of black and a line of white, trees and sand. Finally, the captain said that he could make out a house on the shore. 'That's the house of refuge, sure,' said the cook. 'They'll see us before long, and come out after us.'

The distant lighthouse reared high. 'The keeper ought to be able to make us out now, if he's looking through a glass,' said the captain. 'He'll notify the life-saving people.'

'None of those other boats could have got ashore to give word of the wreck,' said the oiler, in a low voice. 'Else the lifeboat would be out hunting us.'

Slowly and beautifully the land loomed out of the sea. The wind came again. It had veered from the north-east to the south-east. Finally, a new sound struck the ears of the men in the boat. It was the low thunder of the surf on the shore. 'We'll never be able to make the lighthouse now,' said the captain. 'Swing her head a little more north, Billie,' said he.

'A little more north, sir,' said the oiler.

Whereupon the little boat turned her nose once more down the wind, and all but the oarsman watched the shore grow. Under the influence of this expansion doubt and direful apprehension was leaving the minds of the men. The management of the boat was still most absorbing, but it could not prevent a quiet cheerfulness. In an hour, perhaps, they would be ashore.

Their backbones had become thoroughly used to balancing in the boat, and they now rode this wild colt of a dinghy like circus men. The correspondent thought that he had been drenched to the skin, but happening to feel in the top pocket of his coat, he found therein eight cigars. Four of them were soaked with sea-water; four were perfectly scathless. After a search, somebody produced three dry matches, and thereupon the four waifs rode impudently in their little boat, and with an assurance of an impending rescue shining in their eyes, puffed at the big cigars and judged well and ill of all men. Everybody took a drink of water.

IV

'Cook,' remarked the captain, 'there don't seem to be any signs of life about your house of refuge.'

'No,' replied the cook. 'Funny they don't see us!'

A broad stretch of lowly coast lay before the eyes of the men. It was of dunes topped with dark vegetation. The roar of the surf was plain, and sometimes they could see the white lip of a wave as it spun up the beach. A tiny house was blocked out black upon the sky. Southward, the slim lighthouse lifted its little grey length.

Tide, wind, and waves were swinging the dinghy northward. 'Funny they don't see us,' said the men.

The surf's roar was here dulled, but its tone was, nevertheless, thunderous and mighty. As the boat swam over the great rollers, the men sat listening to this roar. 'We'll swamp sure,' said everybody.

It is fair to say here that there was not a life-saving station within twenty miles in either direction, but the men did not know this fact, and in consequence they made dark and opprobrious remarks concerning the eyesight of the nation's life-savers. Four scowling men sat in the dinghy and surpassed records in the invention of epithets.

'Funny they don't see us.'

The lightheartedness of a former time had completely faded. To their sharpened minds it was easy to conjure pictures of all kinds of incompetency and blindness and, indeed, cowardice. There was the shore of the populous land, and it was bitter and bitter to them that from it came no sign.

'Well,' said the captain, ultimately, 'I suppose we'll have to make a try for ourselves. If we stay out here too long, we'll none of us have strength left to swim after the boat swamps.'

And so the oiler, who was at the oars, turned the boat straight for the shore. There was a sudden tightening of muscle. There was some thinking.

'If we don't all get ashore—' said the captain. 'If we don't all get ashore, I suppose you fellows know where to send news of my finish?'

They then briefly exchanged some addresses and admonitions. As for the reflections of the men, there was a great deal of rage in them. Perchance they might be formulated thus: 'If I am going to be drowned – if I am going to be drowned – if I am going to be drowned, why, in the name of the seven mad gods who rule the sea, was I allowed to come thus far and contemplate sand and trees? Was I brought here merely to have my nose dragged away as I was about to nibble the sacred cheese of life? It is preposterous. If this old ninny-woman, Fate, cannot do better than this, she should be deprived of the management of men's fortunes. She is an old hen who knows not her intention. If she has decided to drown me, why did she not do it in the beginning and save me all this trouble? The whole affair is absurd . . . But no, she cannot mean to drown me. She dare not drown me. She cannot drown me. Not after all this work.' Afterward the man might have had an impulse to shake his fist at the clouds: 'Just you drown me, now, and then hear what I call you!'

The billows that came at this time were more formidable. They seemed always just about to break and roll over the little boat in a turmoil of foam. There was a preparatory and long growl in the speech of them. No mind unused to the sea would have concluded that the dinghy could ascend these sheer heights in time. The shore was still afar. The oiler was a wily surfman. 'Boys,' he said swiftly, 'she won't live three minutes more, and we're too far out to swim. Shall I take her to sea again, captain?'

'Yes! Go ahead!' said the captain.

This oiler, by a series of quick miracles, and fast and steady oarsmanship, turned the boat in the middle of the surf and took her safely to sea again.

There was a considerable silence as the boat bumped over the furrowed sea to deeper water. Then somebody in gloom spoke. 'Well, anyhow, they must have seen us from the shore by now.'

The gulls went in slanting flight up the wind toward the grey desolate east. A squall, marked by dingy clouds, and clouds brick-red, like smoke from a burning building, appeared from the south-east.

'What do you think of those life-saving people? Ain't they peaches?'

'Funny they haven't seen us.'

'Maybe they think we're out here for sport! Maybe they think we're fishin'. Maybe they think we're damned fools.'

It was a long afternoon. A changed tide tried to force them southward, but the wind and wave said northward. Far ahead, where coastline, sea, and sky formed their mighty angle, there were little dots which seemed to indicate a city on the shore.

'St. Augustine?'

The captain shook his head. 'Too near Mosquito Inlet.'

And the oiler rowed, and then the correspondent rowed. Then the oiler rowed. It was a weary business. The human back can become the seat of more aches and pains than are registered in books for the composite anatomy of a regiment. It is a limited area, but it can become the theatre of innumerable muscular conflicts, tangles, wrenches, knots, and other comforts.

'Did you ever like to row, Billie?' asked the correspondent.

'No,' said the oiler. 'Hang it!'

When one exchanged the rowing-seat for a place in the bottom of the boat, he suffered a bodily depression that caused him to be careless of everything save an obligation to wiggle one finger. There was cold sea-water swashing to and fro in the boat, and he lay in it. His head, pillowed on a thwart, was within an inch of the swirl of a wave crest, and sometimes a particularly obstreperous sea came in-board and drenched him once more. But these matters did not annoy him. It is almost certain that if the boat had capsized he would have tumbled comfortably out upon the ocean as if he felt sure that it was a great soft mattress.

'Look! There's a man on the shore!'

'Where?'

'There! See 'im? See 'im?'

'Yes, sure! He's walking along.'

'Now he's stopped. Look! He's facing us!'

'He's waving at us!'

'So he is! By thunder!'

'Ah, now we're all right! Now we're all right! There'll be a boat out here for us in half-an-hour.'

'He's going on. He's running. He's going up to that house there.'

The remote beach seemed lower than the sea, and it required a searching glance to discern the little black figure. The captain saw a floating stick and they rowed to it. A bath towel was by some weird chance in the boat, and, tying this on the stick, the captain waved it. The oarsman did not dare turn his head, so he was obliged to ask questions.

'What's he doing now?'

'He's standing still again. He's looking, I think . . . There he goes again. Toward the house . . . Now he's stopped again.'

'Is he waving at us?'

'No, not now! he was, though.'

'Look! There comes another man!'

'He's running.'

'Look at him go, would you.'

'Why, he's on a bicycle. Now he's met the other man. They're both waving at us. Look!'

'There comes something up the beach.'

'What the devil is that thing?'

'Why it looks like a boat.'

'Why, certainly it's a boat.'

'No, it's on wheels.'

'Yes, so it is. Well, that must be the life-boat. They drag them along shore on a wagon.'

'That's the life-boat, sure.'

'No, by—, it's – it's an omnibus.'

'I tell you it's a life-boat.'

'It is not! It's an omnibus. I can see it plain. See? One of these big hotel omnibuses.'

'By thunder, you're right. It's an omnibus, sure as fate. What do you suppose they are doing with an omnibus? Maybe they are going around collecting the life-crew, hey?'

'That's it, likely. Look! There's a fellow waving a little black flag. He's standing on the steps of the omnibus. There come those other two fellows. Now they're all talking together. Look at the fellow with the flag. Maybe he ain't waving it.'

'That ain't a flag, is it? That's his coat. Why, certainly, that's his coat.'

'So it is. It's his coat. He's taken it off and is waving it around his head. But would you look at him swing it.'

'Oh, say, there isn't any life-saving station there. That's just a winter resort hotel omnibus that has brought over some of the boarders to see us drown.'

'What's that idiot with the coat mean? What's he signalling, anyhow?'

'It looks as if he were trying to tell us to go north. There must be a life-saving station up there.'

'No! He thinks we're fishing. Just giving us a merry hand. See? Ah, there, Willie!'

'Well, I wish I could make something out of those signals. What do you suppose he means?'

'He don't mean anything. He's just playing.'

'Well, if he'd just signal us to try the surf again, or to go to sea and wait, or go north, or go south, or go to hell – there would be some reason in it. But look at him. He just stands there and keeps his coat revolving like a wheel. The ass!'

'There come more people.'

'Now there's quite a mob. Look! Isn't that a boat?'

'Where? Oh, I see where you mean. No, that's no boat.'

'That fellow is still waving his coat.'

'He must think we like to see him do that. Why don't he quit it? It don't mean anything.'

'I don't know. I think he is trying to make us go north. It must be that there's a life-saving station there somewhere.'

'Say, he ain't tired yet. Look at 'im wave.'

'Wonder how long he can keep that up. He's been revolving his coat ever since he caught sight of us. He's an idiot. Why aren't they getting men to bring a boat out? A fishing boat – one of those big yawls – could come out here all right. Why don't he do something?'

'Oh, it's all right, now.'

'They'll have a boat out here for us in less than no time, now that they've seen us.'

A faint yellow tone came into the sky over the low land. The shadows on the sea slowly deepened. The wind bore coldness with it, and the men began to shiver.

'Holy smoke!' said one, allowing his voice to express his impious mood, 'If we keep on monkeying out here! If we've got to flounder out here all night!'

'Oh, we'll never have to stay here all night! Don't you worry. They've seen us now, and it won't be long before they'll come chasing out after us.'

The shore grew dusky. The man waving a coat blended gradually into this gloom, and it swallowed in the same manner the omnibus and the group of people. The spray, when it dashed uproariously over the side, made the voyagers shrink and swear like men who were being branded.

'I'd like to catch the chump who waved the coat. I feel like soaking him one, just for luck.'

'Why? What did he do?'

'Oh, nothing, but then he seemed so damned cheerful.'

In the meantime the oiler rowed, and then the correspondent rowed, and then the oiler rowed. Grey-faced and bowed forward, they mechanically, turn by turn, plied the leaden oars. The form of the lighthouse had vanished from the southern horizon, but finally a pale star appeared, just lifting from the sea. The streaked saffron in the west passed before the all-merging darkness, and the sea to the east was black. The land had vanished, and was expressed only by the low and drear thunder of the surf.

'If I am going to be drowned – if I am going to be drowned – if I am going to be drowned, why, in the name of the seven mad gods who rule the sea, was I allowed to come thus far and contemplate sand and trees? Was I brought here merely to have my nose dragged away as I was about to nibble the sacred cheese of life?'

The patient captain, drooped over the water-jar, was sometimes obliged to speak to the oarsman.

'Keep her head up! Keep her head up!'

' "Keep her head up," sir.' The voices were weary and low.

This was surely a quiet evening. All save the oarsman lay heavily and listlessly in the boat's bottom. As for him, his eyes were just capable of noting the tall black waves that swept forward in a most sinister silence, save for an occasional subdued growl of a crest.

The cook's head was on a thwart, and he looked without interest at the water under his nose. He was deep in other scenes. Finally he spoke. 'Billie,' he murmured, dreamfully, 'what kind of pie do you like best?'

V

'Pie,' said the oiler and the correspondent, agitatedly. 'Don't talk about those things, blast you!'

'Well,' said the cook, 'I was just thinking about ham sandwiches, and—'

A night on the sea in an open boat is a long night. As darkness settled finally, the shine of the light, lifting from the sea in the south, changed to full gold. On the northern horizon a new light appeared, a small bluish gleam on the edge of the waters. These two lights were the furniture of the world. Otherwise there was nothing but waves.

Two men huddled in the stern, and distances were so magnificent in the dinghy that the rower was enabled to keep his feet partly warmed by thrusting them under his companions. Their legs indeed extended far under the rowing-seat until they touched the feet of the captain forward. Sometimes, despite the efforts of the tired oarsman, a wave came piling into the boat, an icy wave of the night, and the chilling water soaked them anew. They would twist their bodies for a moment and groan, and sleep the dead sleep once more, while the water in the boat gurgled about them as the craft rocked.

The plan of the oiler and the correspondent was for one to row until he lost the ability, and then arouse the other from his sea-water couch in the bottom of the boat.

The oiler plied the oars until his head drooped forward, and the overpowering sleep blinded him. And he rowed yet afterward. Then he touched a man in the bottom of the boat, and called his name. 'Will you spell me for a little while?' he said, meekly.

'Sure, Billie,' said the correspondent, awakening and dragging himself to a sitting position. They exchanged places carefully, and the oiler, cuddling down in the sea-water at the cook's side, seemed to go to sleep instantly.

The particular violence of the sea had ceased. The waves came without snarling. The obligation of the man at the oars was to keep the boat headed so that the tilt of the rollers would not capsize her, and to preserve her from filling when the crests rushed past. The black waves were silent and hard to be seen in the darkness. Often one was almost upon the boat before the oarsman was aware.

In a low voice the correspondent addressed the captain. He was not sure that the captain was awake, although this iron man seemed to be always awake. 'Captain, shall I keep her making for that light north, sir?'

The same steady voice answered him. 'Yes. Keep it about two points off the port bow.'

The cook had tied a life-belt around himself in order to get even the warmth which this clumsy cork contrivance could donate, and he seemed almost stove-like when a rower, whose teeth invariably chattered wildly as soon as he ceased his labour, dropped down to sleep.

The correspondent, as he rowed, looked down at the two men sleeping under-foot. The cook's arm was around the oiler's shoulders, and, with their fragmentary clothing and haggard faces, they were the babes of the sea, a grotesque rendering of the old babes in the wood.

Later he must have grown stupid at his work, for suddenly there was a growling of water, and a crest came with a roar and a swash into the boat, and it was a wonder that it did not set the cook afloat in his life-belt. The cook continued to sleep, but the oiler sat up, blinking his eyes and shaking with the new cold.

'Oh, I'm awful sorry, Billie,' said the correspondent contritely.

'That's all right, old boy,' said the oiler, and lay down again and was asleep.

Presently it seemed that even the captain dozed, and the correspondent thought that he was the one man afloat on all the oceans. The wind had a voice as it came over the waves, and it was sadder than the end.

There was a long, loud swishing astern of the boat, and a gleaming trail of phosphorescence, like blue flame, was furrowed on the black waters. It might have been made by a monstrous knife.

Then there came a stillness, while the correspondent breathed with the open mouth and looked at the sea.

Suddenly there was another swish and another long flash of bluish light, and this time it was alongside the boat, and might almost have been reached with an oar. The correspondent saw an enormous fin speed like a shadow through the water, hurling the crystalline spray and leaving the long glowing trail.

The correspondent looked over his shoulder at the captain. His face was hidden, and he seemed to be asleep. He looked at the babes of the sea. They certainly were asleep. So, being bereft of sympathy, he leaned a little way to one side and swore softly into the sea.

But the thing did not then leave the vicinity of the boat. Ahead or astern, on one side or the other, at intervals long or short, fled the long sparkling streak, and there was to be heard the whirroo of the dark fin. The speed and power of the thing was greatly to be admired. It cut the water like a gigantic and keen projectile.

The presence of this biding thing did not affect the man with the same horror that it would if he had been a picnicker. He simply looked at the sea dully and swore in an undertone.

Nevertheless, it is true that he did not wish to be alone. He wished one of his companions to awaken by chance and keep him company with it. But the captain hung motionless over the water-jar, and the oiler and the cook in the bottom of the boat were plunged in slumber.

VI

'If I am going to be drowned – if I am going to be drowned – if I am going to be drowned, why, in the name of the seven mad gods who rule the sea, was I allowed to come thus far and contemplate sand and trees?'

During this dismal night, it may be remarked that a man would conclude that it was really the intention of the seven mad gods to drown him, despite the abominable injustice of it. For it was certainly an abominable injustice to drown a man who had worked so hard, so hard. The man felt it would be a crime most unnatural. Other people had drowned at sea since galleys swarmed with painted sails, but still —

When it occurs to a man that nature does not regard him as important, and that she feels she would not maim the universe by disposing of him, he at first wishes to throw bricks at the temple, and he hates deeply the fact that there are no bricks and no temples. Any visible expression of nature would surely be pelleted with his jeers.

Then, if there be no tangible thing to hoot he feels, perhaps, the desire to confront a personification and indulge in pleas, bowed to one knee, and with hands supplicant, saying: 'Yes, but I love myself.'

A high cold star on a winter's night is the word he feels that she says to him. Thereaster he knows the pathos of his situation.

The men in the dinghy had not discussed these matters, but each had, no doubt, reflected upon them in silence and according to his mind. There was seldom any expression upon their faces save the general one of complete weariness. Speech was devoted to the business of the boat.

To chime the notes of his emotion, a verse mysteriously entered the correspondent's head. He had even forgotten that he had forgotten this verse, but it suddenly was in his mind.

> A soldier of the Legion lay dying in Algiers,
> There was a lack of woman's nursing, there was dearth of woman's tears;
> But a comrade stood beside him, and he took that comrade's hand,
> And he said: 'I shall never see my own, my native land.'

In his childhood, the correspondent had been made acquainted with the fact that a soldier of the Legion lay dying in Algiers, but he had never regarded the fact as important. Myriads of his school-fellows had informed him of the soldier's plight, but the dinning had naturally ended by making him perfectly indifferent. He had never considered it his affair that a soldier of the Legion lay dying in Algiers, nor had it appeared to him as a matter for sorrow. It was less to him than the breaking of a pencil's point.

Now, however, it quaintly came to him as a human, living thing. It was no longer merely a picture of a few throes in the breast of a poet, meanwhile drinking tea and warming his feet at the grate; it was an actuality – stern, mournful, and fine.

The correspondent plainly saw the soldier. He lay on the sand with his feet out straight and still. While his pale left hand was upon his chest in an attempt to thwart the going of his life, the blood came between his fingers. In the far Algerian distance, a city of low square forms was set against a sky that was faint with the last sunset hues. The correspondent, plying the oars and dreaming of the slow and slower movements of the lips of the soldier, was moved by a profound and perfectly impersonal comprehension. He was sorry for the soldier of the Legion who lay dying in Algiers.

The thing which had followed the boat and waited, had evidently grown bored at the delay. There was no longer to be heard the slash of the cut-water, and there was no longer the flame of the long trail. The light in the north still glimmered, but it was apparently no nearer to the boat. Sometimes the boom of the surf rang in the correspondent's ears, and he turned the craft seaward then and rowed harder. Southward, some one had evidently built a watch-fire on the beach. It was too low and too far to be seen, but it made a shimmering, roseate reflection upon the bluff back of it, and this could be discerned from the boat. The wind came stronger, and sometimes a wave suddenly raged out like a mountain-cat, and there was to be seen the sheen and sparkle of a broken crest.

The captain, in the bow, moved on his water-jar and sat erect. 'Pretty long night,' he observed to the correspondent. He looked at the shore. 'Those life-saving people take their time.'

'Did you see that shark playing around?'

'Yes, I saw him. He was a big fellow, all right.'

'Wish I had known you were awake.'

Later the correspondent spoke into the bottom of the boat.

'Billie!' There was a slow and gradual disentanglement. 'Billie, will you spell me?'

'Sure,' said the oiler.

As soon as the correspondent touched the cold comfortable sea-water in the bottom of the boat, and had huddled close to the cook's life-belt he was deep in sleep, despite the fact that his teeth played all the popular airs. This sleep was so good to him that it was but a moment before he heard a voice call his name in a tone that demonstrated the last stages of exhaustion. 'Will you spell me?'

'Sure, Billie.'

The light in the north had mysteriously vanished, but the correspondent took his course from the wide-awake captain.

Later in the night they took the boat farther out to sea, and the captain directed the cook to take one oar at the stern and keep the boat facing the seas. He was to call out if he should hear the thunder of the surf. This plan enabled the oiler and the correspondent to get respite together. 'We'll give those boys a chance to get into shape again,' said the captain. They curled down and, after a few preliminary chatterings and trembles, slept once more the dead sleep. Neither knew they had bequeathed to the cook the company of another shark, or perhaps the same shark.

As the boat caroused on the waves, spray occasionally bumped over the side and gave them a fresh soaking, but this had no power to break their repose. The ominous slash of the wind and the water affected them as it would have affected mummies.

'Boys,' said the cook, with the notes of every reluctance in his voice, 'she's drifted in pretty close. I guess one of you had better take her to sea again.' The correspondent, aroused, heard the crash of the toppled crests.

As he was rowing, the captain gave him some whisky-and-water, and this steadied the chills out of him. 'If I ever get ashore and anybody shows me even a photograph of an oar—'

At last there was a short conversation.

'Billie . . . Billie, will you spell me?'

'Sure,' said the oiler.

VII

When the correspondent again opened his eyes, the sea and the sky were each of the grey hue of the dawning. Later, carmine and gold was painted upon the waters. The morning appeared finally, in its splendour, with a sky of pure blue, and the sunlight flamed on the tips of the waves.

On the distant dunes were set many little black cottages, and a tall white windmill reared above them. No man, nor dog, nor bicycle appeared on the beach. The cottages might have formed a deserted village.

The voyagers scanned the shore. A conference was held in the boat. 'Well,' said the captain, 'if no help is coming we might better try a run through the surf right away. If we stay out here much longer we will be too weak to do anything for ourselves at all.' The others silently acquiesced in this reasoning. The boat was headed for the beach. The correspondent wondered if none ever ascended the tall wind-tower, and if then they never looked seaward. This tower was a giant, standing with its back to the plight of the ants. It represented in a degree, to the correspondent, the serenity of nature amid the struggles of the individual – nature in the wind, and nature in the vision of men. She did not seem cruel to him then, nor beneficent, nor treacherous, nor wise. But she was indifferent, flatly indifferent. It is, perhaps, plausible that a man in this situation, impressed with the unconcern of the universe, should see the innumerable flaws of his life, and have them taste wickedly in his mind and wish for another chance. A distinction between right and wrong seems absurdly clear to him, then, in this new ignorance of the grave-edge, and he understands that if he were given another opportunity he would mend his conduct and his words, and be better and brighter during an introduction or at a tea.

'Now, boys,' said the captain, 'she is going to swamp, sure. All we can do is to work her in as far as possible, and then when she swamps, pile out and scramble for the beach. Keep cool now, and don't jump until she swamps sure.'

The oiler took the oars. Over his shoulders he scanned the surf. 'Captain,' he said, 'I think I'd better bring her about, and keep her head-on to the seas and back her in.'

'All right, Billie,' said the captain. 'Back her in.' The oiler swung the boat then and, seated in the stern, the cook and the correspondent were obliged to look over their shoulders to contemplate the lonely and indifferent shore.

The monstrous in-shore rollers heaved the boat high until the men were again enabled to see the white sheets of water scudding up the slanted beach. 'We won't get in very close,' said the captain. Each time a man could wrest his attention from the rollers, he turned his glance toward the shore, and in the expression of the eyes during this contemplation there was a singular quality. The correspondent, observing the others, knew that they were not afraid, but the full meaning of their glances was shrouded.

As for himself, he was too tired to grapple fundamentally with the fact. He tried to coerce his mind into thinking of it, but the mind was dominated at this time by the muscles, and the muscles said they did not care. It merely occurred to him that if he should drown it would be a shame.

There were no hurried words, no pallor, no plain agitation. The men simply looked at the shore. 'Now, remember to get well clear of the boat when you jump,' said the captain.

Seaward the crest of a roller suddenly fell with a thunderous crash, and the long white comber came roaring down upon the boat.

'Steady now,' said the captain. The men were silent. They turned their eyes from the shore to the comber and waited. The boat slid up the incline, leaped at the furious top, bounced over it, and swung down the long back of the wave. Some water had been shipped and the cook bailed it out.

But the next crest crashed also. The tumbling, boiling flood of white water caught the boat and whirled it almost perpendicular. Water swarmed in from all sides. The correspondent had his hands on the gunwale at this time, and when the water entered at that place he swiftly withdrew his fingers, as if he objected to wetting them.

The little boat, drunken with this weight of water, reeled and snuggled deeper into the sea.

'Bail her out, cook! Bail her out,' said the captain.

'All right, captain,' said the cook.

'Now, boys, the next one will do for us, sure,' said the oiler. 'Mind to jump clear of the boat.'

The third wave moved forward, huge, furious, implacable. It fairly swallowed the dinghy, and almost simultaneously the men tumbled into the sea. A piece of lifebelt had lain in the bottom of the boat, and as the correspondent went overboard he held this to his chest with his left hand.

The January water was icy, and he reflected immediately that it was colder than he had expected to find it on the coast of Florida. This appeared to his dazed mind as a fact important enough to be noted at the time. The coldness of the water was sad; it was tragic.

This fact was somehow so mixed and confused with his opinion of his own situation that it seemed almost a proper reason for tears. The water was cold.

When he came to the surface he was conscious of little but the noisy water. Afterward he saw his companions in the sea. The oiler was ahead in the race. He was swimming strongly and rapidly. Off to the correspondent's left, the cook's great white and corked back bulged out of the water, and in the rear the captain was hanging with his one good hand to the keel of the overturned dinghy.

There is a certain immovable quality to a shore, and the correspondent wondered at it amid the confusion of the sea.

It seemed also very attractive, but the correspondent knew that it was a long journey, and he paddled leisurely. The piece of life-preserver lay under him, and sometimes he whirled down the incline of a wave as if he were on a handsled.

But finally he arrived at a place in the sea where travel was beset with difficulty. He did not pause swimming to inquire what manner of current had caught him, but there his progress ceased. The shore was set before him like a bit of scenery on a stage, and he looked at it and understood with his eyes each detail of it.

As the cook passed, much farther to the left, the captain was calling to him, 'Turn over on your back, cook! Turn over on your back and use the oar.'

'All right, sir.' The cook turned on his back, and, paddling with an oar, went ahead as if he were a canoe.

Presently the boat also passed to the left of the correspondent with the captain clinging with one hand to the keel. He would have appeared like a man raising himself to look over a board fence, if it were not for the extraordinary gymnastics of the boat. The correspondent marvelled that the captain could still hold to it.

They passed on, nearer to shore – the oiler, the cook, the

captain – and following them went the water-jar, bouncing gaily over the seas.

The correspondent remained in the grip of this strange new enemy – a current. The shore, with its white slope of sand and its green bluff, topped with little silent cottages, was spread like a picture before him. It was very near to him then, but he was impressed as one who in a gallery looks at a scene from Brittany or Holland.

He thought: 'I am going to drown? Can it be possible? Can it be possible? Can it be possible?' Perhaps an individual must consider his own death to be the final phenomenon of nature.

But later a wave perhaps whirled him out of this small, deadly current, for he found suddenly that he could again make progress toward the shore. Later still, he was aware that the captain, clinging with one hand to the keel of the dinghy, had his face turned away from the shore and toward him, and was calling his name. 'Come to the boat! Come to the boat!'

In his struggle to reach the captain and the boat, he reflected that when one gets properly wearied, drowning must really be a comfortable arrangement, a cessation of hostilities accompanied by a large degree of relief, and he was glad of it, for the main thing in his mind for some months had been horror of the temporary agony. He did not wish to be hurt.

Presently he saw a man running along the shore. He was undressing with most remarkable speed. Coat, trousers, shirt, everything flew magically off him.

'Come to the boat,' called the captain.

'All right, captain.' As the correspondent paddled, he saw the captain let himself down to bottom and leave the boat. Then the correspondent performed his one little marvel of the voyage. A large wave caught him and flung him with ease and supreme speed completely over the boat and far beyond it. It struck him even then as an event in gymnastics, and a true miracle of the sea. An over-turned boat in the surf is not a plaything to a swimming man.

The correspondent arrived in water that reached only to his waist, but his condition did not enable him to stand for more than a moment. Each wave knocked him into a heap, and the under-tow pulled at him.

Then he saw the man who had been running and undressing, and undressing and running, come bounding into the water. He dragged ashore the cook, and then waded towards the captain, but the captain waved him away, and sent him to the correspondent. He was naked, naked as a tree in winter, but a halo was about his head, and he shone like a saint. He gave a strong pull, and a long drag, and a bully heave at the correspondent's ha. The correspondent, schooled in the minor formulae, said: 'Thanks, old man.' But suddenly the man cried: 'What's that?' He pointed a swift finger. The correspondent said: 'Go.'

In the shallows, face downward, lay the oiler. His forehead touched sand that was periodically, between each wave, clear of the sea.

The correspondent did not know all that transpired afterward. When he achieved safe ground he fell, striking the sand with each particular part of his body. It was as if he had dropped from a roof, but the thud was grateful to him.

It seems that instantly the beach was populated with men with blankets, clothes, and flasks, and women with coffeepots and all the remedies sacred to their minds. The welcome of the land to the men from the sea was warm and generous, but a still and dripping shape was carried slowly up the beach, and the land's welcome for it could only be the different and sinister hospitality of the grave.

When it came night, the white waves paced to and fro in the moonlight, and the wind brought the sound of the great sea's voice to the men on shore, and they felt that they could then be interpreters.

JONATHAN RABAN

Passage to Juneau

Once upon a time, people made their way across the sea by reading the surface, shapes, and colours of the water. On clear nights, they took their directions from the stars; by day, they sailed by the wind and waves. In the Homeric world there were four reigning winds: Boreas blew from the north, Notus from the south, Eurus from the east, Zephyrus from the west.

Wind made itself most useful for navigational purposes by generating swells. Whatever the fickle gusts of the moment, the prevailing seasonal wind was registered in the stubborn movement of the sea. Swell continues for many days, and sometimes thousands of miles, after the wind that first raised it has blown itself out. Islands, because they deflect the direction of swell, can be 'felt' from a great distance by a sensitive pilot. As the depth of the sea decreases, the swell steepens, warning of imminent landfall.

Sailing by swell entailed an intense concentration on the character of the sea itself. Wave shape was everything. A single wave is likely to be moulded by several forces: the local wind; a dominant, underlying swell; and, often, a weaker swell coming from a third direction. Early navigators had to be in communion with every lift of the bow as the sea swept under the hull in order to sense each component in the wave and deduce from them the existence of unseen masses of land.

David Lewis, a New Zealand-born doctor who gave up his London practice to become a freelance ocean adventurer, sailed in the

1960s with some of the last traditional Polynesian navigators in their outrigger canoes. *We, the Navigators* is his firsthand report, from the Pacific Ocean in the mid-twentieth century, on how sailors like Odysseus crossed the Mediterranean circa 700 BC, before the invention of the magnetic compass. Most importantly, Lewis's book conveys how the open sea could be as intimately known and as friendly to human habitation as a familiar stretch of land to those seamen who lived on its surface, as gulls do, wave by wave.

Sea*men*. For the testicles were, Lewis wrote, the instruments best attuned to picking up slight variations in the rhythm of a swell – a sudden steepening, an interlocking of two opposed wave-trains. Rest your balls lightly on the top of the stempost and feel the jaunting upsurge of the bow, then its sudden, precipitous collapse into the trough . . . As a four-year-old, I keenly anticipated the approach to humpback bridges in my mother's lightly sprung 1938 Ford. Taken a mile or so too fast, each bridge induced a moment of exquisite, unmentionable pleasure; it was like finding a small but energetic tree frog trapped inside one's scrotum. Had I been blindfolded on these car rides in 1946, I believe I could have identified half the humpback bridges in Norfolk by my genitals alone.

So did Lewis's Polynesian friends feel their way across the humpbacked ocean. On these voyages, Lewis – a vastly experienced small-boat sailor – often found himself totally disoriented, as the wind changed direction, the sea got up and the underlying swells became confused or imperceptible. Yet his guides could sense a regular grain in the roughest, most disorderly sea. Time and again they'd sail through fifty or more miles of murky overcast, without sight of the sun, and make a perfect landfall at – in one instance – a narrow passage between islands, breaking into sudden visibility less than two miles off.

Sailing with no instruments, the primitive navigator knew his local sea in the same unselfconscious way that a farmer knows his fields. The stars supplied a grand chart of paths across the known ocean, but there was often little need of these since the water itself was as legible as acreage farmed for generations. Colour, wind, the flight of birds, and telltale variations of swell gave the sea direction, shape, character.

Here, where you feel the intersection of two swells, each deflected by islands far over the horizon, you make your turn . . . Now you search for *toake*, the tropic bird, and follow its homeward flight until the sea begins to brown with sand . . . In *Polynesian Seafaring and Navigation*, Richard Feinberg, an anthropologist, includes a sequence of interviews with navigators from the island of Anuta in the Solomon Islands. One of these, Pu Maevatau, says of sailing under a cloudy sky that 'the expert navigator . . . will make his bearer the ocean.'

That sense of being borne along to your destination by the ocean itself is strong in Homer, whose voyagers are seen as creatures of nature assisted, or impeded, by the gods. When the gods are with you, the winds and the sea conduct you onward, like thistledown blown from wave to wave. For Odysseus, as for the Polynesian navigators in the books of Lewis and Feinberg, the ocean is a place, not a space; its mobile surface full of portents, clues, and meanings. It is as substantial and particular, as crowded with topographical features, as, say, Oxfordshire.

The arrival of the magnetic compass caused a fundamental rift in the relationship between man and sea. Europeans were sailing with compasses in the eleventh century, and may have used them even earlier. Once the compass became established on the quarterdeck, snug in its wooden binnacle, the whole focus of the helmsman shifted, from the sea itself to an instrument eighteen inches or so under his nose. Suddenly he no longer needed to intuit the meaning of the waves; he had become a functionary, whose job was to keep the ship at an unvarying angle to the magnetised pointer with its scrolled N. First he steered by letters, E by S, W by N; later, by numbers assigned him by the officer of the watch. Holding the bow to the sea at a steady 195, the helmsman was performing a task that eventually would be done more efficiently by a machine.

Such a simple invention, or discovery, the compass. One wet Saturday afternoon, I made one for Julia: we rubbed the eye of a sewing needle against a magnet on the fridge door; slipped the needle into a sawn-off drinking straw to make it float; and launched it in a water-filled salad bowl. Breasting the resistance of the surface tension, the needle obediently swung slowly around to align itself with the earth's magnetic field, pointing 21½° east of true north. With the sofa to the south, TV to the north, bookshelves to the east, and dining table to the west, I set Julia to walking the room on a succession of compass courses. Preschool Navigation: Lesson One.

Possibly we were merely replicating, by mechanical means, a piece of equipment that we both already possessed somewhere in our bodies. A recent study conducted at the University of Auckland in New Zealand shows that rainbow trout have built-in sensors composed of magnetite cells, with nerves connecting the sensors to their brains. With a Pavlovian regime of rewards and punishments, the experimenters were able to persuade the fish to swim on any given compass course; for the food pellet, take 195. When last heard of (on the BBC World Service), the scientists were busy dissecting migratory birds, hoping to isolate similar magnetic sensors, and speculating that humans, too, might be born with such navigational devices, at least in vestigial form. If so, the classic description of Columbus, as a man implanted with a compass rose inside his head, will turn out to be a statement of literal fact.

But the external compass – the magic gizmo in a box – put man at a remove from his surroundings. A compass course is a hypothesis.

It has length, but no width. It can't be seen or felt (though once, perhaps, we could feel it, as the rainbow trout appears to). It cannot even be steered. The autopilot on my boat leaves a cleaner, straighter wake than I can manage, yet it keeps 'on course' – as I do – only by making continuous mistakes. Each time the vessel falls sufficiently away from its heading for the autopilot to notice the error, the machine administers a corrective turn of the wheel which points the bow to the far side of the notional course. The wheel, attached to the autopilot's motor by a strop, spins now to port, now to starboard, now to port again, making a monotonous *hee-haw, hee-haw* sound, like an ailing donkey. The real track of the boat through the sea is a weaving zigzag path whose innumerable deviations define the idealized pencil line of the course as it appears on the chart. Steering a compass course, by machine or hand, it is by indirection that one finds direction out.

So the helmsman looked away from the sea, wedding himself instead to a geometrical abstraction that had no tangible reality in nature. Possession of a compass soon rendered obsolete a great body of inherited, instinctual knowledge, and rendered the sea itself – in fair weather, at least – as a void, an empty space to be traversed by a numbered rhumb line.

Too little has been made of this critical moment in the history of navigation. Because the compass has been with us for a thousand years, we've lost sight of the mental revolution it caused. The figure of the helmsman, his eyes glued to the tilting card in its bowl, turning the spokes of the wheel to keep the assigned number on target against the lubberline, is an early avatar of modern man. The compass has turned him into a steering machine. He is the direct ancestor of Thomas MacWhirr, the dim, unimaginative son of an Ulster grocer who captains Nan-Shan in Conrad's *Typhoon*.

Bound for Fu-Chou, on a north-easterly course through the South China Sea, MacWhirr (whose name gives the game away) drives his ship straight through the eye of a hurricane. He has a timetable to meet, and a steam engine with which to meet it; and so he refuses to budge from the course of 040 that leads across the chart to the approaches to Fu-Chou. To MacWhirr, the compass course has become a blind imperative; he cannot deviate from it for a spot of what he calls 'dirty weather'. Jukes, the chief mate, urges him to turn the ship's head to the east, to meet the huge cross-swell that is the first sign of the coming typhoon.

> *'Head to the eastward?' [MacWhirr] said, struggling to sit up. 'That's more than four points off her course.'*
>
> *'Yes, sir. Fifty degrees . . . Would just bring her head far enough round to meet this . . .'*
>
> *Captain MacWhirr was now sitting up. He had not dropped the book, and he had not lost his place.*
>
> *'To the eastward?' he repeated, with dawning astonishment. 'To the . . . Where do you think we are bound to? You want me to haul a full-powered steamship four points off her course to make the Chinamen comfortable! Now, I've heard more than enough of mad things done in the world – but this . . . If I didn't know you, Jukes, I would think you were in liquor.'*

MacWhirr is Conrad's archetype of the modern technological mariner, blithely, ignorantly divorced from nature. His sea is a placid vacancy, its terrors conquered by the compass and the engine. *Nan-Shan*, built of iron at a great industrial shipyard on the Clyde, crosses the globe in inflexible straight lines, reducing the ocean to a neutral medium for the commercial enterprises of men as literal-minded and mechanical as MacWhirr himself. The whirling cyclone that Conrad brews up to engulf the stupid captain and his crew is the ocean's revenge for the hubris of the steam turbine and the ruled line on the chart.

JAMES COOK

Journal of the Second Voyage (1772–1775)

[JANUARY 1773]

SATURDAY *9th. Thermr.* 35. *Winds NW. Lat. in South* 61°36'. *Var. of the Compass* 30°8'. Gentle gales and clowdy. In the PM passed Several Islands of Ice more than we have seen for some days past, and at 9 o'Clock came to one that had a quantity of loose Ice about it, upon which we hauled our Wind with a view to keep to windward in order to take some of it up in the Morn, at Midnight we tacked and stood for the Island, at this time the Wind shifted two or 3 Points to the Northward so that we could not fetch it, we therefore bore away for the next Island to Leeward which we reached by 8 o'Clock and finding loose pieces of Ice about it, we hoisted out three Boats and took up as much as yeilded about 15 Tons of Fresh Water, the Adventure at the same time got about 8 or 9 and all this was done in 5 or 6 hours time; the pieces we took up and which had broke from the Main Island, were very hard and solid, and some of them too large to be handled so that we were obliged to break them with our Ice Axes before they could be taken into the Boats, the Salt Water that adhered to the pieces was so trifleing as not to be tasted and after they had laid on Deck a little while intirely dreaned of, so that the Water which the Ice yeilded was perfectly well tasted, part of the Ice we packed in Casks and the rest we Milted in the Coppers and filled the Casks up with the Water; the Melting of the Ice is a little tideous and takes up some time, otherwise this is the most expeditious way of Watering I ever met with.

SUNDAY 10*th. Winds NWBN to WNW. Course S* 54° *E. Dist. Saild* 37 *Miles. Lat. in South* 61°58'. *Longd. in East pr. Reckg. Corrected* 36°7' *Watch* 35°48'. *Long, made from C.G.H.* 17°44'. Gentle gales, first part fair and Clowdy, remainder hazy with showers of snow. In the PM hoisted in the Boats after having taken up all the loose Ice with which our Decks were full; having got on board this seasonable supply of fresh Water, I did not hesitate one moment whether or no I should steer farther to the South but directed my course South East by South, and as we had once broke the Ice I did not doubt of geting a supply of Water when ever I stood in need. We had not stood above one hour and a half upon the above Course before I found it necessary to keep away more East and before the Swell to prevent the Sloops from rowling occasioned in some measure by the great weight of Ice they had on their Decks which by 9 o'Clock in the Morning was a good deal reduced and the Swell gone down we resumed our former Course.

TUESDAY 12*th. Winds NE, East, & SE. Course S* 18½° *E. Dist. Saild* 63 *Miles. Lat. in South* 64°12'. *Longd. in East pr. Reckg. Corrected* 38°14' *Watch* 37°47'. *Long. made from C.G.H.* 19°51'. *Variation of the Compass* 23°52½'. Gentle gales and Clowdy, at 4 in the AM it was clear and I took 12 observations of the Suns Azimuth with Mr Gregorys Compass which gave 23°39½' West Variation. I also took a like number with two of Dr Knights Compass's, the one gave 23°15' and the other 24°42' West Variation, the Mean of all these Means is 23°52¼'; our Latitude and Longitude was the same as at Noon. At 6 o'Clock, having but little Wind, we brought to a mong some loose Ice, hoisted out the Boats and took up as much as filled all our empty Casks and compleated our Water to 40 Tons, the Adventure at the same time filled all her Empty Casks; while this was doing Mr Forster shott an Albatross whose plumage was of a Dark grey Colour, its head, uper sides of the Wings rather inclining to black with white Eye brows, we first saw of these Birds about the time of our first falling in with these Ice Islands and they have accompanied us ever sence. Some of the Seamen call them Quaker Birds, from their grave Colour. These and a black one with a yellow Bill are our only Companions of the Albatross kind, all the other sorts have quite left us. Some Penguins were seen this morning.

WEDNESDAY 13*th. Winds Southerly – Calm. Course ESE. Dist. Sailed* 16 *Miles. Latitude in South* 64°18'. *Longd. in East Greenwich Reck.g Corrct.* 38°48'. *Longd. made Cape G. Hope* 20°25'. At 4 o'Clock in the PM hoisted in the Boats and made sail to the SE with a gentle gale at SBW attended with Showers of Snow. At 2 am it fell calm, and at 9 hoisted out a Boat to try the Current which we found to set NW near one third of a Mile an hour which is pretty confirmable to what I have before observed in regard to the Currants; this is a point worth inquiring into, for was the direction of the Currants well assertained, we should be no longer at a loss to know from what quarter the Islands of Ice we daily meet with comes from. At the time of trying the Currant Fahrenheits Thermometer was sent down 100 fathom and when it came up the mercury was at 32 which is the freezing point, some little time after, being exposed to the surface of the Sea, it rose to 33½ and in the open air to 36. Some curious and intresting experiments are wanting to know what effect cold has on Sea Water in some of the following instances: does it freeze or does it not? if it does, what degree of cold is necessary and what becomes of the Salt brine? for all the Ice we meet with yeilds Water perfectly sweet and fresh.

SUNDAY 17*th. Thermr.* 34. *Winds EBS. Course South. Dist. Sailed* 125 *Miles. Lot. in South* 66°36½'. *Longde. in E. Greenwich Reck.g* 39°35'. *Long[de] made E. of C.G.H.* 21°12'. In the PM had fresh gales and Clowdy weather. At 6 o'Clock, being then in the Latitude of 64°56' s I found the Variation by Gregorys Compass to be 26°41' West, at this time the Motion of the Ship was so great that I could not observe with Dr Knights Compass. In the AM had hazy weather with Snow Showers and saw but one Island of Ice in the Course of these 24 hours so that we begin to think that we have got into a clear Sea. At about a ¼ past 11 o'Clock we cross'd the Antarctic Circle for at Noon we were by observation four Miles and a half South of it and are undoubtedly the first and only Ship that ever cross'd that line. We now saw several Flocks of the Brown and White Pintadoes which we have named Antarctic Petrels because they seem to be natives of that Region; the White Petrels also appear in greater numbers than of late and some few Dark Grey Albatrosses, our constant companions the Blue Petrels have not forsaken us but the Common Pintadoes have quite disapeared as well as many other sorts which are Common in lower Latitudes.

MONDAY 18*th Winds EBS. Course North. Distce Sailed* 44 *Miles. Lat. in South* 65°52'. *Longde. in East Greenwich Reck.g* 39°35'. *Longde. East Cape G. Hope* 21°12'. In the PM had a Fresh gale and fair Weather. At 4 o'Clock we discoverd from the Mast head thirty eight Islands of Ice extending from the one Bow to the other, that is from the SE to West, and soon after we discovered Feild or Packed Ice in the same Direction and had so many loose pieces about the Ship that we were obliged to loof for one and bear up for another, the number increased so fast upon us that at ¾ past Six, being then in the Latitude of 67°15' s, the Ice was so thick and close that we could proceed no further but were fain to Tack and stand from it. From the mast head I could see nothing to the Southward but Ice, in the Whole extent from East to WSW without the least appearence of any partition, this immence Feild was composed of different kinds of Ice, such as high Hills or Islands, smaller pieces packed close together and what Greenland men properly call field Ice, a piece of this kind, of such extend that I could see no end to it, lay to the SE of us, it was 16 or 18 feet high at least and appeared of a pretty equal height. I did not think it was consistant with the safty of the Sloops or any ways prudent for me to persevere in going farther to the South as the summer was already half spent and it would have taken up some time to have got round this Ice, even supposing this to have been practicable, which however is doubtfull. The Winds Continued at East and EBS and increased to a strong gale attended with a large Sea, hazy weather Sleet and Snow and obliged us to close reef our Topsails.

SATURDAY 30*th. Thermr.* 39½. *Winds North W to North. Course N* 57° *E. Dist. Sailed* 101 *Miles. Lat. in South* 51°34'. *Longde. in East of Greenwich Reck.g* 55°55'. *Long. made from C.G. Hope* 37°32'. Very hard gale and thick hazey weather with drizling rain which obliged us to close reef our Topsails and at 8 o'Clock to hand the Main sail and Fore Topsail. We spent the night, which was dark and stormy, in making a trip to the SW, and in the morning made sail again to the NE under Courses and Double reefed Topsails, the Wind being some thing abated but it yet blew a fresh gale at NW and NNW attended with drizling rain and hazey thick weather. This is the first and only day we have seen no Ice sence we first discovered it.

[FEBRUARY 1773]

MONDAY 1*st. Thermr.* 41½. *Winds WNW. Course N* 18 ° *E. Diste. Sailed* 126 *Miles. Lat. in South* 48°51'. *Longde. in East Reck.g* 57°47'. *Long. made from C.G.H.* 39°24'. Fresh gales and Clowdy, at 2 pm passed two Is^{ds} of Ice. In the AM saw a small piece of rock weed.

TUESDAY 2*nd. Winds NW to WNW. Course N* 78° *E. Diste. Sailed* 73 *Miles. Lat. in South* 48°36'. *Longde. in East* 59°33'. *Watch* 59°33'.

Long. made from C.G.H. 41°12'. Hazey Clowdy weather and a fresh gale at NW with which we stood NEBN till 4 o'Clock in the PM when being in the Latitude of [48°39'] S and nearly in the Meridian of the Isle of Mauritius, where we were to expect to find the Land said lately to have been discovered by the French, but seeing nothing of it we bore away East and made the Signal to the Adventure to keep on our Starboard beam at 4 miles distance.

MONDAY 8*th. North, East to North. Course S* 54° *E. Distce. Saild* 103 *Miles. Lat. In South* 49°51'. *Longde. in East Reck.g* 63°57'. *Longde. East C.G.H.* 45°34'. At 6 o'Clock in the PM, made the Signal for the Adventure to come under our stern and at the same time took several Az[ths] which gave the var[n] 31°28' but the observations were doubtfull on account of the rowling of the sloop occasioned by a very high Westerly swell. Fair weather continued till Midnight when it became Squally with rain and we took a reef in each Topsail. In the Morning saw several Penguins & Divers and some were heard at different times in the night, these signs of land continuing we at 8 o'Clock sounded but found no ground with 210 fathoms. We were now in the Latitude of 49°53' S, Longitude 63°39' East, Steering South by Compass close upon a Wind which was at ESE the Adventure was about a point or two upon our Larb[d] quarter, distant about one Mile and a half or as some thou[gh]t one Mile, about half an hour after a thick Fogg came on so that we could not see her. At 9 o'Clock we fired a gun and repeated it at 10 and at 11 and at Noon made the Signal to Tack and Tacked accordingly, but neither this last Signal or any of the former were answered by the Adventure which gave us too much reason to apprehend that a seperation would take place. I have said that at 8 o'Clock we laid South by Compass with the wind at ESE but by 9 o'Clock or before the Wind veered to NNE so that we laid E½S, this must have brought the Adventure upon our weather beam or directly to Windward of us, provided she kept her Wind which she ought to have done as no signal was made to the contrary. In short we were intirely at a loss, even to guess by what means she got out of the hearing of the first gun we fired.

TUESDAY 9*th. Winds North, North, NBE & NNW. Course S* 66° *W. Distce. Sailed* 5 *Miles. Lat. in South* 49°53'. *Long*[d]*, in East Reck.g* 63°53'. *Longde. made C.C.H.* 45°30'. The thick Foggy Weather continuing and being apprehensive that the Adventure was still on the other Tack, we at 2 pm, after having run 2 Leagues to the West, made the Signal and Tacked, to which we heard no Answer, we now continued to fire a gun every half hour. At 3 o'Clock just after fireing our gun the officer of the Watch and others on Deck heard or thought they heard the report of a gun on the Weather bow, about the same space of time after fireing the next gun no one on deck doubted but what they heard the report of a nother gun on our beam, the different situations of these two sounds induced us to think that the Adventure was on the other Tack and standing to the Westward, and being to Windward of us I thought she might not hear our guns, but was only fireing half hour guns as well as us and that her fireing so soon after us was only chance. I therefore orderd a nother gun to be fire'd a quarter of an hour after to which we heard no answer, being now satisfied that she did not hear us both from her not answering and not bearing down, I prepared to Tack at 4 o'Clock but first ordered a Gun to be fired after which Mr Forster alone thought he heard the report of a nother to Windward nearly in the same situation as the last, this occasioned my standing half an hour longer to the Westward in which time we fired two guns to which no answer was heard, we then Tack'd and stood to the Westward after having stood some thing more than 8 Miles to the East. We still continued to fire half [h]our guns and the Fogg dissipated at times so to admit us to see two or three Miles or more round us, we however could niether hear nor see any thing of her. Being now well assured that a Separation had taken place I had nothing to do but to repair to the place where we last saw her, Captain Furneaux being directed to do the same and there to cruze three days, accordingly I stood on to the Westward till 8 o'Clock, than made a trip to the East till Midnight and then again to the West till Day light and fired half hour guns and Bur[n]t false fires all night, the Weather continued Foggy and hazey and the Wind remained invariable at North and NBW, which, if the Adventure kept her Wind during the Fogg, was very favourable for her to return back to the apointed station. After day light at which time we could only see about 2 or 3 Miles round us, we tacked and stood to the East till 8 o'Clock, then again to the West and at Noon we were about 6 or 7 Miles East of the place were we last saw the Adventure and would see about 3 or 4 Leagues round us: the Wind which was now at NNW had increased in such a manner as to oblige us to take in our Topsails and the Sea at the same time began to rise from the same point. We still continued to see Penguins and Divers which made us conjector that land was not far off, it was for this reason I tacked yesterday at Noon which probably was the occasion of my loosing the Adventure for seeing such signs of the Vicinity of land, I thought it more prudent to make short boards during the Fogg, over that part of the Sea we had already made our selves accquainted with, than to continue standing to the Eastward at a time we could not see a quarter of a mile before us.

WEDNESDAY 10*th. Thermr.* 40½. *Winds NNW to WBN. Course S* 68°. *Distce. Sailed* 38 *Miles. Lat. in South* 50°7'. *Longde. in East Reck.g* 64°53' *Watch* 64°49'. *Longd. C.G.H.* 46°30'. We stood to the Westward till half past 2 o'Clock pm having run 10 Miles sence Noon and neither hearing or seeing any thing of the Adventure, we wore and lay-too under the Mizen Staysail with our head to the Eastward, at 8 o'Clock the gale being somewhat abated we set the Foresail and this Sail we kept under all night; during the height of the gale the Weather was hazey with rain, but towards evening it cleared up so as to see 3 or 4 Leagues round us and in this state it continued all night, we however kept burning false fires at the mast head and fireing guns every hour, but neither the one nor the other had the desired effect for altho we laid too all the morning we could see nothing of the Adventure which if she had been with[in] 4 or 5 Leagues of us must have been seen from the mast head. Having now spent two Days out of the three assign'd to look for each other, I thought it would be to little purpose to wait any longer and still less to attempt to beat back to the appointed station will knowing that the Adventure must have been drove to leeward equally with our selves. I therefore made sail to the SE with a very fresh gale at WBN accompanied with a high Sea, many dark grey Albatrosses, Blue Petrels and Sheerwaters about the Ship but only two or three divers were seen and not one Penguin.

SATURDAY 13*th. Thermr.* 36. *Winds SW* to *WBS. Course S* 50°30' *E. Distce. Sailed* 104 *Miles. Lat. in South* 53°54'. *Longde. in East Reck.g* 72°34' *Watch* 72°24'. *Longd. C.G.H.* 54°11'. *Varn.* 33°8'. Gentle gales and pleasent Weather. In the Evening the Variation was 32°32' W and in the morning 33°8' W. We were now accompanied by a much greater number of Penguins than at any time before and of a different sort, being smaller, with Redish Bills and brown heads, the meeting with so many of these Birds gave us still some hopes of meeting with land and various were the oppinions among the officers of its situation. Some said we should find it to East others to the North, but it was remarkable that not one gave it as his opinion that any was to be found to the South which served to convince me that they had no inclination to proceed any farther that way. I however was resolved to get as far to the South as I conveniently could without looseing too much easting altho I must confess I had little hopes of meeting with land, for the high swell or Sea which we have had for some time from the West came now gradualy round to SSE so that it was not probable any land was near between these two points and it is less probable that land of any extent can lie to the North as we are not above [160] Leagues South of Tasmans track and this space I expect Captain Furneaux will explore, who I expect is to the North of me.

MONDAY 15*th. Thermr.* 36½. *Winds SWBS. Course SEBE. Dist. Sailed* 162 *Miles. Lat. in South* 56°52'. *Longd. in East Reckg.* 78°48'. *Longd. East Cape G.H.* 60°25'. Fresh gales, with now and then Showers of sleet and snow. In the evening the Variation was found to be 34°48' West. The Wind now veered to SWBS or SSW and the Swell followed the same direction. About 6 in the AM a nother Seal was seen. Some petty thefts having lately been commited in the Ship, I made a thro' search to day for the stolen things and punished those in whose custody they were found.

WEDNESDAY 17*th. Winds NBE to EBS. Course S* 36½° *E. Dist. Sailed* 57 *Miles. Lat. in South* 57°54'. *Longd. in East Reckg.* 82°4'. *Longd. East Cape G.H.* 63°41'. Gentle gales and dark clowdy weather with frequent Showers of Sleet and snow. At 9 o'Clock am Saw an Island of Ice to the westward distant 3 or 4 Leagues, which we bore down to with the same intention as we stood for the one yesterday. The Wind was now at East by South but the swell still continued to come from the West. Last night Lights were seen in the Heavens similar to those seen in the Northern Hemisphere commonly called the Northern lights, I do not remember of any Voyagers makeing mention of them being seen in the Southern before.

THURSDAY 18*th. Thermr.* 30 *to* 33. *Winds SBW. Course S* 87° *E. Distce. Sailed* 48 *Miles. Lat. in South* 57°57'. *Longd. in East Reckg.* 83°44' *Watch* 83°0'. *Longd. from the Cape G.H.* 65°21'. *Var.* 39°33' *West.* A little past Noon we brought-to under the Island of Ice which was full half a mile in circuit and two hundred feet high. At this time there were but a few loose pieces about it but while we were hoisting out the Boats to take this up an immence quantity broke from the Island, a convincing proof that these Islands must decrease pretty fast while floating about in the Sea. I observed the loose pieces to drift fast to the Westward, that is it quited the Island in that direction and this I suppose to be occasioned by a current, for the Wind which was at ESE could have little or no effect upon the Ice.

At 8 o'Clock we hoisted in the Boats and made sail to the Eastward with a gentle gale at South, having got on board as much Ice as yeilded nine or ten Tons of Water. In the morning the Variation was 39°33' West. At Noon we had twelve Islands of Ice in sight besides an immence number of loose pieces which had broke off from the Island. The Southern lights were again seen last night. The Thermometer was 2° or 3° below the freezing point in consequence of which the Water in the Scuttle cask was froze. Swell still continues to come from the West. The morning being clear Mr Wales and some of the officers took several Observations of the Sun and Moon which gave the Longitude reduc'd to Noon 83°44' E one degree less then the Logg gives carried on from the last observations, which indicates that there is a current seting to the West.

SUNDAY 21*st. Thermr.* 36. *Winds South, SW to NWBW. Course S* 61° *E. Dist. Sailed* 27 *Miles. Lat. in South* 59°0'. *Longd. in East Reckg.* 92°30'. *Longd. E. of Cape G.H.* 76°27'. Gentle breeze and clowdy Weather. At 1 pm, thinking we saw land to the SW, we Tacked and stood towards it, but at 3 o'Clock found it to be only clowds, which soon dissipated and we again resumed our Course to the SE. In the Evening the Horizon was unusally clear so as to see full 12 or 15 Leagues round. The variation by several Az[ths] taken with Dr Knights Compass was 40°8' and by Gregorys 40°15' West. Had but little wind all night which increased but very little with the Day and as the Sea was smooth and favourable for the Boats to take up Ice we steered for a large Island were we expected to meet with some and were not disapointed when we reached it. In the morning some Penguins were seen.

MONDAY *22nd. Winds NBE to EBN. Course* S 44° *E. Dist. Sailed* 49 *Miles. Lat. in South* 59°35'. *Longd. in East Reckg.* 93°36'. *Longd. E. of Cape G.H.* 77°33' *Varn. West* 40°51'. After Dinner hoisted out two Boats and set them to take up Ice while we stood to and from under the Island which was about half a mile in circuit and three or four hundred feet high, yet this huge body turned nearly bottom up while we were near it. At 6 o'Clock having got aboard as much Ice as we could dispence with, we hoisted in the Boats and made sail to the SE with a gentle breeze at NBE having at this time Eight Ice Islands in sight and increased in such a manner as we run to the SE that in the morn[g] 23 were seen at one time and yet the Weather was generally hazey with snow showers. Variation p[r] Az[ths] 40°51' West.

WEDNESDAY *24th. Winds ESE. Course S* 16° *E. Dist. Sailed* 22 *Miles. Lat. in South* 61°21'. *Longd. in East Reckg.* 95°15'. *Longd. in C.G.H.* 79°12'. Fresh gales & hazey with Snow and sleet. Stood to the South till 8 pm at which time we were in the Latitude of 61°52' s, the Ice Islands were now so numerous that we had passed upwards of Sixty or Seventy sence noon many of them a mile or a mile and a half in circuit, increasing both in number and Magnitude as we advanced to the South, sufficient reasons for us to tack and spend the night making short boards, accordingly we stood to the north under Reefed Topsails and Fore sail till midnight when we tacked and stood South having very thick hazey weather with Sleet & snow together with a very strong gale and a high Sea from the East. Under these circumstances and surrounded on every side with huge pieces of Ice equally as dangerous as so many rocks, it was natural for us to wish for day-light which when it came was so far from lessening the danger that it served to increase our apprehensions thereof by exhibiting to our view those mountains of ice which in the night would have been passed unseen. These obstacles together with dark nights and the advanced season of the year, discouraged me from carrying into execution a resolution I had taken of crossing the Antarctick Circle once more, according at 4 o'Clock in the AM we Tacked and Stood to the North under our two Courses and double reefed Topsails, stormy Weather still continuing which together with a great Sea from the East, made great distruction among the Islands of Ice. This was so far from being of any advantage to us that it served only to increase the number of pieces we had to avoide, for the pieces which break from the large Islands are more dangerous then the Islands themselves, the latter are generally seen at a sufficient distance to give time to steer clear of them, whereas the others cannot be seen in the night or thick weather till they are under the Bows: great as these dangers are, they are now become so very familiar to us that the apprehensions they cause are never of long duration and are in some measure compencated by the very curious and romantick Views many of these Islands exhibit and which are greatly heightned by the foaming and dashing of the waves against them and into the several holes and caverns which are formed in the most of them, in short the whole exhibits a View which can only be discribed by the pencle of an able painter and at once fills the mind with admiration and horror, the first is occasioned by the beautifullniss of the Picture and the latter by the danger attending it, for was a ship to fall aboard one of these large pieces of ice she would be dashed to pieces in a moment.

SUNDAY 28*th. Thermr.* 36½. *Winds South to SW. Course N* 77°15' *E. Dist. Sailed* 135 *Miles. Lat. in South* 59°58'. *Longd. in East Reck.g* 104°44'. *Long[d] East C.G.H.* 86°21'. In the pm the gale abated and the Wind veered to SSW and SWBS. Hazey weather with sleet continued till 8 o'Clock in the am when it became fair and tolerable clear, at day light in the morn got Topg[t] yards across and set all the sail we could, having a fine fresh gale and but few Islands of Ice to impede us, the late gale having probably distroyed great numbers of them. A large hollow sea hath continued to accompany the Wind that is from the East round by the South to SW, so that no land can be hoped for betwixt these two extreme points. We have a breeding sow on board which yesterday morning Farrowed nine Pigs every one of which were killed by the cold before 4 o'Clock in the afternoon notwithstanding all the care we could take of them, from the same cause several People on board have their feet and hands chilblain'd, from the circumstances a judgement may be formed of the summer weather we injoy here.

EMILY BALLOU

The Beach

I

All along the scratched stretches of beach land
when the tide was tugging back, some evenings
the low pools, miniature seas
contained everything
he could imagine, everything he'd ever
want to see, and more. Bounties of being.

In the opal light, the flats left organ-shaped puddles,
sucking quicksand he squelched through.
He liked the deep squelch around his boots, eyes
to the ground, scouring the stony, spongy surface
trying to disappear.

He tried to banish the thought, newly lodged,
that man's consciousness was mere brain, waves
pushing pebbles forward and backwards over the beach.

The grey-blue clouds, milk-topped, spread long and low.
His shadow preceded him, painted proof on the sand
of his existence; the gold engine of sunlight
as inexplicable and shape-shifting as melting glass
streaming behind the humped Edinburgh hills.

The patterning of gulls on the shore
a huddled form
that altered as he edged closer.
Their feet left triangled marks of retreat.

A lone gull, far from the flock, experimented
with the limits of water, pecked at stones.
When he chortled at its waddle, its purpose and poise
it turned its head as if to say:
'What is so very amusing?'
And pattered on, away,
to do some more of what it did so well.

II

You could squat down for an hour near a tidal pool,
and await the tiniest sway of cilia, or a glimpse
of feathery sea-pen, hovering at the edge
to discover a colony of tentacled polyps; or hurl
your chanced find of grey, furred sea mice back
into the sea like stones
and watch them turn into balls, and watch
them roll back up onto the beach with waves
and stand finally, aching, to stretch and let
out a long, deep growl
of pleasure and be struck
by a sudden bolt of light in the eyes
as if you had dissolved back up
into the sun.

III

The beach just was. *Is.*
It had no idea how it got there.
But to dig back the sand with two hands as he had
as a child, searching its beginning
with his particular gift of persistence
with his willingness to understand what it truly was
while his sisters called: *It never ends. It will never end*
still did not seem to be an impossible task.

IV

That day he had dug until the water
streaming through the sand collapsed his hole and sand
was no longer sand – quite – in this form.
It flowed, liquid, becoming something else across his palms.
He dug another and another hole, each time
trying to find the bottom of the world
before the sea caved in.

Stop! You will exhaust yourself!
his sisters cried.

You cannot hold it, just behold it.

It is what it is. And nothing more.

V

When the sun becomes a cross of orange light
and puddles darken the shore and stones
become miniature mountains in the glow
and the sand blows –
stop and stretch again, the gulls are flying.

Soon,
the crags of the old town will be
clawed by birds
flown in with fog from the sea
circling the air above steeple and castle
with their distinct ache, their screams.

His sisters said to put his faith
not in this world
but the next.

But look how it shines.

CHARLES DARWIN

The Voyage of the Beagle

While sailing a little south of the Plata on one very dark night, the sea presented a wonderful and most beautiful spectacle. There was a fresh breeze, and every part of the surface, which during the day is seen as foam, now glowed with a pale light. The vessel drove before her bows two billows of liquid phosphorus, and in her wake she was followed by a milky train. As far as the eye reached, the crest of every wave was bright, and the sky above the horizon, from the reflected glare of these livid flames, was not so utterly obscure as over the vault of the heavens.

As we proceed further southward the sea is seldom phosphorescent; and off Cape Horn I do not recollect more than once having seen it so, and then it was far from being brilliant. This circumstance probably has a close connexion with the scarcity of organic beings in that part of the ocean. After the elaborate paper* by Ehrenberg, on the phosphorescence of the sea, it is almost superfluous on my part to make any observations on the subject. I may however add, that the same torn and irregular particles of gelatinous matter, described by Ehrenberg, seem in the southern as well as in the northern hemisphere, to be the common cause of this phenomenon. The particles were so minute as easily to pass through fine gauze; yet many were distinctly visible by the naked eye. The water when placed in a tumbler and agitated, gave out sparks, but a small portion in a watch-glass

*An abstract is given in No IV of the Magazine of Zoology and Botany.

scarcely ever was luminous. Ehrenberg states that these particles all retain a certain degree of irritability. My observations, some of which were made directly after taking up the water, gave a different result. I may also mention, that having used the net during one night, I allowed it to become partially dry, and having occasion twelve hours afterwards to employ it again, I found the whole surface sparkled as brightly as when first taken out of the water. It does not appear probable in this case, that the particles could have remained so long alive. On one occasion having kept a jelly-fish of the genus Dianæa till it was dead, the water in which it was placed became luminous. When the waves scintillate with bright green sparks, I believe it is generally owing to minute crustacea. But there can be no doubt that very many other pelagic animals, when alive, are phosphorescent.

On two occasions I have observed the sea luminous at considerable depths beneath the surface. Near the mouth of the Plata some circular and oval patches, from two to four yards in diameter, and with defined outlines, shone with a steady but pale light; while the surrounding water only gave out a few sparks. The appearance resembled the reflection of the moon, or some luminous body; for the edges were sinuous from the undulations of the surface. The ship, which drew thirteen feet water, passed over, without disturbing these patches. Therefore we must suppose that some animals were congregated together at a greater depth than the bottom of the vessel.

Near Fernando Noronha the sea gave out light in flashes. The appearance was very similar to that which might be expected from a large fish moving rapidly through a luminous fluid. To this cause the sailors attributed it; at the time, however, I entertained some doubts, on account of the frequency and rapidity of the flashes. I have already remarked that the phenomenon is very much more common in warm than in cold countries; and I have sometimes imagined that a disturbed electrical condition of the atmosphere was most favourable to its production. Certainly I think the sea is most luminous after a few days of more calm weather than ordinary, during which time it has swarmed with various animals. Observing that the water charged with gelatinous particles is in an impure state, and that the luminous appearance in all common cases is produced by the agitation of the fluid in contact with the atmosphere, I am inclined to consider that the phosphorescence is the result of the decomposition of the organic particles, by which process (one is tempted almost to call it a kind of respiration) the ocean becomes purified.

JENNIFER ACKERMAN

Notes from the Shore

Crabs again. Real crabs this time, *Ocypode quadrata*, the swift-footed ghost crab, abundant from Cape Henlopen to Brazil and hunter of night beaches. I've come down to the cape at sunset on an early summer day. The only sounds are those of crickets buzzing in the beach grass and the regular slap of waves. The sun, a ball of fire, sinks and is swallowed by a bank of clouds long before it reaches the horizon. The light warms and reddens. A flock of gulls lets me come very close before they send up an explosion of white wings. They circle about and a few seconds later, settle again behind me, a quarrelsome knot of dim shadows picking through the sea scraps. A squadron of cormorants passes low against the reflected afterlight. The waters darken. For a brief moment Venus shines alone, then stars fill the night.

I flip on my light and startle a ghost crab. It's a good size, two and a half inches of furious activity against the pale sand. I've watched smaller members of its tribe, sidling from dune to sea, halting to dig for mole crabs, then dashing madly back, their camouflage so effective they look like bits of wind-shifted sand or detritus. Their exquisite protective disguise arises from pigmented cells called chromatophores. The pigment migrates in response to light and temperature, causing color changes that help the crab mimic its surroundings.

This one scuttles sideways on the tips of eight legs. When I press in, it raises the last pair of legs off the ground and accelerates, disappearing down a hole at the toe of the foredunes. I flash my light in the opening, but to no avail. The burrow may shoot or spiral down to depths of five or six feet. One morning I saw an adult crab emerge popeyed from a burrow beneath the awning of a horseshoe crab shell, cradling a load of sand in its legs. It paused for a moment, then flung its load down and flashed back into the hole. It was some time later before it reemerged with another load.

Ghost crabs breathe air through narrow, slitlike openings between their third and fourth legs and can live for long periods out of water, but their gills must be kept moist in order to function. At intervals they visit the swash zone to replenish the moisture. Theirs is an evolutionary drama, says Rachel Carson, the coming to land of a sea creature. The larvae begin life as part of the plankton drifting in the open ocean. They become amphibious as they grow, at some point following an urge to pop through the water membrane into the throttling air. They come ashore a rolled-up, fistlike ball of legs and torso, protected from the bruising surf by a tough cuticle. Small immature crabs burrow near the water, just above high-tide line. As they mature, the crabs become more and more independent of the sea, foraying as much as a quarter of a mile inland. Still, they must return to the wash of broken surf to wet their gills and release their eggs.

In their bondage to the sea, ghost crabs resemble their sometime prey, a tiny creature only a half inch in length that leaps about the light of my lamp when I set it on the sand. Beach fleas explode into the air with an agile flexing of legs. Not fleas at all, but crustaceans with flealike powers of jumping, they hop distances of more than fifty times their own length using three pairs of short, stiff rear legs. The fleas also bear three abdominal legs modified for swimming. They, too, hover between land and sea, still possessing gills, though much reduced in size from those of their marine ancestors.

A beach flea lives close to the wrack line, burrowing in the moist sand beneath drift seaweed in the heat of day to avoid desiccating its gills and body and emerging only at night to browse on bits of decaying plant and animal matter. Using the moon and other celestial cues to guide it, the flea moves up and down the beach with the tides, staying within a narrow ribbon of damp sand. It shares with the ghost crab a fear of the full tide. Both crab and flea will drown if kept under water for any period of time, as we ourselves might drown.

I switch off my lamp. A pale moon has risen, spreading its diffuse light across the water's surface. Small waves shower light foam on the shore. Otherwise the night is black. It was 350 million years ago that the first pioneer of land life heaved itself out of the sea: an arthropod, one of the great phyla that later gave rise to crabs and insects, a stumbling, adventurous refugee that lived the strange half-aquatic, half-terrestrial life of the ghost crab and beach flea. The small scuttlings at my feet presage the future, pointing out that life is not fixed like a butterfly pinned on a board, but still brewing, groping on in countless directions.

One hot morning in late June, I set off in a Boston whaler with friends to look for pelagic birds. I sit in the front of the boat where the jolts are hard, bouncing along toward the open blue, the sea disappearing as we mount a rise, then reappearing as we smack down hard. The coast recedes to a thin featureless crust, then disappears altogether. It takes most of the morning to reach Five Fathoms, a fishing spot forty miles out to sea. Alone with our boat, we cut the motor and drift, binoculars trained on the horizon. Not a pelagic bird in sight. A gull swaying gently overhead cocks its head as if to ask what we are doing way out here in such a duckling of a boat.

Our whaler carries some sophisticated electronic navigation equipment, but I can't help wondering what would happen if it failed, along with our motor, and the weather turned bad. Sea nomads in the archipelagoes of Southeast Asia can look up at the sun and clouds, look down at the sea, and accurately read both time and their whereabouts. But for most of us, the open ocean seems mute and lonely when you're out there in the middle of it. A report I read in *The Journal of Navigation* suggested that humans have an innate sense of direction. Laboratory experiments showed that people's ability to pinpoint North gradually improved after multiple challenges. 'Orientation in humans is a latent sense,' said the researchers, though in most of us it seems to have disappeared from lack of use.

I don't know how long we sit. The sun beats down, and the water calms to a sheet of thick, undulating metal. Suddenly, twenty yards from our tiny craft, a great slick-backed blue mass lifts in a rising swell and rolls forward, flashing a sharp hook of fin and a bright turquoise patch. It disappears, then rises again, slow, cloudlike, and blows a jet of white vapor ten feet high, like an upside-down pyramid.

It is a finback whale, the second largest baleen whale after the blue, and Earth's only asymmetrically colored mammal. The left side of its head and jaw is dark, the right side light. The purpose of this asymmetrical pigmentation remains a mystery, although it may be an adaptation for the capture of small schooling fish such as herring. The whale swims around the fish to the right in smaller and smaller circles, showing only its translucent, invisible white side so as not to startle its prey. As the herring clump together, the whale turns into the pack, mouth open, and gobbles them up, its pleated throat bulging, expanding bellowlike to take in the liquid meal. Then it presses the water out of the baleen plates that grow down from its upper gums and swallows what's left behind.

Our boat bobs about in the waves above Five Fathoms. We each face a different direction to cover the scope of ocean, but the giant beast has disappeared. Finbacks can swim underwater for forty minutes without drawing breath. Like other whales, their ancestors were four-legged land animals that hunted in the tidal shallows around river deltas and the edges of the warm, shallow seas. Sometime around fifty or sixty million years ago, they abandoned land life for the ocean, perhaps following their prey farther and farther out to sea. Palaeontologists digging in Pakistan recently unearthed the remains of an ancient whale with legs and long feet like a seal, a missing link in the evolutionary chain. The scientists called the animal *Ambulocetus natans*, 'the walking whale that swam.'

Over the course of millions of years, these creatures slipped protean through many shapes, losing their legs and pelvises and developing a horizontal fluked tail to propel them through the seas. Their bodies eventually assumed a smooth, hairless form perfectly suited to swimming and swept back for speed. Their forelimbs grew into organs for steering and balancing in a liquid environment. Sound became their light and hearing their vision. Deep within the fifty-ton body of a finback is born a pure, radiant booming an octave deeper than the lowest note on a piano. Such sound waves can travel enormous distances in the sea. Trapped in the deep sound channel, a layer of water where sound waves bend back on themselves and retain their energy, whale sounds can carry several hundred miles.

Both whales and dolphins bear traces of their kinship to land creatures. Their flippers have bones similar to those in a human arm and hand, though much reshaped. Whales often retain tiny leg bones. Dolphins catch the same diseases as pigs and cattle. In a throwback to a dim past, dolphins along the coast of the Carolinas briefly revisit the world of their ancestors, having learned once again to feed at low tide on the edge of the land, herding schools of fish toward the mud flats, riding the waves in, and plucking the fish from the shore.

Once, on a visit to a whaling ship, biologist Victor Scheffer acquired a whale fetus only four inches long. 'I took the little creature, packed in ice cubes, to the mainland,' he wrote. 'At my hotel I bought a pint of vodka and a bottle of shaving lotion. I mixed these in a washbasin, slit the belly and chest of the fetus with a razor blade, and embalmed it overnight in the fragrant solution. Later I dissected it in my laboratory . . . . In profile, the little head could . . . have belonged to an infant pig, with eyes shut, lower jaw protruding beyond the snout, and nostrils at the front . . . . The penis protruded; the rudimentary nipples were evident; even the ears were there – tiny ridges of skin, most unfitting for a whale. There were actually traces of whiskers, casting a long shadow from an ancestor dead now forty million years.'

In the course of development, embryonic whales grow rudimentary legs, nostrils, and surface genitals. Then the hindlimbs disappear, the nostrils slide backward to become blowholes, the genitals vanish inside a slit. In the fetus of a finback whale, tooth germs appear in the gums, but by the time the fetus has reached a length of thirteen feet, they have vanished.

I once saw a jar containing a pickled human embryo in the National Museum of Natural History. It had a bulging reptilian head, a tail, and arches like gills just beneath the head. Our own living organs, eyes, backbones, hands, and feet originated in far places and different eras of time. Four hundred million years ago our piscine forebears wiggled over muddy flats, throwing their bodies in an S-curve. As a consequence, our arms swing in opposition to the swing of our legs. Our reflected past and some shadow of the future is paradoxically written in our bodies. We, too, are changelings, made of millions of bits of information strung together from an odd little alphabet and brought into being by an astronomical number of chance events over the long course of evolution. But for this we might be hovering just above the warm mud. As Stephen Jay Gould has written, those stubby, sluggish fins that became weight-bearing limbs – the necessary prerequisite to terrestrial life – evolved in an uncommon group of fishes off the main line. They were a fluke.

In some way all creatures bear traces of their past: ghost crabs their gills, whales their vestigial limbs, humans our liquid cells, the salt water running in our veins, our feeling for the sea. 'Why upon your first voyage as a passenger,' wrote Melville, 'did you yourself feel such a mystical vibration when first told that you and your ship were out of sight of land?'

ERNEST HEMINGWAY

The Old Man and the Sea

Sometimes someone would speak in a boat. But most of the boats were silent except for the dip of the oars. They spread apart after they were out of the mouth of the harbour and each one headed for the part of the ocean where he hoped to find fish. The old man knew he was going far out and he left the smell of the land behind and rowed out into the clean early morning smell of the ocean. He saw the phosphorescence of the Gulf weed in the water as he rowed over the part of the ocean that the fishermen called the great well because there was a sudden deep of seven hundred fathoms where all sorts of fish congregated because of the swirl the current made against the steep walls of the floor of the ocean. Here there were concentrations of shrimp and bait fish and sometimes schools of squid in the deepest holes and these rose close to the surface at night where all the wandering fish fed on them.

In the dark the old man could feel the morning coming and as he rowed he heard the trembling sound as flying fish left the water and the hissing that their stiff set wings made as they soared away in the darkness. He was very fond of flying fish as they were his principal friends on the ocean. He was sorry for the birds, especially the small delicate dark terns that were always flying and looking and almost never finding, and he thought, 'The birds have a harder life than we do except for the robber birds and the heavy strong ones. Why did they make birds so delicate and fine as those sea swallows when the ocean can be so cruel? She is kind and very beautiful. But she can be so cruel and it comes so suddenly and such birds that fly, dipping and hunting, with their small sad voices are made too delicately for the sea.'

He always thought of the sea as *la mar* which is what people call her in Spanish when they love her. Sometimes those who love her say bad things of her but they are always said as though she were a woman. Some of the younger fishermen, those who used buoys as floats for their lines and had motor-boats, bought when the shark livers had brought much money, spoke of her as *el mar* which is masculine. They spoke of her as a contestant or a place or even an enemy. But the old man always thought of her as feminine and as something that gave or withheld great favours, and if she did wild or wicked things it was because she could not help them. The moon affects her as it does a woman, he thought.

He was rowing steadily and it was no effort for him since he kept well within his speed and the surface of the ocean was flat except for the occasional swirls of the current. He was letting the current do a third of the work and as it started to be light he saw he was already further out than he had hoped to be at this hour.

I worked the deep wells for a week and did nothing, he thought. Today I'll work out where the schools of bonito and albacore are and maybe there will be a big one with them.

Before it was really light he had his baits out and was drifting with the current. One bait was down forty fathoms. The second was at seventy-five and the third and fourth were down in the blue water at one hundred and one hundred and twenty-five fathoms. Each bait hung head down with the shank of the hook inside the bait fish, tied and sewed solid, and all the projecting part of the hook, the curve and the point, was covered with fresh sardines. Each sardine was hooked through both eyes so that they made a half-garland on the projecting steel. There was no part of the hook that a great fish could feel which was not sweet-smelling and good-tasting.

The boy had given him two fresh small tunas, or albacores, which hung on the two deepest lines like plummets and, on the others, he had a big blue runner and a yellow jack that had been used before; but they were in good condition still and had the excellent sardines to give them scent and attractiveness. Each line, as thick around as a big pencil, was looped onto a green-sapped stick so that any pull or touch on the bait would make the stick dip and each line had two forty-fathom coils which could be made fast to the other spare coils so that, if it were necessary, a fish could take out over three hundred fathoms of line.

Now the man watched the dip of the three sticks over the side of the skiff and rowed gently to keep the lines straight up and down and at their proper depths. It was quite light and any moment now the sun would rise.

The sun rose thinly from the sea and the old man could see the other boats, low on the water and well in toward the shore, spread out across the current. Then the sun was brighter and the glare came on the water and then, as it rose clear, the flat sea sent it back at his eyes so that it hurt sharply and he rowed without looking into it. He looked down into the water and watched the lines that went straight down into the dark of the water. He kept them straighter than anyone did, so that at each level in the darkness of the stream there would be a bait waiting exactly where he wished it to be for any fish that swam there. Others let them drift with the current and sometimes they were at sixty fathoms when the fishermen thought they were at a hundred.

But, he thought, I keep them with precision. Only I have no luck any more. But who knows? Maybe today. Every day is a new day. It is better to be lucky. But I would rather be exact. Then when luck comes you are ready.

The sun was two hours higher now and it did not hurt his eyes so much to look into the east. There were only three boats in sight now and they showed very low and far inshore.

All my life the early sun has hurt my eyes, he thought. Yet they are still good. In the evening I can look straight into it without getting the blackness. It has more force in the evening too. But in the morning it is painful.

Just then he saw a man-of-war bird with his long black wings circling in the sky ahead of him. He made a quick drop, slanting down on his backswept wings, and then circled again.

'He's got something,' the old man said aloud. 'He's not just looking.'

He rowed slowly and steadily toward where the bird was circling. He did not hurry and he kept his lines straight up and down. But he crowded the current a little so that he was still fishing correctly though faster than he would have fished if he was not trying to use the bird.

The bird went higher in the air and circled again, his wings motionless. Then he dove suddenly and the old man saw flying fish spurt out of the water and sail desperately over the surface.

'Dolphin,' the old man said aloud. 'Big dolphin.'

He shipped his oars and brought a small line from under the bow. It had a wire leader and a medium-sized hook and he baited it with one of the sardines. He let it go over the side and then made it fast to a ring bolt in the stern. Then he baited another line and left it coiled in the shade of the bow. He went back to rowing and to watching the long-winged black bird who was working, now, low over the water.

As he watched the bird dipped again slanting his wings for the dive and then swinging them wildly and ineffectually as he followed the flying fish. The old man could see the slight bulge in the water that the big dolphin raised as they followed the escaping fish. The dolphin were cutting through the water below the flight of the fish and would be in the water, driving at speed, when the fish dropped. It is a big school of dolphin, he thought. They are wide spread and the flying fish have little chance. The bird has no chance. The flying fish are too big for him and they go too fast.

He watched the flying fish burst out again and again and the ineffectual movements of the bird. That school has gotten away from me, he thought. They are moving out too fast and too far. But perhaps I will pick up a stray and perhaps my big fish is around them. My big fish must be somewhere.

The clouds over the land now rose like mountains and the coast was only a long green line with the grey-blue hills behind it. The water was a dark blue now, so dark that it was almost purple. As he looked down into it he saw the red sifting of the plankton in the dark water and the strange light the sun made now. He watched his lines to see them go straight down out of sight into the water and he was happy to see so much plankton because it meant fish. The strange light the sun made in the water, now that the sun was higher, meant good weather and so did the shape of the clouds over the land. But the bird was almost out of sight now and nothing showed on the surface of the water but some patches of yellow, sun-bleached Sargasso weed and the purple, formalized, iridescent, gelatinous bladder of a Portuguese man-of-war floating close beside the boat. It turned on its side and then righted itself. It floated cheerfully as a bubble with its long deadly purple filaments trailing a yard behind it in the water.

'*Agua mala*,' the man said. 'You whore.'

From where he swung lightly against his oars he looked down into the water and saw the tiny fish that were coloured like the trailing filaments and swam between them and under the small shade the bubble made as it drifted. They were immune to its poison. But men were not and when some of the filaments would catch on a line and rest there slimy and purple while the old man was working a fish, he would have welts and sores on his arms and hands of the sort that poison ivy or poison oak can give. But these poisonings from the *agua mala* came quickly and struck like a whiplash.

The iridescent bubbles were beautiful. But they were the falsest thing in the sea and the old man loved to see the big sea turtles eating them. The turtles saw them, approached them from the front, then shut their eyes so they were completely carapaced and ate them filaments and all. The old man loved to see the turtles eat them and he loved to walk on them on the beach after a storm and hear them pop when he stepped on them with the horny soles of his feet.

He loved green turtles and hawks-bills with their elegance and speed and their great value and he had a friendly contempt for the huge, stupid logger-heads, yellow in their armour-plating, strange in their love-making, and happily eating the Portuguese men-of-war with their eyes shut.

He had no mysticism about turtles although he had gone in turtle boats for many years. He was sorry for them all, even the great trunk-backs that were as long as the skiff and weighed a ton. Most people are heartless about turtles because a turtle's heart will beat for hours after he has been cut up and butchered. But the old man thought, I have such a heart too and my feet and hands are like theirs. He ate the white eggs to give himself strength. He ate them all through May to be strong in September and October for the truly big fish.

He also drank a cup of shark liver oil each day from the big drum in the shack where many of the fishermen kept their gear. It was there for all fishermen who wanted it. Most fishermen hated the taste. But it was no worse than getting up at the hours that they rose and it was very good against all colds and grippes and it was good for the eyes.

Now the old man looked up and saw that the bird was circling again.

'He's found fish,' he said aloud. No flying fish broke the surface and there was no scattering of bait fish. But as the old man watched, a small tuna rose in the air, turned and dropped head first into the water. The tuna shone silver in the sun and after he had dropped back into the water another and another rose and they were jumping in all directions, churning the water and leaping in long jumps after the bait. They were circling it and driving it.

If they don't travel too fast I will get into them, the old man thought, and he watched the school working the water white and the bird now dropping and dipping into the bait fish that were forced to the surface in their panic.

'The bird is a great help,' the old man said. Just then the stern line came taut under his foot, where he had kept a loop of the line, and he dropped his oars and felt the weight of the small tuna's shivering pull as he held the line firm and commenced to haul it in. The shivering increased as he pulled in and he could see the blue back of the fish in the water and the gold of his sides before he swung him over the side and into the boat. He lay in the stern in the sun, compact and bullet-shaped, his big, unintelligent eyes staring as he thumped his life out against the planking of the boat with the quick shivering strokes of his neat, fast-moving tail. The old man hit him on the head for kindness and kicked him, his body still shuddering, under the shade of the stern.

'Albacore,' he said aloud. 'He'll make a beautiful bait. He'll weigh ten pounds.'

He did not remember when he had first started to talk aloud when he was by himself. He had sung when he was by himself in the old days and he had sung at night sometimes when he was alone steering on his watch in the smacks or in the turtle boats. He had probably started to talk aloud, when alone, when the boy had left. But he did not remember. When he and the boy fished together they usually spoke only when it was necessary. They talked at night or when they were storm-bound by bad weather. It was considered a virtue not to talk unnecessarily at sea and the old man had always considered it so and respected it. But now he said his thoughts aloud many times since there was no one that they could annoy.

'If the others heard me talking out loud they would think that I am crazy,' he said aloud. 'But since I am not crazy, I do not care. And the rich have radios to talk to them in their boats and to bring them the baseball.'

Now is no time to think of baseball, he thought. Now is the time to think of only one thing. That which I was born for. There might be a big one around that school, he thought. I picked up only a straggler from the albacore that were feeding. But they are working far out and fast. Everything that shows on the surface today travels very fast and to the north-east. Can that be the time of day? Or is it some sign of weather that I do not know?

He could not see the green of the shore now but only the tops of the blue hills that showed white as though they were snow-capped and the clouds that looked like high snow mountains above them. The sea was very dark and the light made prisms in the water. The myriad flecks of the plankton were annulled now by the high sun and it was only the great deep prisms in the blue water that the old man saw now with his lines going straight down into the water that was a mile deep.

The tuna, the fishermen called all the fish of that species tuna and only distinguished among them by their proper names when they came to sell them or to trade them for baits, were down again. The sun was hot now and the old man felt it on the back of his neck and felt the sweat trickle down his back as he rowed.

I could just drift, he thought, and sleep and put a bight of line around my toe to wake me. But today is eighty-five days and I should fish the day well.

Just then, watching his lines, he saw one of the projecting green sticks dip sharply.

'Yes,' he said. 'Yes,' and shipped his oars without bumping the boat. He reached out for the line and held it softly between the thumb and forefinger of his right hand. He felt no strain nor weight and he held the line lightly. Then it came again. This time it was a tentative pull, not solid nor heavy, and he knew exactly what it was. One hundred fathoms down a marlin was eating the sardines that covered the point and the shank of the hook where the hand-forged hook projected from the head of the small tuna.

The old man held the line delicately, and softly, with his left hand, unleashed it from the stick. Now he could let it run through his fingers without the fish feeling any tension.

This far out, he must be huge in this month, he thought. Eat them, fish. Eat them. Please eat them. How fresh they are and you down there six hundred feet in that cold water in the dark. Make another turn in the dark and come back and eat them.

He felt the light delicate pulling and then a harder pull when a sardine's head must have been more difficult to break from the hook. Then there was nothing.

'Come on,' the old man said aloud. 'Make another turn. Just smell them. Aren't they lovely? Eat them good now and then there is the tuna. Hard and cold and lovely. Don't be shy, fish. Eat them.'

He waited with the line between his thumb and his finger, watching it and the other lines at the same time for the fish might have swum up or down. Then came the same delicate pulling touch again.

'He'll take it,' the old man said aloud. 'God help him to take it.'

He did not take it though. He was gone and the old man felt nothing.

'He can't have gone,' he said. 'Christ knows he can't have gone.

He's making a turn. Maybe he has been hooked before and he remembers something of it.'

Then he felt the gentle touch on the line and he was happy.

'It was only his turn,' he said. 'He'll take it.'

He was happy feeling the gentle pulling and then he felt something hard and unbelievably heavy. It was the weight of the fish and he let the line slip down, down, down, unrolling off the first of the two reserve coils. As it went down, slipping lightly through the old man's fingers, he still could feel the great weight, though the pressure of his thumb and finger were almost imperceptible.

'What a fish,' he said. 'He has it sideways in his mouth now and he is moving off with it.'

Then he will turn and swallow it, he thought. He did not say that because he knew that if you said a good thing it might not happen. He knew what a huge fish this was and he thought of him moving away in the darkness with the tuna held crosswise in his mouth. At that moment he felt him stop moving but the weight was still there. Then the weight increased and he gave more line. He tightened the pressure of his thumb and finger for a moment and the weight increased and was going straight down.

'He's taken it,' he said. 'Now I'll let him eat it well.'

JOHN STEINBECK

The Log from the Sea of Cortez

In the evening we came back restlessly to the top of the deckhouse, and we discussed the Old Man of the Sea, who might well be a myth, except that too many people have seen him. There is some quality in man which makes him people the ocean with monsters and one wonders whether they are there or not. In one sense they are, for we continue to see them. One afternoon in the laboratory ashore we sat drinking coffee and talking with Jimmy Costello, who is a reporter on the Monterey *Herald*. The telephone rang and his city editor said that the decomposed body of a sea-serpent was washed up on the beach at Moss Landing, half-way around the Bay. Jimmy was to rush over and get pictures of it. He rushed, approached the evil-smelling monster from which the flesh was dropping. There was a note pinned to its head which said, 'Don't worry about it, it's a basking shark. [Signed] Dr. Rolph Bolin of the Hopkins Marine Station.' No doubt that Dr. Bolin acted kindly, for he loves true things; but his kindness was a blow to the people of Monterey. They so wanted it to be a sea-serpent. Even we hoped it would be. When sometimes a true sea-serpent, complete and undecayed, is found or caught, a shout of triumph will go through the world. 'There, you see,' men will say, 'I knew they were there all the time. I just had a feeling they were there.' Men really need sea-monsters in their personal oceans. And the Old Man of the Sea is one of these. In Monterey you can find many people who have seen him. Tiny Colletto has seen him close up and can draw a crabbed sketch of him. He is very large. He stands up in the water, three or four feet emerged above the waves, and watches an approaching boat until it comes too close, and then he sinks slowly out of sight. He looks somewhat like a tremendous diver, with large eyes and fur shaggily hanging from him. So far, he has not been photographed. When he is, probably Dr. Bolin will identify him and another beautiful story will be shattered. For this reason we rather hope he is never photographed, for if the Old Man of the Sea should turn out to be some great malformed sea-lion, a lot of people would feel a sharp personal loss – a Santa Claus loss. And the ocean would be none the better for it. For the ocean, deep and black in the depths, is like the low dark levels of our minds in which the dream symbols incubate and sometimes rise up to sight like the Old Man of the Sea. And even if the symbol vision be horrible, it is there and it is ours. An ocean without its unnamed monsters would be like a completely dreamless sleep. Sparky and Tiny do not question the Old Man of the Sea, for they have looked at him. Nor do we question him because we know he is there. We would accept the testimony of these boys sufficiently to send a man to his death for murder, and we know they saw this monster and that they described him as they saw him.

We have thought often of this mass of sea-memory, or sea-thought, which lives deep in the mind. If one ask for a description of the unconscious, even the answer-symbol will usually be in terms of a dark water into which the light descends only a short distance. And we have thought how the human fetus has, at one stage of its development, vestigial gill-slits. If the gills are a component of the developing human, it is not unreasonable to suppose a parallel or concurrent mind or psyche development. If there be a life-memory strong enough to leave its symbol in vestigial gills, the preponderantly aquatic symbols in the individual unconscious might well be indications of a group psyche-memory which is the foundation of the whole unconscious. And what things must be there, what monsters, what enemies, what fear of dark and pressure, and of prey! There are numbers of examples wherein even invertebrates seem to remember and to react to stimuli no longer violent enough to cause the reaction. Perhaps, next to that of the sea, the strongest memory in us is that of the moon. But moon and sea and tide are one. Even now, the tide establishes a measurable, although minute, weight differential. For example, the steamship *Majestic* loses about fifteen pounds of its weight under a full moon.[1] According to a theory of George Darwin (son of Charles Darwin), in pre-Cambrian times, more than a thousand million years ago, the tides were tremendous; and the weight differential would have been correspondingly large. The moon-pull must have been the most important single environmental factor of littoral animals. Displacement and body weight then must certainly have decreased and increased tremendously with the rotation and phases of the moon, particularly if the orbit was at that time elliptic. The sun's reinforcement was probably slighter, relatively.

Consider, then, the effect of a decrease in pressure on gonads turgid with eggs or sperm, already almost bursting and awaiting the slight extra pull to discharge. (Note also the dehiscence of ova through the body walls of the polychaete worms. These ancient worms have their ancestry rooted in the Cambrian and they are little changed.) Now if we admit for the moment the potency of this tidal effect, we have only to add the concept of inherited psychic pattern we call 'instinct' to get an inkling of the force of the lunar rhythm so deeply rooted in marine animals and even in higher animals and in man.

When the fishermen find the Old Man rising in the pathways of their boats, they may be experiencing a reality of past and present. This may not be a hallucination; in fact, it is little likely that it is. The interrelations are too delicate and too complicated. Tidal effects are

[1] Marmer, *The Tide*, 1926, p. 26.

mysterious and dark in the soul, and it may well be noted that even today the effect of the tides is more valid and strong and widespread than is generally supposed. For instance, it has been reported that radio reception is related to the rise and fall of Labrador tides[2], and that there may be a relation between tidal rhythms and the recently observed fluctuations in the speed of light[3]. One could safely predict that all physiological processes correspondingly might be shown to be influenced by the tides, could we but read the indices with sufficient delicacy.

It appears that the physical evidence for this theory of George Darwin is more or less hypothetical, not in fact, but by interpretation, and that critical reasoning could conceivably throw out the whole process and with it the biologic connotations, because of unknown links and factors. Perhaps it should read the other way around. The animals themselves would seem to offer a striking confirmation to the tidal theory of cosmogony. One is almost forced to postulate some such theory if he would account causally for this primitive impress. It would seem far-fetched to attribute the strong lunar effects actually observable in breeding animals to the present fairly weak tidal forces only, or to coincidence. There is tied up to the most primitive and powerful racial or collective instinct a rhythm sense or 'memory' which affects everything and which in the past was probably more potent than it is now. It would at least be more plausible to attribute these profound effects to devastating and instinct-searing tidal influences active during the formative times of the early race history of organisms; and whether or not any mechanism has been discovered or is discoverable to carry on this imprint through the germ plasms, the fact remains that the imprint is there. The imprint is in us and in Sparky and in the ship's master, in the palolo worm, in mussel worms,

[2] *Science Supplement,* Vol. 80, No. 2069, p. 7, Aug. 24, 1934.
[3] *Science,* Vol. 81, No. 2091, p. 101, Jan. 25, 1935.

in chitons, and in the menstrual cycle of women. The imprint lies heavily on our dreams and on the delicate threads of our nerves, and if this seems to come a long way from sea-serpents and the Old Man of the Sea, actually it has not come far at all. The harvest of symbols in our minds seems to have been planted in the soft rich soil of our pre-humanity. Symbol, the serpent, the sea, and the moon might well be only the signal light that the psycho-physiologic warp exists.

THOR HEYERDAHL

The Kon-Tiki Expedition

The very first day we were left alone on the sea we had noticed fish round the raft, but were too much occupied with the steering to think of fishing. The second day we went right into a thick shoal of sardines, and soon afterwards an eight-foot blue shark came along and rolled over with its white belly uppermost as it rubbed against the raft's stern, where Herman and Bengt stood barelegged in the seas, steering. It played round us for a while, but disappeared when we got the hand harpoon ready for action.

Next day we were visited by tunnies, bonitos and dolphins, and when a big flying fish thudded on board we used it as bait and at once pulled in two large dolphins (dorados) weighing from 20 to 35 lbs. each. This was food for several days. On steering watch we could see many fish we did not even know, and one day we came into a school of porpoises which seemed quite endless. The black backs tumbled about, packed close together, right in to the side of the raft, and sprang up here and there all over the sea as far as we could see from the masthead. And the nearer we came to the equator, and the farther from the coast, the commoner flying fish became. When at last we came out into the blue water where the sea rolled by majestically, sunlit and sedate, ruffled by gusts of wind, we could see them glittering like a rain of projectiles, shooting from the water and flying in a straight line till their power of flight was exhausted and they vanished beneath the surface.

If we set the little paraffin lamp out at night flying fish were attracted by the light and, large and small, shot over the raft. They often struck the bamboo cabin or the sail and tumbled helpless on the deck. For, unable to get a take-off by swimming through the water, they just remained lying and kicking helplessly, like large-eyed herrings with long breast fins. It sometimes happened that we heard an outburst of strong language from a man on deck when a cold flying fish came unexpectedly at a good speed slap into his face. They always came at a good pace and snout first, and if they caught one full in the face they made it burn and tingle. But the unprovoked attack was quickly forgiven by the injured party, for this, with all its drawbacks, was a maritime land of enchantment where delicious fish dishes came hurtling through the air. We used to fry them for breakfast, and whether it was the fish, the cook, or our appetites, they reminded us of fried troutlings once we had scraped the scales off.

The cook's first duty when he got up in the morning was to go out on deck and collect all the flying fish that had landed on board in the course of the night. There were usually half a dozen or more, and one morning we found twenty-six fat flying fish on the raft. Knut was much upset one morning because, when he was standing operating with the frying pan, a flying fish struck him on the hand instead of landing right in the cooking fat.

Our neighbourly intimacy with the sea was not fully realised by Torstein till he woke one morning and found a sardine on his pillow. There was so little room in the cabin that Torstein was lying with his head in the doorway and, if anyone inadvertently trod on his face when going out at night, he bit him in the leg. He grasped the sardine by the tail and confided to it understandingly that all sardines had his entire sympathy. We conscientiously drew in our legs so that Torstein should have more room the next night, but then something happened which caused Torstein to find himself a sleeping-place on the top of all the kitchen utensils in the wireless corner.

It was a few nights later. It was overcast and pitch dark, and Torstein had placed the paraffin lamp just by his head, so that the night watches should see where they were treading when they crept in and out over his head . . . About four o'clock Torstein was woken by the lamp tumbling over and something cold and wet flapping about his ears. 'Flying fish,' he thought, and felt for it in the darkness to throw it away. He caught hold of something long and wet that wriggled like a snake, and let go as if he had burned himself. The unseen visitor twisted itself away and over to Herman, while Torstein tried to get the lamp alight. Herman started up too, and this made me wake thinking of the octopus which came up at night in these waters. When we got the lamp alight, Herman was sitting in triumph with his hand gripping the neck of a long thin fish which wriggled in his hands like an eel. The fish was over three feet long, as slender as a snake, with dull black eyes and a long snout with a greedy jaw full of long sharp teeth. The teeth were as sharp as knives and could be folded back into the roof of the mouth to make way for what it swallowed. Under Herman's grip a large-eyed white fish, about eight inches long, was suddenly thrown up from the stomach and out of the mouth of the predatory fish, and soon after up came another like it. These were clearly two deep water fish, much torn by the snake-fish's teeth. The snake-fish's thin skin was bluish violet on the back and steel blue underneath, and it came loose in flakes when we took hold of it.

Bengt too was woken at last by all the noise, and we held the lamp and the long fish under his nose. He sat up drowsily in his sleeping bag and said solemnly:

'No, fish like that don't exist.'

With which he turned over quietly and fell asleep again.

Bengt was not far wrong. It appeared later that we six sitting round the lamp in the bamboo cabin were the first men to have seen this fish alive. Only the skeleton of a fish like this one had been found a few times on the coast of South America and the Galapagos Islands; ichthyologists called it *Gempylus*, or snake mackerel, and thought it lived at the bottom of the sea at a great depth, because no one had ever seen it alive. But if it lived at a great depth, this must at any rate be by day, when the sun blinded the big eyes. For on dark nights *Gempylus* was abroad high over the surface of the seas; we on the raft had experience of that.

A week after the rare fish had landed in Torstein's sleeping bag, we had another visit. Again it was four in the morning, and the new moon had set, so that it was dark, but the stars were shining. The raft was steering easily, and when my watch was over I took a turn along the edge of the raft to see if everything was ship-shape for the new watch. I had a rope round my waist, as the watch always had, and, with the paraffin lamp in my hand, I was walking carefully along the outermost log to get round the mast. The log was wet and slippery, and I was furious when someone quite unexpectedly caught hold of the rope behind me and jerked till I nearly lost my balance. I turned round wrathfully with the lantern, but not a soul was to be seen. There came a new tug at the rope, and I saw something shiny lying writhing on the deck. It was a fresh *Gempylus*, and this time it had got its teeth so deep into the rope that several of them broke before I got the rope loose. Presumably the light of the lantern had flashed along the curving white rope, and our visitor from the depths of the sea had caught hold in the hope of jumping up and snatching an extra long and tasty tit-bit. It ended its days in a jar of formalin.

The sea contains many surprises for him who has his floor on a level with the surface, and drifts along slowly and noiselessly. A sportsman who breaks his way through the woods may come back and say that no wild life is to be seen. Another may sit down on a stump and wait, and often rustlings and cracklings will begin, and curious eyes peer out. So it is on the sea too. We usually plough across it with roaring engines and piston strokes, with the water foaming round our bows. Then we come back and say that there is nothing to see far out on the ocean.

Not a day passed but we, as we sat floating on the surface of the sea, were visited by inquisitive guests which wriggled and waggled about us, and a few of them, such as dolphins and pilot fish, grew so familiar that they accompanied the raft across the sea and kept round us day and night.

When night had fallen, and the stars were twinkling in the dark tropical sky, the phosphorescence flashed around us in rivalry with the stars, and single glowing plankton resembled round live coals so vividly that we involuntarily drew in our bare legs when the glowing pellets were washed up round our feet at the raft's stern. When we caught them we saw that they were little brightly shining species of shrimp. On such nights we were sometimes scared when two round shining eyes suddenly rose out of the sea right alongside the raft and glared at us with an unblinking hypnotic stare – it might have been the Old Man of the Sea himself. These were often big squids which came up and floated on the surface with their devilish green eyes shining in the dark like phosphorus. But sometimes they were the shining eyes of deep water fish which only came up at night and lay staring, fascinated by the glimmer of light before them. Several times, when the sea was calm, the black water round the raft was suddenly full of round heads two or three feet in diameter, lying motionless and staring at us with great glowing eyes. On other nights balls of light three feet and more in diameter would be visible down in the water, flashing at irregular intervals like electric lights turned on for a moment.

We gradually grew accustomed to having these subterranean or submarine creatures under the floor, but nevertheless we were just as surprised every time a new version appeared. About two o'clock on a cloudy night, on which the man at the helm had difficulty in distinguishing black water from black sky, he caught sight of a faint illumination down in the water which slowly took the shape of a large animal. It was impossible to say whether it was plankton shining on its body, or if the animal itself had a phosphorescent surface, but the glimmer down in the black water gave the ghostly creature obscure, wavering outlines. Sometimes it was roundish, sometimes oval or triangular, and suddenly it split into two parts which swam to and fro under the raft independently of one another. Finally there were three of these large shining phantoms wandering round in slow circles under us. They were real monsters, for the visible parts alone were some five fathoms long, and we all quickly collected on deck and followed the ghost dance. It went on for hour after hour, following the course of the raft. Mysterious and noiseless, our shining companions kept a good way beneath the surface, mostly on the starboard side, where the light was, but often they were right under the raft or appeared on the port side. The glimmer of light on their backs revealed that the beasts were bigger than elephants, but they were not whales, for they never came up to breathe. Were they giant ray-fish which changed shape when they turned over on their sides? They took no notice at all if we held the light right down on the surface to lure them up, so that we might see what kind of creatures they were. And like all proper goblins and ghosts, they had sunk into the depths when the dawn began to break.

We never got a proper explanation of this nocturnal visit from the three shining monsters, unless the solution was afforded by another visit we received a day and a half later in the full midday sunshine. It was May 24, and we were lying drifting on a leisurely swell in exactly 95° west by 7° south. It was about noon, and we had thrown overboard the guts of two big dolphins we had caught early in the morning. I was having a refreshing plunge overboard at the bows, lying in the water, keeping a good look out and hanging on to a rope-end, when I caught sight of a thick brown fish, six feet long, which came swimming inquisitively towards me through the crystal-clear sea water. I hopped quickly up on to the edge of the raft and sat in the hot sun looking at the fish as it passed quietly, when I heard a wild war-whoop from Knut, who was sitting aft behind the bamboo cabin. He bellowed 'Shark!' till his voice cracked in a falsetto, and as we had sharks swimming alongside the raft almost daily without creating such excitement, we all realised that this must be something extra special, and flocked astern to Knut's assistance.

Knut had been squatting there, washing his pants in the swell, and when he looked up for a moment he was staring straight into the biggest and ugliest face any of us had ever seen in the whole of our lives. It was the head of a veritable sea monster, so huge and so hideous that if the Old Man of the Sea himself had come up he could not have made such an impression on us. The head was broad and flat like a frog's, with two small eyes right at the sides, and a toadlike jaw which was four or five feet wide and had long fringes hanging drooping from the corners of the mouth. Behind the head was an enormous body ending in a long thin tail with a pointed tail fin which stood straight up and showed that this sea monster was not any kind of whale. The body looked brownish under the water, but both head and body were thickly covered with small white spots. The monster came quietly, lazily swimming after us from astern. It grinned like a bulldog and lashed gently with its tail. The large round dorsal fin projected clear of the water and sometimes the tail fin as well, and when the creature was in the trough of the swell the water flowed about the broad back as though washing round a submerged reef. In front of the broad jaws swam a whole crowd of zebra-striped pilot fish in fan formation, and large remora fish and other parasites sat firmly attached to the huge body and travelled with it through the water, so that the whole thing looked like a curious zoological collection crowded round something that resembled a floating deep water reef.

A 25 lbs. dolphin, attached to six of our largest fish-hooks, was hanging behind the raft as bait for sharks, and a swarm of pilot fish shot straight off, nosed the dolphin without touching it, and then hurried back to their lord and master, the sea king. Like a mechanical monster it set its machinery going and came gliding at leisure towards the dolphin which lay, a beggarly trifle, before its jaws. We tried to pull the dolphin in, and the sea monster followed slowly, right up to the side of the raft. It did not open its mouth, but just let the dolphin bump against it, as if to throw open the whole door for such an insignificant scrap was not worth while. When the giant came right up to the raft, it rubbed its back against the heavy steering oar, which was just lifted up out of the water, and now we had ample opportunity of studying the monster at the closest quarters – at such close quarters that I thought we had all gone mad, for we roared stupidly with laughter and shouted over-excitedly at the completely fantastic sight we saw. Walt Disney himself, with all his powers of imagination, could not have created a more hair-raising sea monster than that which thus suddenly lay with its terrific jaws along the raft's side.

The monster was a whale shark, the largest shark and the largest fish known in the world today. It is exceedingly rare, but scattered specimens are observed here and there in the tropical oceans. The whale shark has an average length of 50 feet, and according to zoologists it weighs 15 tons. It is said that large specimens can attain a length of 65 feet, and a harpooned baby had a liver weighing 600 lbs., and a collection of three thousand teeth in each of its broad jaws.

The monster was so large that when it began to swim in circles round us and under the raft its head was visible on one side while the whole of its tail stuck out on the other. And so incredibly grotesque, inert and stupid did it appear when seen full-face that we could not help shouting with laughter, although we realised that it had strength enough in its tail to smash both balsa logs and ropes to pieces if it attacked us. Again and again it described narrower and narrower circles just under the raft, while all we could do was to wait and see what might happen. When out on the other side it glided amiably under the steering oar and lifted it up in the air, while the oar-blade slid along the creature's back. We stood round the raft with hand harpoons ready for action, but they seemed to us like toothpicks in relation to the heavy beast we had to deal with. There was no indication that the whale shark ever thought of leaving us again; it circled round us and followed like a faithful dog, close to the raft. None of us had ever experienced or thought we should experience anything like it; the whole adventure, with the sea monster swimming behind and under the raft, seemed to us so completely unnatural that we could not really take it seriously.

In reality the whale shark went on encircling us for barely an hour, but to us the visit seemed to last a whole day. At last it became too exciting for Erik, who was standing at a corner of the raft with an eight-foot hand harpoon, and encouraged by ill-considered shouts, he raised the harpoon above his head. As the whale shark came gliding slowly towards him, and had got its broad head right under the corner of the raft, Erik thrust the harpoon with all his giant strength down between his legs and deep into the whale shark's gristly head. It was a second or two before the giant understood properly what was happening. Then in a flash the placid half-wit was transformed into a mountain of steel muscles. We heard a swishing noise as the harpoon line rushed over the edge of the raft, and saw a cascade of water as the giant stood on its head and plunged down into the depths. The three men who were standing nearest were flung about the place head over heels, and two of them were flayed and burnt by the line as it rushed through the air. The thick line, strong enough to hold a boat, was caught up on the side of the raft but snapped at once like a piece of twine, and a few seconds later a broken-off harpoon shaft came up to the surface two hundred yards away. A shoal of frightened pilot fish shot off through the water in a desperate attempt to keep up with their old lord and master, and we waited a long time for the monster to come racing back like an infuriated submarine; but we never saw anything more of the whale shark.

We were now in the South Equatorial Current and moving in a westerly direction just 400 sea miles south of the Galapagos. There was no longer any danger of drifting into the Galapagos currents, and the only contacts we had with this group of islands were greetings from big sea turtles which no doubt had strayed far out to sea from the islands. One day we saw a thumping big sea turtle lying struggling with its head and one great fin above the surface of the water. As the swell rose we saw a shimmer of green and blue and gold in the water under the turtle, and we discovered that it was engaged in a life and death struggle with dolphins. The fight was apparently quite one-sided; it consisted in twelve to fifteen big-headed, brilliantly coloured dolphins attacking the turtle's neck and fins and apparently trying to tire it out, for the turtle could not lie for days on end with its head and paddles drawn inside its shell.

When the turtle caught sight of the raft, it dived and made straight for us, pursued by the glittering fish. It came close up to the side of the raft and was showing signs of wanting to climb up on to the timber when it caught sight of us already standing there. If we had been more practised we could have captured it with ropes without difficulty as the huge carapace paddled quietly along the side of the raft. But we spent the time that mattered in staring, and when we had the lasso ready the giant turtle had already passed our bows. We flung the little rubber dinghy into the water, and Herman, Bengt and Torstein went in pursuit of the sea turtle in the round nutshell, which was not a great deal bigger than what swam ahead of them. Bengt as steward saw in his mind's eye endless meat dishes and the most delicious turtle soup. But the faster they rowed, the faster the turtle slipped through the water just below the surface, and they were not more than a hundred yards from the raft when the turtle suddenly disappeared without trace. But they had done one good deed at any rate. For when the little yellow rubber dinghy came dancing back over the water, it had the whole glittering school of dolphins after it. They circled round the new turtle, and the boldest snapped at the oar-blades which dipped in the water like fins; meanwhile the peaceful turtle escaped successfully from all its ignoble persecutors.

THOMAS FARBER

On Water

'Water remains a chaos,' Ivan Illich writes, 'until a creative story interprets its seeming equivocation . . . Most myths of creation have as one of their main tasks the conjuring of water. This conjuring always seems to be a division . . . the creator, by dividing the waters, makes space for creation.' Illich also writes that 'to keep one's bearing when exploring water, one must not lose sight of its dual nature.' Deep and shallow, life-giving and murderous, etc. etc.

Melville, for instance, reading the mystery of both sea and terra firma: 'consider them . . . and do you not find a strange analogy to something in yourself? For as this appalling ocean surrounds the verdant land, so in the soul of man there lies an insular Tahiti, full of peace and joy, but encompassed by all the horrors of the half known life. God keep thee! Push not off from that isle, thou canst never return!'

Reading water. 'Water represents the unconscious,' says the psychologist at the party. The surface the boundary between consciousness and unconsciousness. Water in this view thus a form of seductive regression, representing the purely instinctive. As Heraclitus cautioned, 'It is fatal for the soul to dissolve in water.'

For instance, naturalist Ann Zwinger, entering the 'peaceful, cradling ocean,' then looking up from below at the surface, 'a silken tent in the underwater wind, gray blue with moving ovals.' Dazzled by the 'interlocking lozenges of light,' she writes: 'how simple it is for those who pivot or rasp, supported and fed, adrift in an infinite womb, bathed in a life-support medium full of needed nutrients, needing less to eat than their land-based counterparts, less plagued by temperature fluctuations and desiccation, no need to heat their blood, filled with body fluids with the same osmotic pressure, so needing less protective covering to separate them from an alien atmosphere, suspended easily in this friendly bath without having to battle the incessant pull of gravity.'

Or consider the vision of author Joan Ocean, who in seeking a name that conveyed 'no male lineage', took on the first word that came to her. As she notes in *Dolphin Connection*, it 'was easy to spell, easy for other people to remember, unusual.' What's in a name? It was not until seven years later that she first experienced a 'symbiotic love for the ocean . . . I accepted completely my personal connection to it, and my responsibility to preserve and respect all of the vast life forms that resided within it.' This epiphany came 'by "absorption" ', after she had for the first time been in the water with (captive) dolphins. Soon, Joan Ocean wanted to be a dolphin, which for her meant 'to be weightless and buoyant', to 'move within the changing currents of water, to feel the earth turn, to slide on the waves, and be surrounded by the varied and unique concerts of sea animals. To play in the sparkling shafts of sunlight and bubbles, to feel the pull of the moon and the stars . . . to have seventy percent of the planet as your intimate and cosy home.'

Ocean does not mention, however, that dolphins are predators – of squid, lantern fish, shrimp, etc. – and that they have their own predators, including the eighteen-inch 'cookie cutter shark', which uses suction to extract large discs of dolphin flesh. More, deep-water dolphins school because it increases their capacity to protect themselves and/or to feed successfully. Lone Ranger dolphins, James Dean dolphins? Apparently not in the Pacific. Nonetheless, Joan Ocean has experienced 'commingling with particles of the water', has come to believe that telepathic interspecies messages can be 'conveyed on this air-water conduit.' (In Rodman's 'The Dolphin Papers', it is recorded in cetacean lore that a few human beings will one day make contact with the dolphins, but in one version of this cetacean legend, apparently, the humans never return to their own kind. In others, they return to be imprisoned as lunatics, or, alternatively, to implement either a bloodless coup d'etat or an 'insurrection of all the beasts', which eliminates humans as 'irredeemably depraved and dangerous to the planet.')

Liquid eyes. For Claudel, 'that unexplored pool of liquid light which God put in the depths of our being.' And, by extension: 'water is the gaze of the earth, its instrument for looking at time.' Water perhaps also being the earth's instrument for looking at us. And/or, our instrument for us to look at ourselves. As Melville wrote, Narcissus could not grab hold of the 'tormenting, mild image he saw in the fountain, plunged into it and was drowned. But that same image we ourselves see in all rivers and oceans. It is the image of the ungraspable phantom of life.' Put another way, the story of Narcissus suggests that the images of the self we see in water are not only seductive, but lethal.

Reading water. Dylan Thomas' 'carol-singing sea', for example. Or that sentient, conscious ocean in Lem's science fiction novel *Solaris*, capable of diagnosing the hearts of the scientists observing it and, even, of creating for them incarnations of their hungers. One studies the ocean on the planet Solaris, then, only at enormous risk. And always, of course, there are the limits of perception: the ocean on Solaris, Lem seems to be saying, is beyond human ken, whatever the human yearning for 'Contact.' (Poet Robinson Jeffers, looking west from Carmel, for years studied 'the hill of water' which is 'half the / planet: this dome, this half-globe, this bulging / Eyeball of water' with 'eyelids that / never close', what Jeffers termed 'the staring unsleeping / Eye of the earth.' And, he concluded, 'what it watches is not our wars.')

Reading water. 'To describe a wave analytically,' Calvino wrote, 'to translate its every movement into words, one would have to invent a new vocabulary and perhaps also a new grammar and a new syntax, or else employ a system of notation like a musical score.' Making a start, Pablo Neruda says, 'Teach us to see the sea wave by wave.' Not a bad aspiration, though Neruda himself is quick to figurative language: the ocean's 'gifts and dooms', the 'spent comet' of the wave's 'scorn and desire.' The need of the poet, like Lem's scientists, to make Contact. To name the qualities of even Earth's ocean, Lem seems to be arguing, thus reveals our hungers. Takes us to the limits of our capacities. And beyond.

JACQUES COUSTEAU

The Silent World

The sea is a most silent world. I say this deliberately on long accumulated evidence and aware that wide publicity has recently been given to the noises of the sea. Hydrophones have recorded clamours that have been sold as gramophone curiosities, but the recordings have been grossly amplified. It is not the reality of the sea as we have known it with naked ears. There are noises under water, very interesting ones that the sea transmits exceptionally well, but a diver does not hear boiler factories.

An undersea sound is so rare that one attaches great importance to it. The creatures of the sea express fear, pain and joy without audible comment. The old round of life and death passes silently, save among the mammals – whales and porpoises. The sea is unaffected by man's occasional uproars of dynamite and ships' engines. It is a silent jungle, in which the diver's sounds are keenly heard – the soft roar of exhalations, the lisp of incoming air, and the hoots of a comrade. One's hunting companion may be hundreds of yards away out of sight, but his harpoon may be clearly heard clanging on the rocks when he misses a shot, and when he returns one can taunt him by holding up a finger for each unsuccessful stroke.

Attentive ears may occasionally perceive a remote creaking sound, especially if the breath is held for a moment. The hydrophone can, of course, swell this faint sound to a din, which is helpful for analysis but not how it sounds to the submerged ear. We have not been able to adduce a theory to explain the creaking sounds. Syrian fishermen select fishing grounds by putting their heads down into their boats to the focal point of the sound shell that is formed by the hull. Where they hear creaking sounds they cast nets. They believe that the sound somehow emanates from rocks below, and rocks mean fish pasturage. Some marine biologists suppose the creaking sound comes from thick thousands of tiny shrimps, scraping pincers in concert. Such a shrimp in a specimen jar will transmit audible snaps. But the Syrians net fish, not shrimps. When we have dived into creaking areas we have never found a single shrimp. The distant rustle seems stronger in calm seas after a storm, but this is not always the case. The more we experience the sea, the less certain we are of conclusions.

Some fish can croak like frogs. At Dakar I swam in a loud orchestration of these monotonous animals. Whales, porpoises, croakers, and whatever makes the creaking noise are the only exceptions we know to the silence of the sea.

Some fish have internal ears with otoliths, or ear stones, which make attractive necklaces called 'lucky stones'. But fish show little or no reaction to noises. The evidence is that they are much more responsive to non-audible vibrations. They have a sensitive lateral line along their flanks which is, in effect, the organ of a sixth sense. As a fish undulates, the lateral receiver probably establishes its main sense of being. We think the lateral line can detect pressure waves, such as those generated by a struggling creature at a great distance. We have noticed that shouting at fish does not perturb them, but pressure waves generated by rubber foot fins seem to have a distinct influence. To approach fish we move our legs in a liquid sluggish stroke, expressing a peaceful intention. A nervous or rapid kick will empty the area of fish, even those behind rocks which cannot see us. The alarm spreads in successive explosions; one small fleeing creature is enough to frighten the others. The water trembles with emergency and fish far from sight receive the silent warning.

It has become second nature to swim unobtrusively among them. We will pass casually through a landscape where all sorts of fish are placidly enjoying life and showing us the full measure of acceptance. Then, without an untoward move on our part, the area will be deserted of all fish. What portent removes hundreds of creatures, silently and at once? Were porpoises beating up pressure waves out of sight, or were hungry dentex marauding off in the mist? All we know, hanging in the abandoned space, is that some inaudible raid siren has sent all but us to shelter. We feel like deaf men. With all senses attuned to the sea, we are still without the sixth sense, perhaps the most important of all in undersea existence.

At Dakar I was diving in water where sharks ranged peacefully among hundreds of tempting red sea-bream unwary of the predators. I returned to the boat and threw in a fishing line and hooked several bream. The sharks snapped them in two before I could get them into the boat. I think perhaps the struggle of the hooked fish transmitted vibrations that told the sharks there was easy prey available, animals in distress. In tropical waters we have used dynamite to rally sharks. I doubt whether the explosion is anything more than a dull, insignificant noise to them, but they answer the pressure waves of the fluttering fish that have been injured near the burst.

JAMES HAMILTON-PATERSON

Seven-Tenths: The Sea and its Thresholds

On the coast of a Philippine province there is a small town. On the landward side of the road, set well back among coconut palms and jasmine, is a whitewashed church with a green tin roof, only one of several civic buildings including an elementary school and an abandoned health centre. A legend surrounds this church, one known to every fisherman in town and to every boy who ever jumped off the little coral pier clutching a speargun. The legend *underlies* the church rather than surrounds it, for the story goes that there is a passage leading from the sea to a cave deep beneath it which is the lair of a giant octopus. There certainly is a fissure in the thick cap of fossil coral which covers much of the volcanic basalt of the island's coasts. Its mouth lies about 25 feet below the surface at high tide and at night a powerful underwater flashlight shone nervously in reveals no end to its interior.

In the absence of scuba gear there is nothing to be done, since only a madman attempts to explore a submarine cave with a chestful of air. There would be nothing to be done in any case. Whether or not one has been worked on by the legend, this particular depthless black slit does exude a peculiar aura of menace. The water around the mouth is always several degrees colder than elsewhere and very few fish appear ever to venture in. If there really were some great monster lying tucked away inside, a good deal of food would need to swim unwarily in for it to survive. On the other hand it may be that one way or another much of the town's drains seep into this crack and the creature survives on ordure. It might even have grown so fat it could no longer leave if it wanted to and is bottled up in its coral crypt. The thought of swimming up a sewer to confront a trapped monster is another good reason for not making the attempt.

These are all excuses, of course. The fact is, I am afraid of the place and so is everybody else. From the sunlit surface above its opening one can look across the road through the palms and see the church's corrugated roof. If the fissure really does extend that far it must be at least 100 metres long. Up in the brilliant daylight the whole legend looks different. The idea of a demon Kraken lying in its lair beneath a church is too naively Filipino, too redolent of Christian mythmaking to be more than the embroidery with which the credulous have ornamented a freak of local geology. Nevertheless, I am not going inside.

Only a mile or two down the coast and not far out to sea the reefs drop suddenly into ultramarine depths. By swimming out over the shallow corals it is possible to pretend one is low flying, hedgehopping above rough coral terrain, an illusion strengthened by soaring out over this great abyss. So abrupt and powerful is the effect one may even feel one's stomach drop. It was here, over the years, that I would practise seeing how far down I could swim, to set my own private record. Soon I knew every ledge on this cliff-face, each downward step of the agonising but exhilarating journey. I knew each level by its peculiar feature – a coral outcrop or eccentric sponge – and also by its ambient light. I knew as well as Beebe where red became grey, where my blood looked dark green in the water. This series of terraces now seems a lunatic vertical *via crucis*, every step gained representing pain, but it also stands as a chill measure of ageing. I can no longer reach the very deepest of the shelves I once touched, raising a confirmatory plume of silt still visible from the surface like a triumphant smoke signal impressively far below. I might again, I tell myself, but only if I lost weight since fat is buoyant and means one has to burn up more oxygen to drive it all downwards. Since I can no longer measure my own record, I have to estimate it as between 85 and 95 feet, rather less than the average local teenager can manage when harvesting big white sea cucumbers.

It was here, perhaps to spite myself, I tried 'riding the rock' instead, or using a weight to pull me down. This is how world free-diving records are set. Nobody labours to swim down; they ride the rock and somewhere beyond 300 feet let go and hope their still uncollapsed lungs contain enough air for them to claw their way back to the surface. Taken to such depths it is a dangerous sport, but I had no intention of going that far. In the event, the whole business felt faintly embarrassing. There is something foolish about loading rocks into a dinghy, rowing out, and jumping overboard clutching them to one's chest. The first was too heavy and took me down so fast I could not 'clear' in time and the pain in my ears made me let go at about 30 feet. The next took me down rather languidly, and it was a pleasure to see ledges I had fought to reach drifting upwards past me like floors in a descending lift. With a subsequent rock I passed my own record and was pulled onwards into unknown territory. I do not believe I ever went further than about 160 feet. There was something disagreeably inexorable about the downward tug. It was not as if one doubted for a moment that one's arms would release the weight before it was too late, but it had something to do with increasing pressure and deepening gloom which I had never experienced without breathing apparatus. Perhaps because the motion was entirely vertical and swift one imagined dissolving like a meteor, leaving a trail of silver bubbles, soon to be worn away to nothing by the rasping caress of the sea.

Beebe had written: 'The only other place comparable to these marvellous nether regions must surely be naked space itself . . . where sunlight has no grip upon the dust and rubbish of planetary air.' The exploration of space and of the deep sea have obvious things in common. Both require venturers to be supplied with complex life-support systems and defended against extreme ambient pressure, whether positive or negative. At a mythic level, however, there are important differences, many of which – in the case of deeps – have to do with the dark.

The famous and fatuous opposition of light and darkness is pre-Socratic in origin, only one pair of many made up of a 'noble' element (right, above, hot, male, dry, etc) and an 'ignoble' (left, below, cold, female, wet). By the sixth and fifth centuries BC the faculty of vision and the attributes of knowledge had run together in the Greek word *theorein*, meaning both 'to see' and 'to know'. Knowledge was henceforth a register of vision. Ignorance therefore becomes a lack of knowledge predicated on objects not being visible, so darkness equals ignorance. In turn, the dark becomes a source of fear as if a knowledge of visible objects were the only defence against terror and anxiety. By the eighteenth century the light of reason stood for the banishing of primordial fear: literally, enlightenment. Superstition as a concept is a product of eighteenth-century topology.[1] Where the ocean's deeps are concerned several other dualities operate as well, such as up/down, lightness/pressure, outwards/inwards and future/past. To go into space is in some sense to go forwards; to go down into the depths is at a psychic level to regress.

Why should this be? Space travel is 'going forwards' in the obvious sense that it involves technological 'progress', but so does deep sea exploration. It is as though *Homo* viewed himself in spatial rather than temporal terms, as if his history had been one not of eras and dynasties so much as of steady territorial expansion. Maybe the whole of human history might be rewritten, leaving out dates and

[1] This passage originates in a lecture given by Mark Cousins at the Architectural Association in London on 23 November 1990.

measuring instead the boundaries pushed outwards by tribes on their way to becoming nations, by earthlings as they stake out their claims to colonise the solar system. Yet even with nations claiming EEZs and seabed rights it never feels an appropriate choice of cliché when journalists call the ocean depths 'the last frontier'. As always, the sea is really less connected with space than with *time*, as if there were a correlation between going deep and going back. Thus the deeper one went the more primitive would be the life forms encountered, the more prehistoric and inchoate.

This must be a comparatively recent idea, post-Darwinian, at any rate, and taking into account a popular version of Victorian scientists' excitement on learning that the deeps were not azoic. The finding of the first coelacanth would have strengthened this, as does every fresh 'sighting' of the Loch Ness Monster. Legends of monsters and sea serpents are at least as ancient as the written word, but presumably it is only after the mid-nineteenth century that they begin to be depicted as prehistoric and corresponding loosely to fossil forms. The Loch Ness Monster is almost invariably spoken of nowadays not as some unknown species of sea snake or eel but as a saurian of prehistoric type. Since this is what people wish to be there it is faithfully confirmed by all the 'sightings'. It is thus a true remnant of a misapprehension by nineteenth-century science.

Myths of space travel do include visits to worlds at an earlier stage of evolution than our own. Yet even these are often in 'obscure' backwaters of space as if in the scriptwriters' imaginations space did correspond to a vast ocean in which the most developed regions tended to be those appearing from Earth most brightly lit. ('Rigel Concourse' in Jack Vance's stories is a good example, Rigel being a pure white, first-magnitude star. This is exactly where one would expect to find our outwardly bound pioneering descendants rather than huddled around some dismal cepheid variable out in the galactic sticks.) All this apart, the creatures most commonly associated with space operas as well as with UFOlogy are of an intelligence superior to ours: and with the waning of American paranoia about communism they tend to be less and less bent on kidnapping and brainwashing. Nowadays space aliens may well incline towards the godlike, beings from whom we might acquire knowledge, enlightenment, light itself, before it is too late.

The mythology of our own Planet's oceans is the polar reverse of all this, so much so that the nether world sometimes seems hardly part of the Earth at all. It is worth examining this from the popular standpoint for a moment because it shows how the concept of 'the deeps' relies on a jumble of associative ideas. Far from being likely to find enlightenment the further down we go, then, we expect to meet ever-dumber creatures. Moreover – exactly opposite to actuality – we envisage them near the bottom as still bigger, more terrifying in their mindless strength, and *uglier* . . . In fact, monsters. To this extent they are remarkably similar to the nightmare creatures of the unconscious: tentacular horrors which enwrap and bear their victims down and down to lairs where, in due time, begins the business of the hideous rending beak and saucerlike eyes. The very gradations of sleep itself seem to suggest a vertical descent into annihilating depths, the deepest levels of sleep being those of oblivion. The levels of dreaming, like the layers of the ocean which can support the biggest life forms, lie nearer the surface. In any case, by descending into the sea we would expect to meet the monstrous rather than the divine. Gods are the last things we would imagine finding in the deeps. It is no accident that even the men we encounter tend to be people like Captain Nemo, ominous whichever way we read his name. Astronauts have claimed close encounters with a Supreme Being, but never deep sea divers. Nor should we be surprised. Superior beings are by definition on top, while only the inferior can lurk below. The deeps also remind us of where we suppose we originally came from, what we have left behind. Going back thus to our genetic roots rather than to the sunlit idyll of Eden is a disquieting affair. Did we not abandon our ancestral dark by crawling towards the light?

No; we did not. The sea, to its dwellers, is not a dark place. With exceedingly acute eyes perceiving low levels of light and complex codes of bioluminescence; with sensitivity to sounds, smells and minute pressure differentials far beyond the spectrum of our own senses, it is as pointless to speak in crude human terms of 'light' and 'dark' as it would be when speculating about what a bat sees. A bat 'sees' with its ears with great precision and at speed. In short, there *is* no such thing as darkness. It exists only in the perception of the beholder. Vision does not depend on light.

To these 'oppositions' and their associations (up/down, above/below, superior/inferior, heaven/hell) should be added striving/sinking, where the first generally implies upward aspiration and self-betterment and the second is redolent of slummocking on a downward path, of Jack finding his own level while still undrowned. 'Sinking' is also used to describe wretched people glistening with sickness on their deathbeds, as if their problem were only one of weakness and they could no longer resist the force of gravity tugging them down towards their graves. That there might be something subtler at work than these pairs of opposites is suggested by the Latin word *altus* which can mean both high and deep (as it does still in Italian and where *l'alto* means 'the Deep' in an oceanic sense which also lingers in the English phrase 'the high seas'). In the Freudian unconscious, at least, such an idea would not embody a contradiction because there are no contradictions in the unconscious. Entirely antithetical and mutually cancelling propositions can exist simultaneously with not the slightest difficulty.

Perhaps, then, the least strange thing about the Deep is the degree to which it has retained its psychic force, its sonorous and chilling stateliness, its amalgamation of height and depth, of gulfs of space and of time. Almost no matter what is done to the oceans, however much they are explored and exploited, even ravaged and polluted, the Deep surprises us by its resistance to contamination. In this respect it resembles the Moon, which still feels much the same even though we know its dust bears the frivolous prints of cleated boots playing golf. The fact is, it was a different moon on which the astronauts landed, just as it is a different deep which GLORIA deafens with its sonar signals and whose silt is scarred by remotely controlled sleds gathering the sort of things a sled would gather. Neither Beebe nor Piccard nor Ballard ever visited the Deep. They reached various depths, even the ocean bed, but they carried the Deep within them. It is not a space to which there is physical access. Yet an air of mystery, no matter how slight, still surrounds objects retrieved from the depths, even beer bottles and polystyrene cups lowered by the curious. People like to touch things brought up, such as hoppers full of nodules. They like to feel the chill of aeons before it fades, just as they like to handle meteorites and moon rocks. If the ocean vanished tomorrow its mystery would not be found in the sum of its creatures flopping and dying and rotting on its bed. It exists elsewhere altogether, as Tennyson well knew when he capitalised on its high melancholy to express his grief over Arthur Hallam's death, hidden and heightened in a transition: 'From the great deep to the great deep he goes.'[2]

[2] Tennyson, *Idylls of the King*, 'The Coming of Arthur'. I.140.

WILLIAM BEEBE

Half Mile Down

Barton and I were screwed down and bolted in at ten o'clock, and four minutes later touched water. The surface was crossed with small wavelets, and three times before we were completely submerged the distant *Gladisfen* and the level horizon were etched clearly on the glass, and as instantly erased by a green and white smother. We sank slowly and I peered upward and watched the under side of the surface rise above me. When the rush of silvery bubble-smoke imprisoned beneath the sphere had passed, the surface showed clear. From the point of view of a submarine creature, I should by rights call it the floor of the air, and not the ceiling of the water. Even when diving in the helmet I am always conscious of the falsity of calling the water wet when I am once immersed in it. Spray blows in one's face and leaves it wet, but down below, the imprisoned air sailing upward, slips through one's fingers like balloon pearls, dry, mobile beauty, leaving only a pleasant sensation.

And now I looked up at our vertical wake of thousands of iridescent swimming bits of air, and, for a moment, forgot whither we were bound.

The boundary of air and water above me appeared perfectly solid, and like a slowly waving, pale green canopy, quilted everywhere with deep, pale puckers – the sharp apexes of the wavelets above showing as smooth, rounded indentations below. The sunlight sifted down in long, oblique rays as if through some unearthly beautiful cathedral window.

The host of motes of dust had their exact counterpart in mid-water, only the general feeling of colour was cool green, not yellow. The water was so clear that I could see dimly the distant keel of the *Gladisfen*, rolling gently. And here and there, like bunches of mistletoe hanging from a chandelier, were clusters of golden sargassum weed, with only their upper tips hidden, breaking through into the air. A stray berry went past my window and I saw an amusing likeness between its diminutive air-filled sphere and that which was at present my home.

The last thing in focus, of the upper world, was a long, undulating sea serpent of a rope dangling down from the side of the *Ready*.

We had asked to be lowered slowly. When less than 50 feet beneath the surface I happened to glance at a large, deep-sea prawn which I had taken for colour experiment. To my astonishment it was no longer scarlet, but a deep velvety black. I opened my copy of *Depths of the Ocean* and the plate of bright red shrimps was dark as night: No wonder I thought of the light as cool.

On this and other dives I carefully studied the changing colours, both by direct observation and by means of the spectroscope (Plate IV). Just beneath the surface the red diminished to one-half its normal width. At 20 feet there was only a thread of red and at 50 the orange was dominant. This in turn vanished at 150 feet. 300 feet found the whole spectrum dimmed, the yellow almost gone and the blue appreciably narrowed. At 350 I should give as a rough summary of the spectrum fifty per cent blue violet, twenty-five per cent green, and an equal amount of colourless pale light. At 450 feet no blue remained, only violet, and green too faint for naming. At 800 feet there was nothing visible but a narrow line of pale greyish-white in the green-blue area, due of course to the small amount of light reaching my eye. Yet when I looked outside I saw only the deepest, blackest-blue imaginable. On every dive this unearthly colour brought excitement to our eyes and minds.

A few familiar aurelia jellyfish drifted past while we were sinking to 50 feet, and at 100 feet a cloud of brown thimble jellies vibrated by the window. These were identical with those which we had observed in vast swarms in Haiti.[1] They are supposed to be surface forms, but here they were pushing against my window 20 fathoms down. They were the first organisms which showed that the fused quartz did away with all distortion. Full 20 feet away I could see them coming, and the knowledge of their actual size – that of a thimble – gave me a gauge of comparison which helped in estimating distance, size, and speed of unknown organisms.

I found that little things could change my whole mental outlook in the bathysphere. Up to this moment I had been watching the surface or seeing surface organisms, and I had focused so intensely upward that what was beneath had not yet become vivid. As the last thimble jelly passed, an air bubble broke loose from some hidden corner of the sphere, and writhing from the impetus of its wrenching free, rose swiftly, breaking into three just overhead, and the trio vanished. Now I felt the isolation and the awe which increased with the dimming of the light; the bubble seemed the last link with my upper world, and I wondered whether any of the watchers saw it coming, silver at first, then clothing itself in orange and red iridescence as it reached the surface – to break and merge and be lost for ever.

At 200 feet there occurred my first real deep-sea experience on this dive, something which could never be duplicated on the surface of the water. A six-inch fish suddenly appeared, nosed the bag of ancient squid and then took up its position close to the glass of my window, less than a foot away from my face. Something about it seemed familiar, yet it was strange. In size, shape and general pattern it was very like a pilot-fish, *Naucrates ductor*. Twice it swam back to the delectable bait and three times returned to where it almost

[1] *Beneath Tropic Seas*, Linuche jellies, pp. 20–23.

diametered my circular outlook. Then I knew what the trouble was – it was the ghost of a pilot-fish – pure white with eight wide, black, upright bands. At 200 feet a pilot-fish could not be the colour he is at the surface, and, like Einstein's half-sized world, here was a case where only the faulty, transient memory of man sealed up in a steel sphere had any right to assert that under different conditions the fish would show any colours other than the dark upright bands. I am certain that the fish itself aided this pale appearance, for it has considerable power of colour change, but this was very different from the mere expansion and contraction of dermal chromatophores. At 250 feet I saw the pilot-fish going *upward.*

There was a similarity between two- and three-hundred-foot levels in that most of the fish seen were carangids, such as pilot-fish and *Psenes* (this has no human or Christian name, but its technical one is so interesting to pronounce that this can be excused!). Long strings of siphonophores drifted past, lovely as the finest lace, and schools of jellyfish throbbed on their directionless but energetic road through life. Small vibrating motes passed in clouds, wholly mysterious until I could focus exactly and knew them for pteropods, or flying snails, each of which lived within a delicate, tissue shell, and flew through life with a pair of flapping, fleshy wings.

At 400 feet there came into view the first real deep-sea fish – *Cyclothones* or round-mouths, lanternfish, and bronze eels. The former meant nothing at first; I took them for dark-coloured worms or shrimps. Only when I saw them at greater depths in the searchlight did I recognize them. Of all the many thousands of these fish which I have netted, I never saw one alive until now. The lanternfish (Myctophids) came close to the glass and were easy to call by name. Instead of having only a half-dozen scales left, like those caught in the nets, these fish were ablaze with their full armour of iridescence. Twice I caught the flash of their light organs, but only for an instant.

An absurdly small and rotund puffer appeared quite out of place at this depth, but with much more reason he probably thought the same of me.

Big silvery bronze eels came nosing about the bait, although what they expected to accomplish with their exceedingly slender and delicate jaws is hard to imagine. Their transparent larva also appeared, swimming by itself, a waving sheet of watery tissue. Pale shrimps drifted by, their transparency almost removing them from vision. Now and then came a flash as from an opal, probably the strange, flat crustacean, well-named *Sapphirina*. Ghosts of pilot-fish swam into view again at this level.

Here, at 400 feet, we found that we could just read ordinary print with an effort, and yet to the unfocussed eye the illumination seemed very brilliant. I found that the two hours' difference between 10 a.m. and noon, marking the two dives, Numbers Four and Seven, although both were made in full sunlight, resulted in fifty per cent less illumination at 10 a.m. than at noon.

At 500 feet I had fleeting glimpses of fish nearly two feet long, perhaps surface forms, and here for the first time I saw strange, ghostly, dark forms hovering in the distance – forms which never came nearer, but reappeared at deeper, darker depths. Flying snails passed in companies of fifty or more, looking like brown bubbles. I had seen them alive in the net hauls, but here they were at home in thousands. As they perished from old age or accident or what-not, their shells drifted slowly to the bottom, a mile and a half down, and several times when my net had accidentally touched bottom it had brought up quarts of the empty, tinkling shells.

Small, ordinary-looking squids balanced in mid-water. I hoped to see some of the larger ones, those with orange bull's-eye lights at the tips of their arms, or the ones which glow with blue, yellow, and red light organs. None came close enough, however, or it may be

I must wait until I can descend a mile and still live, before I can come to their haunts.

A four-inch fish came into view and nosed the baited hook. It was almost transparent, the vertebrae and body organs being plainly visible, the eyes and the food-filled stomach the only opaque parts. Since making the dive I have twice captured this fish, the pinkish, semi-transparent young of the scarlet, big-eyed snapper.

At 550 feet I found the temperature inside the sphere was seventy-six degrees, twelve degrees lower than on deck. Near here a big leptocephalus undulated past, a pale ribbon of transparent gelatine with only the two iridescent eyes to indicate its arrival. As it moved I could see the outline faintly – ten inches long at least, and as it passed close, even the parted jaws were visible. This was the larva of some great sea eel.

As 600 feet came and passed I saw flashes of light in the distance and at once turned on the searchlight, but although the blue outside seemed dark, yet the electric glare had no visible effect, and we turned it off. The sparks of light and the distant flashes kept on from time to time showing the power of these animal illuminations.

A pale blue fish appeared, yet the blue of the pilot-fish does not exist at this depth. Several seriola-like forms nosed toward me. They must have drifted down from the surface waters into these great pressures without injury. Dark jellyfish twice came to my eyes, and the silvery eels again. The flying snails looked dull gold and I saw my first shrimps with minute but very distinct port-holes where the lights must be. Again a great cloud of a body moved in the distance – this time pale, much lighter than the water. How I longed for a single near view, or telescopic eyes which could pierce the murk. I felt as if some astonishing discovery lay just beyond the power of my eyes.

At another hundred feet a dozen fish passed the sphere swimming almost straight upright, yet they were not unduly elongate like the trumpetfish which occasionally assume this position in shallow waters near shore. I had a flash only of the biggest fish yet – dark, with long, tapering tail and quite a foot in length. Shrimps and snails drifted past like flakes of unheard-of storms. Also a large transparent jellyfish bumped against the glass, its stomach filled with a glowing mass of luminous food.

Here and at 800 feet a human being was permitted for the first time the sight of living, silver hatchet-fish, heliographing their silver sides. I made Barton look quickly out so he could verify the unexpected sight.

Here is an excerpt, of a very full seventeen minutes, direct from the transcription which Miss Hollister took of my notes telephoned up from 800 feet on Dive Number Eleven:

June 19, 1930. 1.24 p.m. Depth 800 feet: 2 black fish, 8 inches long going by, rat-tailed, probably *Idiacanthus*. 2 long, silver, eel-like fish, probably *Serrivomer*. Fish and invertebrates go up and down the shaft of light like insects. 3 Myctophids with headlights; *Diaphus*. (Work with a mirror next time.) 2 more different Myctophids. The same 3 Myctophids with headlights. 20 Pteropods and 6 or 8 *Argyropelecus* together. 3 more Pteropods. Little twinkling lights in the distance all the time, pale greenish in colour. Eels, 1 dark and 1 light. Big *Argyropelecus* coming; looks like a worm head on. *Eustomias*-like fish 5 inches long. 30 *Cyclothones*, greyish white.

We had left the deck at ten o'clock, and it was twenty-five minutes later that we had again reached our record floor – 800 feet. This time I had no hunch – reasonable or unreasonable – and three minutes later we were passing through a mist of crustaceans and flapping snails at 900 feet. We both agreed that the light was quite bright enough to read by and then we tried pica type and found that our eyes showed nothing definite whatever. With the utmost straining I could just distinguish a plate of figures from a page of type. Again the word 'brilliant' slipped wholly free of its usual meaning, and we looked up from our effort to see a real deep-sea eel undulating close to the glass – a slender-jawed *Serrivomer*, bronzy red, as I knew in the dimly-remembered upper world, but here black and white.

At 1000 feet we had a moment's excitement when a loop of black, sea-serpenty hose swung around before us, a jet-black line against blackish-blue.

Almost at once the sparks we had seen higher up became more abundant and larger. At 1050 feet I saw a series of luminous, coloured dots moving along slowly, or jerking unsteadily past, similar and yet independent. I turned on the searchlight and found it effective at last. At 600 feet it could not be distinguished; here it cut a swath almost material, across my field of vision, and for the first time, as far as I know, in the history of scientific inquiry, the life of these depths was visible. The searing beams revealed my coloured lights to be a school of silver hatchet-fish, *Argyropelecus*, from a half to two inches in length and gleaming like tinsel (*Frontispiece*). The marvel of the searchlight was that up to its sharp-cut border the blue-blackness revealed nothing but the lights of the fish. In this species these burned steadily, and each showed a colourful swath directed downward – the little iridescent channels of glowing reflections beneath the source of the actual light. These jerked and jogged along until they reached the sharp-edged border-line of the searchlight's beam, and as they entered it, every light was quenched, at least to my vision, and they showed as spots of shining silver, revealing every detail of fin and eye and utterly absurd outline. When I switched off the electricity or the fish moved out of its path, their pyrotechnics again rushed into visibility. The only effect of the yellow rays was to deflect the path of each fish slightly away from their course. Like active little rays of light entering a new medium, the Argyros passed into the searchlight at right angles to my eye and left it headed slightly away. With them was a mist of jerking pteropods with their delicate shields, frisking in and out among the hatchet-fish like a pack of dogs round the horses.

My hand turned the switch and I looked out into a world of inky blueness where constellations formed and reformed and passed without ceasing. At this moment I heard Miss Hollister's voice faintly seeping through Barton's head-phones, and it seemed as if the sun-drenched deck of the *Ready* must surely be hundreds of miles away.

I used the searchlight intermittently, and by waiting until I saw some striking illumination I could suddenly turn it on and catch sight of the author before it dashed away.

At 1100 feet we surveyed our sphere carefully. There was no evidence of the hose coming inside, the door was dry as a bone, the oxygen tanks were working well and by occasional use of our palm-leaf fans, the air was kept sweet. The walls of the bathysphere were dripping with moisture, probably sweating from the heat of our bodies condensing on the cold steel. The chemicals were working well, and we had a grand shifting of legs and feet, and settled down for what was ahead of us.

In the darkness of these levels I had not been able to see the actual forms of the hatchet-fish, yet a glance out of the window now showed distinctly several rat-tailed macrourid-like fish twisting around the bend of the hose. They were distinct, and were wholly new to me. Their profiles were of no macrourid I had ever seen. As I watched, from the sides of at least two, there flashed six or more dull greenish lights, and the effect on my eyes was such that the fish vanished as if dissolved into water, and the searchlight showed not a trace. I have no idea of what they were.

At 1200 feet there dashed into the searchlight, without any previous hint of illumination, what I identified as *Idiacanthus*, or golden-tailed serpent dragon, a long, slender, eel-like form, which twisted and turned about in the glare, excited by some form of emotion. Twice it touched the edge and turned back as if in a hollow cylinder of light. I saw it when at last it left, and I could see no hint of its own light, although it possesses at least three hundred light organs. The great advantage of the electric light was that even transparent fins – as in the present case – reflected a sheen and were momentarily visible.

From this point on I tied a handkerchief about my face just below the eyes, thus shunting my breath downward and keeping the glass clear, for I was watching with every available rod and cone of both eyes, at what was going on outside the six-inch circle of the quartz.

At 1250 feet several more of the silver hatchets passed, going upward, and shrimps became abundant. Between this depth and 1300 feet not a light or an organism was seen: it was 50 feet of terrible emptiness, with the blue mostly of some wholly new colour term – a term quite absent from any human language. It was probably sheer imagination but the characteristic most vivid was its transparency. As I looked out I never thought of feet or yards of visibility, but of the hundreds of miles of this colour stretching over so much of the world. And with this I will try to leave colour alone for a space.

Life again became evident around 1300 feet and mostly luminous. After watching a dozen or more firefly-like flashes I turned on the searchlight and saw nothing whatever. These sparks, brilliant though they were, were kindled into conflagration and quenched in the same instant upon invisible bodies. Whatever made them was too small to reach my eyes, as was almost the host of copepods or tiny crustaceans through which we passed now and then. At one time I kept the electric light going for a full minute while we were descending, and I distinctly observed two zones of abundance and a wide interval of very scanty, mote-like life. When they were very close to the glass

I could clearly make out the jerking movements of copepods, but they were too small to show anything more. The milky sagitta, or arrow worms, were more easily detected, the eye catching their swift dart and then focussing on their quiet forms. While still near 1300 feet a group of eight large shrimps passed, showing an indeterminate colouration. We never took large shrimps at these comparatively shallow levels in the trawling nets.

Barton had just read the thermometer as seventy-two degrees when I dragged him over to the window to see two more hatchet-fish and what I had at last recognized as round-mouths. These are the most abundant of deep-sea fish and we take them in our nets by the thousand. Flickering forms had been bothering me for some time, giving out no light that I could detect, and twisting and wriggling more than any shrimp should be able to do. Just as my eyes had at first refused to recognize pteropods by their right names, I now knew that several times in the last few hundred feet I had seen *Cyclothones*, or round-mouths. In the searchlight they invariably headed uplight, so that only their thin-lipped mouths and tiny eyes were turned toward me.

Before Barton went back to his instruments, three squids shot into the light, out and in again, changing from black to barred to white as they moved. They showed no luminescence.

At 10.44 we were sitting in absolute silence, our faces reflecting a faint bluish sheen. I became conscious of the pulse-throb in my temples and remember that I kept time to it with my fingers on the cold, damp steel of the window ledge. I shifted the handkerchief on my face and carefully wiped the glass, and at this moment we felt the sphere check in its course – we felt ourselves press slightly more heavily on the floor and the telephone said '1400 feet.' I had the feeling of a few more metres' descent and then we swung quietly at our lowest floor, over a quarter of a mile beneath the surface.

I pressed my face against the glass and looked upward and in the slight segment which I could manage I saw a faint paling of the blue. I peered down and again I felt the old longing to go further, although it looked like the black pit-mouth of hell itself – yet still showed blue. I thought I saw a new fish flapping close to the sphere, but it proved to be the waving edge of the Explorers' Club flag – black as jet at this depth.

My window was clear as crystal, in fact clearer, for, as I have said before and want to emphasize, fused quartz is one of the most transparent of all substances and transmits all wave-lengths of sunlight. The outside world I now saw through it was, however, a solid, blue-black world one which seemed born of a single vibration – blue, blue, for ever and for ever blue.

Once, in a tropical jungle, I had a mighty tree felled. Indians and convicts worked for many days before its downfall was accomplished, and after the cloud of branches, leaves, and dust had settled, a small, white moth fluttered up from the very heart of the wreckage. As I looked out of my window now I saw a tiny, semi-transparent jellyfish throbbing slowly past. I had seen numerous jellyfish during my descent and this one aroused only a mental note that this particular species was found at a greater depth than I expected. Barton's voice was droning out something, and when it was repeated I found that he had casually informed me that on every square inch of glass on my window there was a pressure of slightly more than six hundred and fifty pounds. The little moth flying unharmed from the terrific tangle, and the jellyfish drifting gently past seemed to have something in common. After this I breathed rather more gently in front of my window and wiped the glass with a softer touch, having in mind the nine tons of pressure on its outer surface!

However, it was not until I had ascended that the further information was vouchsafed me that the pressure of the water at our greatest depth, upon the bathysphere from all directions, was more than six and a half million pounds, or more concisely, 3366.2 tons. So far from bringing about an anticlimax of worry, this meant hardly more than the statement that the spiral nebula in Andromeda is 900,000 light years away. Nevertheless I am rather glad that this bit of information was withheld until I had returned to the surface. If I had known it at the time I think the two-tenths of a ton might have distracted my attention – that 400 pounds being fraught with rather a last-straw-on-the-camel's-back significance!

Like making oneself speak of earthrise instead of sunset, there was nothing but continued mental reassertion which made the pressure believable. A six-inch dragon-fish, or *Stomias*, passed – lights first visible, then three seconds of searchlight for identification, then lights alone – and there seemed no reason why we should not swing the door open and swim out. The baited hooks waved to and fro, and the edge of one of the flags flapped idly and I had to call upon all my imagination to realize that instant, unthinkably instant, death would result from the least fracture of glass or collapse of metal. There was no possible chance of being drowned, for the first few drops would have shot through flesh and bone like steel bullets.

The duration of all this rather maudlin comment and unnecessary philosophizing occupied possibly ten seconds of the time we spent at 1426 feet.

When, at any time in our earthly life, we come to a moment or place of tremendous interest it often happens that we realize the full significance only after it is all over. In the present instance the opposite was true and this very fact makes any vivid record of feelings and emotions a very difficult thing. At the very deepest point we reached I deliberately took stock of the interior of the bathysphere; I was curled up in a ball on the cold, damp steel, Barton's voice relayed my observations and assurances of our safety, a fan swished back and forth through the air and the ticking of my wrist-watch came as a strange sound of another world.

Soon after this there came a moment which stands out clearly, unpunctuated by any word of ours, with no fish or other creature visible outside. I sat crouched with mouth and nose wrapped in a handkerchief, and my forehead pressed close to the cold glass – that transparent bit of old earth which so sturdily held back nine tons of water from my face. There came to me at that instant a tremendous wave of emotion, a real appreciation of what was momentarily almost superhuman, cosmic, of the whole situation; our barge slowly rolling high overhead in the blazing sunlight, like the merest chip in the midst of ocean, the long cobweb of cable leading down through the spectrum to our lonely sphere, where, sealed tight, two conscious human beings sat and peered into the abyssal darkness as we dangled in mid-water, isolated as a lost planet in outermost space. Here, under a pressure which, if loosened, in a fraction of a second would make amorphous tissue of our bodies, breathing our own homemade atmosphere, sending a few comforting words chasing up and down a string of hose – here I was privileged to peer out and actually see the creatures which had evolved in the blackness of a blue midnight which, since the ocean was born, had known no following day; here I was privileged to sit and try to crystallize what I observed through inadequate eyes and interpret with a mind wholly unequal to the task. To the ever-recurring question, 'How did it feel?' etc., I can only quote the words of Herbert Spencer, I felt like 'an infinitesimal atom floating in illimitable space.' No wonder my sole written contribution to science and literature at the time was, 'Am writing at a depth of a quarter of a mile. A luminous fish is outside the window.'

DANIEL DUANE

Caught Inside

When Krakatoa erupted in 1883 in Indonesia, creating a blast heard three thousand miles away in Madagascar, a cataclysmic wave – a tsunami – ranging from 60 to 120 feet high, destroyed hundreds of towns and killed over 36,000 people. Seismic, benthic disturbances can do the same: in 1960 an 8.5 earthquake in Chile generated a tsunami that crossed the Pacific and levelled the city of Hilo in Hawaii. Then, of course, there are rogue waves, the occasional out-of-scale monsters formed by converging swells, the kind that are supposed to keep you from walking out exposed rock spits on otherwise flat days. Like the 112-footer encountered February 7, 1933, by the USS *Ramapo* in the North Pacific; a sailor on the bow apparently had the sangfroid to measure the wave's height by triangulating off a high point at the ship's middle. But as delicious as such tales are, the surfer lives around and for more mundane phenomena, wind waves and even tides, which are themselves waves half the circumference of the earth and half a day apart. The high tide, then, is a global crest, while the low is a global trough. And one rides the decaying pulses of far-off storms: solar heat and global winds create high and low pressure systems, areas of storm and areas of calm; a ridge over Oregon, a trough off Anchorage. Wind from the warmer high fills in the colder low and in transit makes a 'fetch' over a stretch of sea, which makes waves: wind friction first ripples the water, then, pushing on the ripples, makes chop. Pushing on the chop makes a constant sea, which leaves the storm and fetch and rolls across open ocean, consolidating, compressing, combining into cleaner lines of swell, trains of waves. And in trains, the foremost wave constantly yields to the ones behind, and new ones appear in the back: a cycling through of shapes to produce a macro ripple, a revolving chunk of sine curve.

So the swell that rolls into harbour, so clean and so beyond storm, has travelled calm water for hundreds, even thousands of miles. When the pulse hits shallow water it slows and bends to the shape of the shoaling bottom; wave energy deflects off the bottom at an angle, steepening until breaking. One floats, then, waiting for a pulse from the sun, just so processed: the shape of the wave ridden is an expression of that faraway storm, of the shape of the seafloor over which one floats, of the precise direction from which a swell hits the seafloor, and of local winds. Onshore wind against the backs of the waves breaks them too soon; offshore into their faces holds them up beautifully steep and fast. And that ocean floor, rarely seen but intimately known: the shape and foam of breaking waves stain a white map of the bottom in liquid relief and motion. A surfer gets to know the benthic topography through its expression in the wave, through holes into which the wave vanishes and shelves over which it stands up. In fact, over the past few weeks, the rains had brought a near-constant audience to the far end of the cliff by my house: the overlook at the mouth of the San Lorenzo River. Every evening, guys gathered by the wooden rail to watch tons upon tons of silt flowing into the bay, sloshing through opaque brown breakers. Pulling their hats low in the downpour, they stood around chatting, speculating on whether this could be the year everyone's been waiting for, the year when the river-mouth sandbar finally regained the shape it had in the legendary winter of 1983. A year when similar, very wet 'Pineapple Express' Pacific storms drew enough mud and sand out of the mountains in just the right formation to make a flawless tubular wave break for nearly a hundred yards. It hadn't come close since, but every year – especially in a wet year, like this one – there was hope.

And still, a certain *something* one wonders about in the wave, some hint of the 'source,' some fundamental means of energy transfer suggesting the essentially *energetic* nature of matter. So, call up an old-friend-cum-physicist:

'Our data says we've done it,' the physicist pronounced, when I asked about his thesis experiment.

Done what?

'Built a clock that will count backwards if time reverses in the universe.' He sounded dismayed, irritated. 'And our data says we've done it.'

Reversed time, or built the clock?

'The clock, but it can't be true, because if it is, I win the Nobel Prize.'

I pictured his earnest blue eyes and sculpted, boyish face, tortoise-shell glasses. Then changed the subject, asked the big question: What the hell *are* waves, anyway?

He thought for a moment – an awkward silence in the delayed relay between Munich and California – then said, 'Waves of radiation or light or sound – it's all elongated sinusoidal oscillations. You really want to get into this? Get a pen. Frequency's the distance between peaks; the peak's the sine and the trough's the cosine; think of amplitude as the wave height, which is basically the vertical distance from trough to crest, and wavelength as the horizontal distance from trough to crest, and the period's the time it takes a crest to travel a wavelength. So, when you're sitting out there, a wave moves toward you but the water doesn't: you probably move in a pretty circular orbit when that wave passes under you.

Energy, though?

'Certainly. Same as sound waves,' he said, "cause energy's energy.

And, uhh . . . I think there's about four hundred joules of energy in a cc of water at room temperature, so . . . that means one quantum for a water wave's about . . .' He stopped talking for a moment, and I heard tapping noises. 'Zero point, oh, about thirty-three zeros and six joules. But a wave would have a lot of quanta. Joules per second is power, right? As in watts in your light bulb? So waves have power.'

Indeed. I'm somewhat mealy-minded at things quantitative; in fact, the physicist had always embarrassed me in such discussions, made me cling to my belief in the great variety of human intelligence.

'Anyway,' he said, 'you got to understand some things about the water molecule – it's unbelievably hard, and hates to bend or stretch. So picture water like this. Get a pen.' He talked me through a little sketch, mark at a time:

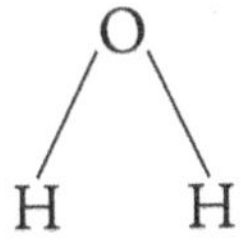

'And the bond's really strong,' he said, 'so water molecules like to endlessly stack on top of each other – sort of full-on group sex for most of the ocean. These stacks move in two ways: first, as a collective, and second, individually. In other words, the molecules can either bump into each other, or just vibrate alone.' He coughed into the line, complained about the lousy weather in northern Europe. 'So the ocean's frozen,' he said, 'rock-hard and nothing you'd call a wave. You hit it with a hammer and it rings a bit, right? So there's some sounds ice supports, basically really high wave frequencies. But melt it into a liquid, and the molecule bonds get looser. So now they're farther apart and take longer to bang into each other, giving you a lower resonant frequency – big, full-on waves. So what you're surfing,' he said, suddenly changing tone, 'is a giant collective undulation of stacked water molecules . . . Is that what you wanted to know?'

I supposed it was, and we chatted a little longer about the North America Wall, a gargantuan, five-day rock climb he'd done the month before on Yosemite's El Capitan. His hands had yet to heal, he said, still couldn't quite make a fist. He rambled a bit about his growing anxiety that he wasn't a true physics genius after all; they had usually peaked by his age, he said, solved some profound problem or redefined certain terms. A call came from an engaged friend of mine, inviting me to a wedding shower up in the city, and then I stepped outside. To corroborate the physicist's interpretation, I strolled down to the beach for a look: our walk in the park, our local copse or lane. It was a path I followed nightly now as well, even found a few familiar faces. Looked forward to the upcoming Christmas parade in the yacht harbour, in which sailboats motored along the docks at night, decked out in colored lights and electrified Santas and Nativity scenes. People with dogs, brooders, barefoot waders, and stumbling, inseparable couples. The sun setting a deep orange behind the west side of town. Above black silhouettes of eucalyptus and palm, a rich flaming band gave way to tropical purple, then to cobalt in the night behind. The orderly west swell of the last few days had faded to a murmuring jibberish, a pointless mumbling. One often hears surfers talk of waves in these terms, particularly on days when the bands are broken up and the interval too short, days when the sea resembles so much sloshing water rather than a field of rolling energy. 'It's all confused,' a guy will say, by way of visual description. I once even heard myself tell Willie of the water before us, without a trace of irony, 'Dude, it's making *no* sense.' I loved the implication that, at times of clean swell, the sea actually did make sense, as if it had some message to communicate, some language to speak. And at times of truly remarkable surf, when all the swell and wind vectors coincide with precisely the right tide to make your home break charge with life, it's as though the friend you've tolerated through weeks or months of doggerel has finally learned to sing the very pulse of the world. As if a speech normally hard to decipher, somehow *inarticulate*, has suddenly been uttered not just well, but in a language unmistakably meant for you; chaos reduced, at last, to a single idea. So when Vince rode his bike to the cliff each morning to calculate the day's optimal conflation of elements, he was, in a sense, looking for that moment when the world would intone just what he – and I – most wanted to hear.

Of course, different moments will sing different songs: stately and warlike Wagner during grand northwesters or, when the light breaks through storm clouds and a south swell catches the point perfectly square, the wit and mystical clarity of a Wallace Stevens lyric. (The evidence then for a divine hand seems awfully good.) I'd had a similar feeling once on the final headwall of a long route on El Capitan; we turned the lip of a big roof to see several hundred feet of polished, overhanging granite. Splitting the middle of that otherwise unclimbable expanse was a single perfect crack. The world then seemed to have a humane design; or, to take the other tack, my own relationship with the world seemed finally clear enough to recognize nature's capacity for precise expression. Waves do, after all, have a staggering complexity. Skinny had teased that I wouldn't be a real surfer until I bought myself a weather radio, and since I have quite a hang-up with authenticity, I'd scrambled out and bought one. Kept it on low volume at all times. As we all know, too much information can be a bad thing. The way nonstop TV coverage of a war or murder trial can tyrannize a person, so did these hourly wind updates tyrannize me, three hourly buoy reports complete with interval. This last figure, the average time elapsing between the passage of distinct wave peaks beneath the buoy, becomes in many ways the primary obsession: a nine-foot swell with a nine-second interval might register at the beach as only a feeble two or three feet; a nine-foot swell with a nineteen-second interval might register as nine feet and very powerful. So, whenever sitting there reading like a good boy, I occasionally heard a jump in average wave height or a lengthening of interval – hints of new swell – productivity plummeted. My quasi girlfriend had begun to find all this talk painfully tedious as she suffered through much of it on the phone, the unfortunate fabric of a long-distance relationship. She was quite alive to the birds and otters, even started keeping an Audubon bird book in her Toyota Celica in a sweet gesture of solidarity with me. But waves? Categorically uninteresting to her, although she'd even bought me a beautiful coffee-table book called *The Book of Waves* that Christmas. She just found them inert somehow. Not alive. About as interesting as rocks. Gendered concerns? Maybe.

A painter on disability who lived down the street had her evening Scotch on a concrete bench as I walked past; six feet and lanky, with an attractive, angular face. A woman I never got to know well, but with whom I exchanged regular, wearily complex smiles. The surly man with a big red beard sat on his usual bench and read; never with a friend or lover, just a book and a bag of walnuts. He also appeared every morning before dawn to, as he put it, 'read the paper and watch the fishing boats head out.' I'd noticed the next day that while he certainly read the paper, he always chose the one bench that did not allow a view of the departing boats. Two paddleboarders were out near the pier, but by and large swimmers held fast to the *Pequod*, stayed by the sand. At the southern end of the beach, a sandbar threw up a wedging collective undulation; four skimboarders in short-sleeved wetsuits held their boards, watching, not looking at one another, not talking. Then one trotted toward the water, finless plywood board held to one side and forward. He turned once, adjusted, then sprinted. At the inch-deep film of lingering water, he dropped the board, danced on, and skimmed into the muddy green face. He banked against it, ducked under its lip, and shot toward the beach, tubed. One second of glory before getting body-slammed. The other three guys stared away, bored.

A schooner sailed slowly past, visitor from the past surrounded by gawking modern sailboats; I tried to conjure Juan Cabrillo, George Vancouver, or Sir Francis Drake, wanted to imagine my beach pre-contact, but it wasn't here; only an Army Corps of Engineers harbor jetty had stopped up enough littoral drift to fill it in. To the west, one could see Lighthouse Point jutting into the bay and the great surf break at Steamer Lane. Mostly for this view, a procession of men passed here daily, parked their pickups on the way to work, sat a moment with the motor running. A few drove big Ford works trucks, clearly the general contractors; most drove the standard-issue surfer vehicle, the Japanese light pickup with fibreglass camper shell, four-wheel-drive only for those with Baja ambitions. They watched the Lane for swell, the outer bay waters for wind. A few came every day at lunch, rolled down a window and ate their sandwiches quietly – the ocean holding down a part of their lives, maintaining a self always renewing.

The lighthouse beacon turned pointlessly, and rills washed along the pilings of the tawdry pier just north, only occasionally thumping, never booming. The three-quarter moon hung as a flattened circle over a dusty Salinas sky to the southeast, and the Monterey ring of lights appeared due south, scribing the other side of the bay into the evening. A white sailboat drifted in the onshore, land-scented breeze and a big fishing boat sat near the pier. Earth and dry-grass smells, making the ocean seem less oppressively wild, more contained. Swimmers in no hurry to get out – a late trace of Indian summer without the cruel contrasts of high summer, without cold morning fog or blistering noon heat. I sat on the bench in a warm, gentle breeze with no impending storm, no unnerving drought or brooding fog: just a lovely evening in a lovely place. A silhouette dog chased a stick against anodized blue water scintillating with ambient crimson; a wave lifted dark over the sand, its back blocking out the dusk, and in the moment before collapsing – a pause like the death between breaths, the muteness between words – shot a streak of rising moonlight along its black face.

Later, sea lions barking under the pier and night fully come, the breeze reversed, blew cool off the Pacific and brought marine smells through my open window. I read once about a woman who closed her seaward windows at night to keep out the lost souls that wander the ocean's spaces.

WAYNE LEVIN

Resident Spirits

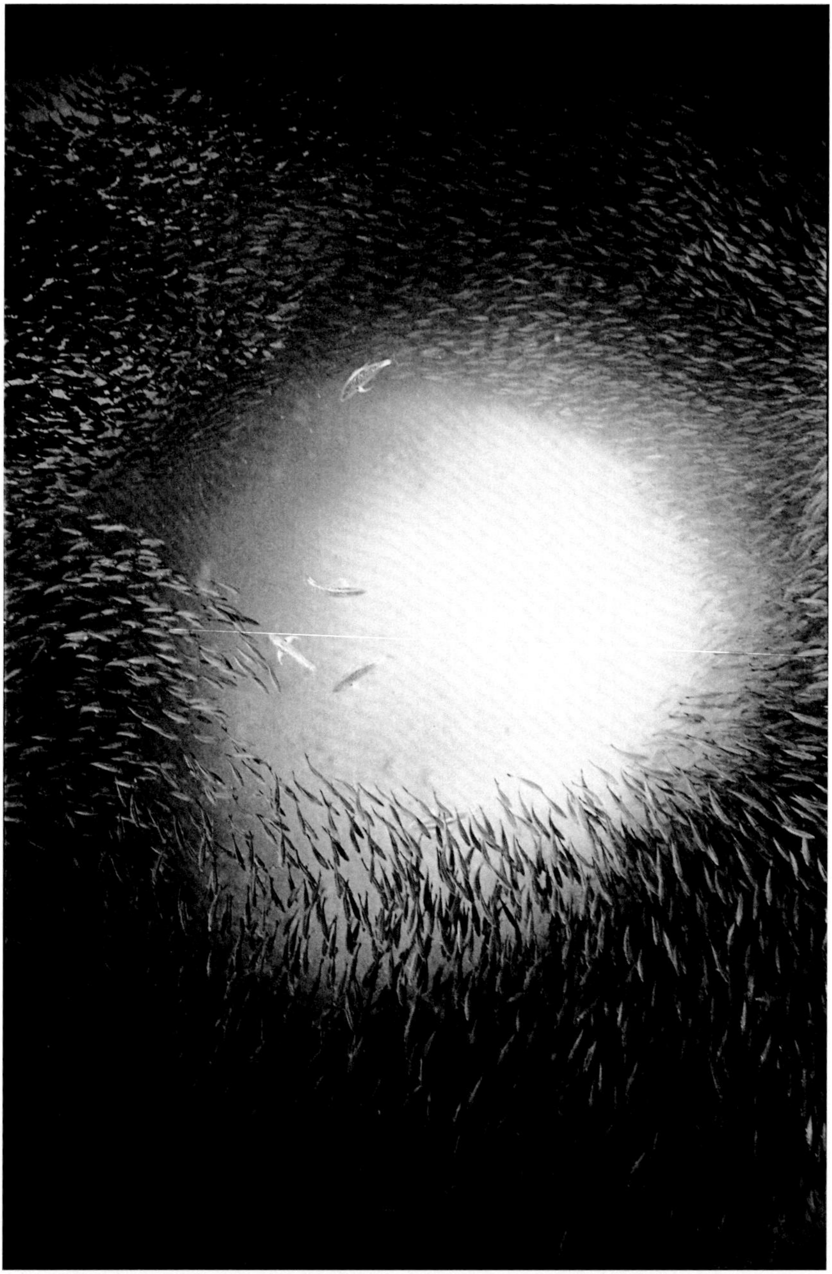

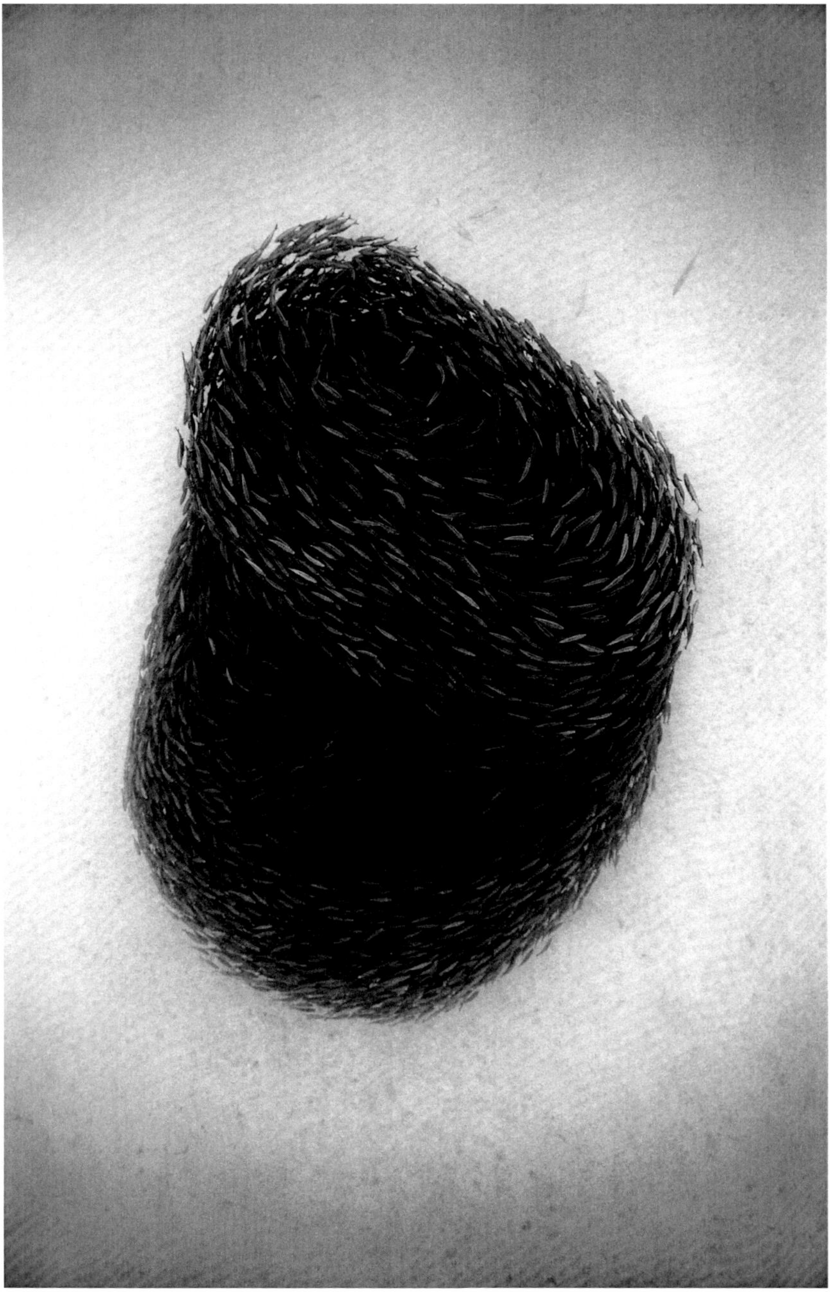

KEVIN HART

Facing the Pacific at Night

Driving east, in the darkness between two stars
Or between two thoughts, you reach the greatest ocean,
That cold expanse the rain can never net,

And driving east, you are a child again –
The web of names is brushed aside from things.
The ocean's name is quietly washed away

Revealing the thing itself, an energy,
An elemental life flashing in starlight.
No word can shrink it down to fit the mind,

It is already there, between two thoughts,
The darkness in which you travel and arrive,
The nameless one, the surname of all things.

The ocean slowly rocks from side to side,
A child itself, asleep in its bed of rocks,
No parent there to wake it from a dream,

To draw the ancient gods between the stars.
You stand upon the cliff, no longer cold,
And you are weightless, back before the thrust

And rush of birth when beards of blood are grown;
Or outside time, as though you had just died
To birth and death, no name to hide behind,

No name to splay the world or burn it whole.
The ocean quietly moves within your ear
And flashes in your eyes: the silent place

Outside the world we know is here and now,
Between two thoughts, a child that does not grow,
A silence undressing words, a nameless love.

JAMES BRADLEY

The Turtle's Graveyard

It has been fifteen hours since the boat left Koh Tao. Not long perhaps, in the greater scheme of things, but long enough. Long enough for the sun to wheel across the cerulean immensity of the sky. Long enough for the massing light of the constellations to rise and begin the long arc of their nightly motion. Long enough for this water, once warm as blood, to turn cool, then cold against my body. Long enough for my arms to tire, fall motionless to my side. Long enough for my skin to swell and pucker.

Have you ever raised your head, gazed upwards into rain? Seen that strange perspectiveless world of cloud and tumbling water? If you have you will know there is a moment, as the world recedes, when the rain will seem to pause in its descent, like the motion of a stone's parabola at apogee. I say a moment, but that is not quite the word, for time itself becomes elastic, slowing finally to a point, before it begins again, and there is reversal, slow at first, but hastening like the stone falling earthwards and all of a sudden the rain is no longer falling, it is you who is falling, or rather rising, upwards, against gravity, like flight, like breathing.

I fly like that now. Once this was water, but now I float in stars. Far beneath me the plankton darts and burns, phosphor bright, cold meteors against the dark. Around me the light from the stars dapples the water's surface, motes of light that move with the water like leaves across the limpid surface of a lake. And all the while I am rising into the sky, falling upwards into the ancient deep of night, the stars passing me like rain as I rise.

Beneath the water's surface my watch's face floats, ghostly green. One hand on three, the other at four: 3.20 a.m. Eight hours since Jenny left, swimming steadily away across the shimmering water, growing smaller and smaller until her body vanished into the haze of light and all I could hear was the faint sound of her stroke, growing fainter, and fainter. Then nothing. Only the fire of the setting sun on the lapping sea.

At first I wondered whether I should have gone with her, tried swimming for the shore, but I saw how low the sun had fallen, and knew, despite Jenny's protestations, that once it set we would no longer know which way to swim across the featureless sea. I cannot read the stars, nor can she. There are no stars in the city of light. And besides, I still hoped that someone might realise we were missing, seek Gerhard or Chas out in their cluttered office and demand to know who had counted the heads when they climbed back on board. That Gerhard and Chas would feel the yawning space of their error open within them as they glanced one to the other, realising what they had done, and come racing for the boat. For us.

But Jenny was frightened, too scared and angry to hear these possibilities. She knew we would not last the night, and wanted to swim for the shore. When I refused she swore and wept and then finally kissed me, her lips cold with the seawater, so I knew at the end she did not mean her words of anger.

I wonder where she is. Perhaps she can see the land, almost hear the music from the beachside discos as it echoes across the water. Maybe it is not stars that play around her but the reflected neon of the tow, slithering silkily across the waves. Or maybe she is lost like me, floating somewhere in the vastness of the night, too far from land to see the faint glow of the lights, needing the dawn and the sun to gauge which way to swim now. Or maybe she is dead, pulled beneath the surface by a shark, or just drowned, her corpse already puffy and bloating.

I know it is morbid to think like this. But I am not frightened. Instead I feel a strange calmness, almost a euphoria, which has grown in me as the night has passed. This is not how it was at the beginning, when the two of us broke the surface, excitedly laughing and talking one over the other as we pulled at our regulators, eager to share the delights we had just witnessed. Our first reef dive. The culmination of two weeks of training in that grass-walled hut with Gerhard and Chas, all for this. Not realising at first, as we turned slowly, looking for the boat, that they were gone, not quite believing our eyes as we turned once more, faster this time, only to see that it was no deception, no trick of the light, but that they had gone without us, back to the island, to the bars and lights and restaurants, leaving us out here. Alone.

Then, all of a sudden, we saw them, a receding speck in the immense blueness of sea, and although we knew it was futile we screamed and waved, using our fins to propel our bodies upwards, launching ourselves out of the water like dolphins performing in some wretched show, hoping that someone might turn and see us. But it was too late. Then, for a time, we waited here, telling each other that someone would notice, that they would realise there were too few of them, that some of the equipment was missing, that we had not been seen since they weighed anchor, but as the long day wore on that possibility receded too, and we were left, just waiting. Alone.

How strange that this should happen now. After Jenny and I saved and planned for this holiday, lured by the falling baht, by the promise of the beaches and the discos. For weeks we spent our lunch-times planning where we would go, what we would do, until I could think of little else, my excitement as consuming as a child's. I do not remember when the diving course became part of our plans, but once it appeared it lodged, like a burr, and everything else receded, the discos of Koh Samui replaced by the more austere pleasures of Koh Tao, the sightseeing by depth charts and equipment tests. Nor do I remember whose idea it was. I would like to think that it was Jenny's but it might just as well have been mine. Not that it matters now.

This holiday was special to me. I'm sure it was special to Jenny as well, but for her it did not carry the weight of symbolism. Instead it was just another holiday, three weeks of sun and drinking and sex. Sex. Even now it makes me laugh to think of Jenny and sex. The men she took to her bed, without shame or guilt. I cannot do that, not yet, although I hoped that tonight, maybe Peter and I, we could have done it, high on beer and sun and maybe just enough grass to make me feel loose and distant from my stern, lonely flesh. For it frightens me, the idea of it, of a man I do not know taking his member and placing it into me. Would it have been like Roger? His face, pale and cruel and somehow earnest as he hammered away on top of me. I think it was the earnestness I hated most, more even than his cruelty, which was real, particularly since that afternoon at the doctor's. More even than his hypocrisy; his false tenderness before, the way he would roll off without a word when he was done. I am not even sure how other people, *normal* people actually do it. I saw a movie, at a party that Jenny took me to. I remember the way she grinned at the way I stared at it, open-mouthed, but I was surprised, there's no sense denying it. The size of his cock, the things those two women did with it. However joyless it seemed I could not help but be startled by the inventiveness of their frenzied panting. The joylessness I have experience in, the inventiveness I will have to improvise. Or would have.

I wonder where Roger is now. I imagine he is asleep, alone in that bed we shared for six years. I can almost hear his breath, the choking rattle of his snore. It will be dawn there soon, and he will wake to the alarm, shower, dress for church. It is a while since I've been, but I imagine the services are still at six each weekday, seven on Sundays. I wonder whether he has met someone else yet. Maybe Hanny Crawford, or that new girl from Perth. He's not bad looking, not like most of the creeps in the church, so he shouldn't have too much trouble. I ran into his brother Mikey on George Street a month ago, and he told me the Elders had annulled our marriage because of my Godlessness, that they had absolved him of all blame. I think he thought I would be ashamed, but I only laughed, told him that I was pleased, maybe now Roger would get a new girlfriend and stop driving past my flat in the middle of the night. He didn't like that much, but I didn't care. Mikey's a jerk.

Roger blamed my job, said if I had never gone to work in the city I would never have met Jenny or Bill or Jodie, and I suppose he is right, at least partly. It was strange, my days spent with people who were not in the Church, men and women with lives that revolved around things other than prayer. People who fornicated, who smoked and drank and took drugs, men who loved other men, like Bill. Sweet, gentle Bill. Sinners, Roger called them. And worse things.

But it wasn't Jenny or Bill or Jodie. Nor even was it Roger. It was God. That afternoon in the doctor's office, when the doctor told me the infection had affected something within me, that I would never conceive. And Roger, white faced, his sweaty palm wrapped around mine, so tight my hand turned white, then blue. That can't be, he kept saying, You're wrong, but the doctor just looked at him over his glasses, his fat face exuding sorrow, although in his eyes you could see he was bored, bored with us, with all these desperate women and men, bored with everything. I'm afraid not, he said, The tests are conclusive, and then he looked at me and said, There's always in vitro, and I said, What? and he said it again, In vitro, and Roger looked from him to me and said, No, we can't, it's against our beliefs, and

I remember thinking, Our beliefs? I don't remember believing babies were wrong.

And afterwards, in the car, as the miles and miles of suburbs passed by outside, I remember thinking that this couldn't be right, not in the way that Roger thought it couldn't be right, but that it couldn't be right at all, not if God was in the world, not if God truly loved us. I had done everything I could, I had been good, and now God took away my right to bear a child, and then forbade me to conceive one outside my body. How could He be so cruel?

For a month I carried these thoughts with me, too afraid to tell anyone. And then there was the night of Jenny's party. Roger didn't want to go, but I made him, and grudgingly, he went. We parked in the street at 8.00, winding our way up the stairs to Jenny's apartment, only to find ourselves the first ones there. Roger wouldn't take a drink, his eyes black as murder when I sipped at the beer that Jenny put in my hand. I didn't miss the way she watched his face as she closed my fingers around the cold glass of the stubbie. Then later, Roger grabbing me in front of everyone, telling me I had to come, he was leaving, his voice booming out into sudden silence, everyone turning. Now, he was saying, pulling at me, Now! and then Bill was there, between us, and Jenny was pulling me back, her arms around me. I thought Roger would strike Bill, as he did me, as he had that night after the doctor's, but he didn't, he didn't do anything, just turned and walked away without a word.

I saw him a week later, and he told me I was a whore in the eyes of God, then begged me to come back, and I said no, so he called me a whore again, and seeing the way his fist was tightening I looked him in the eye and told him that if he hit me I'd have a restraining order put on him. Jenny had told me to say it, and it worked, he went quiet. I love you, he said, and I shrugged. Don't you love me? he asked and I shrugged again. Is this about you and children? he asked, and I nodded, although I knew he didn't understand how it all connected.

On my wrist my watch still glows, the silent gleam of the hands reminding me that dawn is still two hours away. I seem weightless, as if I have slipped free of all that bound me, as if my life has no weight. As if I am become light, and I fall through the water in shafts. This morning, when we made the first dive, I saw the fish dancing, their myriad quicksilvered bodies rising as one in a spinning cylinder of bodies, upwards, through a column of light. Back in the boat between dives I told Peter what I had seen, and he just smiled, told me it was like birds in flocks, that strange unity of movements, the way their motion flickers and changes as if they were one creature in many bodies. I didn't understand, and he said it was all a matter of maths, the fish keeping a certain distance between themselves and any other fish, so that when one moves they all move, but in unison. Nothing is truly random, he said, It's all a matter of discerning deeper patterns, and I said What? like God? and as soon as it was said I regretted the sharpness of my tone, but Peter only smiled and shook his head. I don't believe in God, he said, but I do believe the world has shape and meaning if you only know where to look.

And then he told me about flowers which have sums in the structure of their petals and patterns in smoke and the way the spots of a leopard can be drawn by a computer, and while he was talking I thought I understood what he was saying, although afterwards, when I tried to explain it to Jenny, all I could remember were the sums in the petals and the flocking birds. It seemed strange to hear a mathematician talk this way, but somehow, when I listened to him, and when I thought of those fish, I felt less alone, less adrift.

We were diving here so we could see the turtles' graveyard. A hole in the coral ten metres deep, five metres round, the sand at the bottom gleaming white. Swimming downwards the water grew colder, darker, until scattered across its bottom like some primeval ossuary we saw them; the bones of the turtles. Their great bony heads and jaws half swallowed by the sand, surrounded by the protruding serrations of their ribs and shells, the folded knuckles of their fins, all picked clean and white by the darting coral fish, the tiny chomping plankton.

Gerhard told us that no-one knows what draws turtles to these holes, what invisible lines of force guide them onwards, but somehow they feel it when their time arrives, an impulse in their blood that pulls them here, and so they swim, sometimes thousands of kilometres, steady, determined, unwavering, their heavy bodies graceful against the shifting light of the surface, like great birds across a sapphire sky. Maybe that is where my bones will settle, deep amongst theirs, my flesh devoured by the fish and crabs. Will I know that then, will I remember me? As my body passes outwards, through the teeth and bellies of the fish, will I be like the schooling mackerel, one mind in many bodies? My matter part of some greater fertility, some greater whole.

On the horizon I think I see a light, the fading dark before the dawn. Maybe a boat will come, maybe not, but either way it will be too late. Am I sky? Or am I water? I no longer know. Maybe this is how the turtles feel it, the coming of their time. Maybe they too feel it like a waking, like breathing.

Like flight.

SEBASTIAN JUNGER

The Perfect Storm

All waves, no matter how huge, start as rough spots – cats' paws – on the surface of the water. The cats' paws are filled with diamond-shaped ripples, called capillary waves, that are weaker than the surface tension of water and die out as soon as the wind stops. They give the wind some purchase on an otherwise glassy sea, and at winds over six knots, actual waves start to build. The harder the wind blows, the bigger the waves get and the more wind they are able to 'catch.' It's a feedback loop that has wave height rising exponentially with wind speed.

Such waves are augmented by the wind but not dependent on it; were the wind to stop, the waves would continue to propagate by endlessly falling into the trough that precedes them. Such waves are called gravity waves, or swells; in cross-section they are symmetrical sine curves that undulate along the surface with almost no energy loss. A cork floating on the surface moves up and down but not laterally when a swell passes beneath it. The higher the swells, the farther apart the crests and the faster they move. Antarctic storms have generated swells that are half a mile or more between crests and travel thirty or forty miles an hour; they hit the Hawaiian islands as breakers forty feet high.

Unfortunately for mariners, the total amount of wave energy in a storm doesn't rise linearly with wind speed, but to its fourth power. The seas generated by a forty-knot wind aren't twice as violent as those from a twenty-knot wind, they're seventeen times as violent.

A ship's crew watching the anemometer climb even ten knots could well be watching their death sentence. Moreover, high winds tend to shorten the distance between wave crests and steepen their faces. The waves are no longer symmetrical sine curves, they're sharp peaks that rise farther above sea level than the troughs fall below it. If the height of the wave is more than one-seventh the distance between the crests – the 'wavelength' – the waves become too steep to support themselves and start to break. In shallow water, waves break because the underwater turbulence drags on the bottom and slows the waves down, shortening the wavelength and changing the ratio of height to length. In open ocean the opposite happens: wind builds the waves up so fast that the distance between crests can't keep up, and they collapse under their own mass. Now, instead of propagating with near-zero energy loss, the breaking wave is suddenly transporting a huge amount of water. It's cashing in its chips, as it were, and converting all its potential and kinetic energy into water displacement.

A general rule of fluid dynamics holds that an object in the water tends to do whatever the water it replaces would have done. In the case of a boat in a breaking wave, the boat will effectively become part of the curl. It will either be flipped end over end or shoved backward and broken on. Instantaneous pressures of up to six tons per square foot have been measured in breaking waves. Breaking waves have lifted a 2700-ton breakwater, *en masse*, and deposited it inside the harbour at Wick, Scotland. They have blasted open a steel door 195 feet above sea level at Unst Light in the Shetland Islands. They have heaved a half-ton boulder ninety-one feet into the air at Tillamook Rock, Oregon.

There is some evidence that average wave heights are slowly rising, and that freak waves of eighty or ninety feet are becoming more common. Wave heights off the coast of England have risen an average of 25 percent over the past couple of decades, which converts to a twenty-foot increase in the highest waves over the next half-century. One cause may be the tightening of environmental laws, which has reduced the amount of oil flushed into the oceans by oil tankers. Oil spreads across water in a film several molecules thick and inhibits the generation of capillary waves, which in turn prevent the wind from getting a 'grip' on the sea. Plankton releases a chemical that has the same effect, and plankton levels in the North Atlantic have dropped dramatically. Another explanation is that the recent warming trend – some call it the greenhouse effect – has made storms more frequent and severe. Waves have destroyed docks and buildings in Newfoundland, for example, that haven't been damaged for decades.

As a result, stresses on ships have been rising. The standard practice is to build ships to withstand what is called a twenty-five-year stress – the most violent condition the ship is likely to experience in twenty-five years. The wave that flooded the wheelhouse of the *Queen Mary*, ninety feet up, must have nearly exceeded her twenty-five-year stress. North Sea oil platforms are built to accommodate a 111-foot wave beneath their decks, which is calculated to be a one-hundred-year stress. Unfortunately, the twenty-five-year stress is just a statistical concept that offers no guarantee about what will happen next year, or next week. A ship could encounter several twenty-five-year waves in a month or never encounter any at all. Naval architects simply decide what level of stress she's likely to encounter in her lifetime and then hope for the best. It's economically and structurally impractical to construct every boat to hundred-year specifications.

Inevitably, then, ships encounter waves that exceed their stress rating. In the dry terminology of naval architecture, these are called 'nonnegotiable waves.' Mariners call them 'rogue waves' or 'freak seas.' Typically they are very steep and have an equally steep trough in front of them – a 'hole in the ocean' as some witnesses have described it. Ships cannot get their bows up fast enough, and the ensuing wave breaks their back. Maritime history is full of encounters with such waves. When Sir Ernest Shackleton was forced to cross the South Polar Sea in a twenty-two-foot open life boat, he saw a wave so big that he mistook its foaming crest for a moonlit cloud. He only had time to yell, 'Hang on, boys, it's got us!' before the wave broke over his boat. Miraculously, they didn't sink. In February 1883, the 320-foot steamship *Glamorgan* was swept bow-to-stern by an enormous wave that ripped the wheelhouse right off the deck, taking all the ship's officers with it. She later sank. In 1966, the 44,000-ton *Michelangelo*, an Italian steamship carrying 775 passengers, encountered a single massive wave in an otherwise unremarkable sea. Her bow fell into a trough and the wave stove in her bow, flooded her wheelhouse, and killed a crewman and two passengers. In 1976, the oil tanker *Cretan Star* radioed, '. . . vessel was struck by a huge wave that went over the deck . . .' and was never heard from again. The only sign of her fate was a four-mile oil slick off Bombay.

South Africa's 'wild coast,' between Durban and East London, is home to a disproportionate number of these monsters. The four-knot Agulhas Current runs along the continental shelf a few miles offshore and plays havoc with swells arriving from Antarctic gales. The current shortens their wavelengths, making the swells steeper and more dangerous, and bends them into the fastwater the way swells are bent along a beach. Wave energy gets concentrated in the center of the current and overwhelms ships that are there to catch a free ride. In 1973 the 12,000-ton cargo ship *Bencruachan* was cracked by an enormous wave off Durban and had to be towed into port, barely afloat. Several weeks later the 12,000-ton *Neptune Sapphire* broke in half on her maiden voyage after encountering a freak sea in the same area. The crew were hoisted off the stern section by helicopter. In 1974, the 132,000-ton Norwegian tanker *Wilstar* fell into a huge trough ('There was no sea in front of the ship, only a hole,' said one crew member) and then took an equally huge wave over her bow. The impact crumpled inch-thick steel plate like sheetmetal and twisted railroad-gauge I-beams into knots. The entire bow bulb was torn off.

The biggest rogue on record was during a Pacific gale in 1933, when the 478-foot Navy tanker *Ramapo* was on her way from Manila to San Diego. She encountered a massive low-pressure system that blew up to sixty-eight knots for a week straight and resulted in a fully developed sea that the *Ramapo* had no choice but to take on her stern. (Unlike today's tankers, the *Ramapo*'s wheel-house was slightly forward of amidships.) Early on the morning of February 7th, the watch officer glanced to stern and saw a freak wave rising up behind him that lined up perfectly with a crow's nest above and behind the bridge. Simple geometry later showed the wave to be 112 feet high.

Rogue waves such as that are thought to be several ordinary waves that happen to get 'in step', forming highly unstable piles of water. Others are waves that overlay long-distance swells from earlier storms. Such accumulations of energy can travel in threes – a phenomenon called 'the three sisters' – and are so huge that they can be tracked by radar. There are cases of the three sisters crossing the Atlantic Ocean and starting to shoal along the 100-fathom curve off the coast of France. One hundred fathoms is six hundred feet, which means that freak waves are breaking over the continental shelf as if it were a shoreline sandbar. Most people don't survive encounters with such waves, and so firsthand accounts are hard to come by, but they do exist. An Englishwoman named Beryl Smeeton was rounding Cape Horn with her husband in the 1960s when she saw a shoaling wave behind her that stretched away in a straight line as far as she could see. 'The whole horizon was blotted out by a huge grey wall,' she writes in her journal. 'It had no curling crest, just a thin white line along the whole length, and its face was unlike the sloping face of a normal wave. This was a wall of water with a completely vertical face, down which ran white ripples, like a waterfall.'

The wave flipped the forty-six-foot boat end over end, snapped Smeeton's harness, and threw her overboard.

APSLEY CHERRY-GARRARD

The Worst Journey in the World

'Evening.—Loom of land and Cape Saunders Light blinking.'[1]

The ponies and dogs were the first consideration. Even in quite ordinary weather the dogs had a wretched time. 'The seas continually break on the weather bulwarks and scatter clouds of heavy spray over the backs of all who must venture into the waist of the ship. The dogs sit with their tails to this invading water, their coats wet and dripping. It is a pathetic attitude deeply significant of cold and misery; occasionally some poor beast emits a long pathetic whine. The group forms a picture of wretched dejection; such a life is truly hard for these poor creatures.'[2]

The ponies were better off. Four of them were on deck amidships and they were well boarded round. It is significant that these ponies had a much easier time in rough weather than those in the bows of the ship. 'Under the forecastle fifteen ponies close side by side, seven one side, eight the other, heads together, and groom between – swaying, swaying continually to the plunging, irregular motion.

'One takes a look through a hole in the bulkhead and sees a row of heads with sad, patient eyes come swinging up together from the starboard side, whilst those on the port swing back; then up come the port heads, while the starboard recede. It seems a terrible ordeal for these poor beasts to stand this day after day for weeks together,

[1] *Scott's Last Expedition*, vol. i, p. 7.
[2] *Ibid*, p. 9.

and indeed though they continue to feed well the strain quickly drags down their weight and condition; but nevertheless the trial cannot be gauged from human standards.'[1]

The seas through which we had to pass to reach the pack-ice must be the most stormy in the world. Dante tells us that those who have committed carnal sin are tossed about ceaselessly by the most furious winds in the second circle of Hell. The corresponding hell on earth is found in the southern oceans, which encircle the world without break, tempest-tossed by the gales which follow one another round and round the world from West to East. You will find albatross there – great Wanderers, and Sooties, and Mollymawks – sailing as lightly before these furious winds as ever do Paolo and Francesca. Round the world they go. I doubt whether they land more than once a year, and then they come to the islands of these seas to breed.

There are many other beautiful sea-birds, but most beautiful of all are the Snowy petrels, which approach nearer to the fairies than anything else on earth. They are quite white, and seemingly transparent. They are the familiar spirits of the pack, which, except to nest, they seldom if ever leave, flying 'here and there independently in a mazy fashion, glittering against the blue sky like so many white moths, or shining snowflakes.'[2] And then there are the Giant petrels, whose coloration is a puzzle. Some are nearly white, others brown, and they exhibit every variation between the one and the other. And, on the whole, the white forms become more general the farther south you go. But the usual theory of protective coloration will not fit in, for there are no enemies against which this bird must protect itself. Is it something to do with radiation of heat from the body?

A ship which sets out upon this journey generally has a bad time, and for this reason the overladen state of the Terra Nova was a cause

[1] *Ibid*, p. 8.
[2] Wilson in the *Discovery Natural History Reports*.

of anxiety. The Australasian meteorologists had done their best to forecast the weather we must expect. Everything which was not absolutely necessary had been ruthlessly scrapped. Yet there was not a square inch of the hold and between-decks which was not crammed almost to bursting, and there was as much on the deck as could be expected to stay there. Officers and men could hardly move in their living quarters when standing up, and certainly they could not all sit down. To say that we were heavy laden is a very moderate statement of the facts.

Thursday, December 1, we ran into a gale. We shortened sail in the afternoon to lower topsails, jib and staysail. Both wind and sea rose with great rapidity, and before the night came our deck cargo had begun to work loose. 'You know how carefully everything had been lashed, but no lashings could have withstood the onslaught of these coal sacks for long. There was nothing for it but to grapple with the evil, and nearly all hands were labouring for hours in the waist of the ship, heaving coal sacks overboard and re-lashing the petrol cases, etc., in the best manner possible under such difficult and dangerous circumstances. The seas were continually breaking over these people and now and again they would be completely submerged. At such times they had to cling for dear life to some fixture to prevent themselves being washed overboard, and with coal bags and loose cases washing about, there was every risk of such hold being torn away.

'No sooner was some semblance of order restored than some exceptionally heavy wave would tear away the lashing, and the work had to be done all over again.'[1]

The conditions became much worse during the night and things were complicated for some of us by seasickness. I have lively recollections of being aloft for two hours in the morning watch on Friday and being sick at intervals all the time. For sheer downright misery give

[1] *Scott's Last Expedition*, vol. i, pp. 11–12.

me a hurricane, not too warm, the yard of a sailing ship, a wet sail and a bout of seasickness.

It must have been about this time that orders were given to clew up the jib and then to furl it. Bowers and four others went out on the bowsprit, being buried deep in the enormous seas every time the ship plunged her nose into them with great force. It was an education to see him lead those men out into that roaring inferno. He has left his own vivid impression of this gale in a letter home. His tendency was always to underestimate difficulties, whether the force of wind in a blizzard, or the troubles of a polar traveller. This should be remembered when reading the vivid accounts which his mother has so kindly given me permission to use:

'We got through the forties with splendid speed and were just over the fifties when one of those tremendous gales got us. Our Lat. was about 52° S., a part of the world absolutely unfrequented by shipping of any sort, and as we had already been blown off Campbell Island we had nothing but a clear sweep to Cape Horn to leeward. One realised then how in the Nimrod – in spite of the weather – they always had the security of a big steamer to look to if things came to the worst. We were indeed alone, by many hundreds of miles, and never having felt anxious about a ship before, the old whaler was to give me a new experience.

'In the afternoon of the beginning of the gale I helped make fast the T.G. sails, upper topsails and foresail, and was horrified on arrival on deck to find that the heavy water we continued to ship, was starting the coal bags floating in places. These, acting as battering-rams, tore adrift some of my carefully stowed petrol cases and endangered the lot. I had started to make sail fast at 3 P.M. and it was 9.30 P.M. when I had finished putting on additional lashings to everything I could. So rapidly did the sea get up that one was continually afloat and swimming about. I turned in for 2 hours and lay awake hearing the crash of the seas and thinking how long those cases would stand it, till my watch came at midnight as a relief. We were under 2 lower topsails and hove to, the engines going dead slow to assist keeping head to wind. At another time I should have been easy in my mind; now the water that came aboard was simply fearful, and the wrenching on the old ship was enough to worry any sailor called upon to fill his decks with garbage fore and aft. Still "Risk nothing and do nothing," if funds could not supply another ship, we simply had to overload the one we had, or suffer worse things down south. The watch was eventful as the shaking up got the fine coal into the bilges, and this mixing with the oil from the engines formed balls of coal and grease which, ordinarily, went up the pumps easily; now however with the great strains, and hundreds of tons on deck, as she continually filled, the water started to come in too fast for the half-clogged pumps to cope with. An alternative was offered to me in going faster so as to shake up the big pump on the main engines, and this I did – in spite of myself – and in defiance of the first principles of seamanship. Of course, we shipped water more and more, and only to save a clean breach of the decks did I slow down again and let the water gain. My next card was to get the watch on the hand-pumps as well, and these were choked, too, or nearly so.

'Anyhow with every pump, – hand and steam, – going, the water continued to rise in the stokehold. At 4 A.M. all hands took in the fore lower topsail, leaving us under a minimum of sail. The gale increased to storm force (force 11 out of 12) and such a sea got up as only the Southern Fifties can produce. All the afterguard turned out and the pumps were vigorously shaken up, – sickening work as only a dribble came out. We had to throw some coal overboard to clear the after deck round the pumps, and I set to work to rescue cases of petrol which were smashed adrift. I broke away a plank or two of the lee bulwarks to give the seas some outlet as they were right over the level of the rail, and one was constantly on the verge of floating clean over the side with the cataract force of the backwash. I had all the swimming I wanted that day. Every case I rescued was put on the weather side of the poop to help get us on a more even keel. She sagged horribly and the unfortunate ponies, – though under cover, – were so jerked about that the weather ones could not keep their feet in their stalls, so great was the slope and strain on their forelegs. Oates and Atkinson worked among them like Trojans, but morning saw the death of one, and the loss of one dog overboard. The dogs, made fast on deck, were washed to and fro, chained by the neck, and often submerged for a considerable time. Though we did everything in our power to get them up as high as possible, the sea went everywhere. The wardroom was a swamp and so were our bunks with all our nice clothing, books, etc. However, of this we cared little, when the water had crept up to the furnaces and put the fires out, and we realized for the first time that the ship had met her match and was slowly filling. Without a pump to suck we started the forlorn hope of buckets and began to bale her out. Had we been able to open a hatch we could have cleared the main pump well at once, but with those appalling seas literally covering her, it would have meant less than 10 minutes to float, had we uncovered a hatch.

'The Chief Engineer (Williams) and carpenter (Davies), after we had all put our heads together, started cutting a hole in the engine room bulkhead, to enable us to get into the pump-well from the engine room; it was iron and, therefore, at least a 12 hours job. Captain Scott was simply splendid, he might have been at Cowes, and to do him and Teddy Evans credit, at our worst strait none of our landsmen who were working so hard knew how serious things were. Capt. Scott said to me quietly – "I am afraid it's a bad business for us – What do you think?" I said we were by no means dead yet, though at that moment, Oates, at peril of his life, got aft to report another horse dead; and more down. And then an awful sea swept away our lee bulwarks clean, between the fore and main riggings, – only our chain lashings saved the lee motor sledge then, and I was soon diving after petrol cases. Captain Scott calmly told me that they "did not matter" – This was our great project for getting to the Pole – the much advertised motors that "did not matter"; our dogs looked finished, and horses were finishing, and I went to bale with a strenuous prayer in my heart, and "Yip-i-addy" on my lips, and so we pulled through that day. We sang and re-sang every silly song we ever knew, and then everybody in the ship later on was put on 2-hour reliefs to bale, as it was impossible for flesh to keep heart with no food or rest. Even the fresh-water pump had gone wrong so we drank neat lime juice, or anything that came along, and sat in our saturated state awaiting our next spell. My dressing gown was my great comfort as it was not very wet, and it is a lovely warm thing.

'To make a long yarn short, we found later in the day that the storm was easing a bit and that though there was a terrible lot of water in the ship, which, try as we could, we could not reduce, it certainly had ceased to rise to any great extent. We had reason to hope then that we might keep her afloat till the pump wells could be cleared. Had the storm lasted another day, God knows what our state would have been, if we had been above water at all. You cannot imagine how utterly helpless we felt in such a sea with a tiny ship, – the great expedition with all its hopes thrown aside for its life. God had shown us the weakness of man's hand and it was enough for the best of us, – the people who had been made such a lot of lately – the whole scene was one of pathos really. However, at 11 P.M. Evans and I with the carpenter were able to crawl through a tiny hole in the bulkhead, burrow over the coal to the pump-well cofferdam, where, another hole having been easily made in the wood, we got down below with Davy lamps and set to work. The water was so deep that you had to continually dive to get your hand on to the suction. After 2 hours or so it was cleared for the time being and the pumps worked merrily. I went in again at 4.30 A.M. and had another lap at clearing it. Not till the afternoon of the following day, though, did we see the last of the water and the last of the great gale. During the time the pumps were working, we continued the baling till the water got below the furnaces. As soon as we could light up, we did, and got the other pumps under weigh, and, once the ship was empty, clearing away the suction was a simple matter. I was pleased to find that after all I had only lost about 100 gallons of the petrol and bad as things had been they might have been worse . . .

'You will ask where all the water came from seeing our forward leak had been stopped. Thank God we did not have that to cope with as well. The water came chiefly through the deck where the tremendous strain, – not only of the deck load, but of the smashing seas, – was beyond conception. She was caught at a tremendous disadvantage and we were dependent for our lives on each plank standing its own strain. Had one gone we would all have gone, and the great anxiety was not so much the existing water as what was going to open up if the storm continued. We might have dumped the deck cargo, a difficult job at best, but were too busy baling to do anything else . . .

'That Captain Scott's account will be moderate you may be sure. Still, take my word for it, he is one of the best, and behaved up to our best traditions at a time when his own outlook must have been the blackness of darkness . . .'

Characteristically Bowers ends his account:

'Under its worst conditions this earth is a good place to live in.'

Priestley wrote in his diary:

'If Dante had seen our ship as she was at her worst, I fancy he would have got a good idea for another Circle of Hell, though he would have been at a loss to account for such a cheerful and ribald lot of Souls.'

The situation narrowed down to a fight between the incoming water and the men who were trying to keep it in check by baling her out. The Terra Nova will never be more full of water, nearly up to the furnaces, than she was that Friday morning, when we were told to go and do our damndest with three iron buckets. The constructors had not allowed for baling, only for the passage of one man at a time up and down the two iron ladders which connected the engine-room floor plates with the deck. If we used more than three buckets the business of passing them rapidly up, emptying them out of the hatchway, and returning them empty, became unprofitable. We were divided into two gangs, and all Friday and Friday night we worked two hours on and two hours off, like fiends.

Wilson's Journal describes the scene:

'It was a weird night's work with the howling gale and the darkness and the immense seas running over the ship every few minutes and no engines and no sail, and we all in the engine-room oil and bilge water, singing chanties as we passed up slopping buckets full of bilge, each man above slopping a little over the heads of all below him; wet through to the skin, so much so that some of the party worked altogether naked like Chinese coolies; and the rush of the wave backwards and forwards at the bottom grew hourly less in the dim light of a couple of engine-room oil lamps whose light just made the darkness visible, the ship all the time rolling like a sodden lifeless log, her lee gunwale under water every time.

'There was one thrilling moment in the midst of the worst hour on Friday when we were realizing that the fires must be drawn, and when every pump had failed to act, and when the bulwarks began to go to pieces and the petrol cases were all afloat and going overboard, and the word was suddenly passed in a shout from the hands at work in the waist of the ship trying to save petrol cases that smoke was coming up through the seams in the afterhold. As this was full of coal and patent fuel and was next the engine-room, and as it had not been opened for the airing it required to get rid of gas, on account of the flood of water on deck making it impossible to open the hatchway, the possibility of a fire there was patent to every one, and it could not possibly have been dealt with in any way short of opening the hatches and flooding the ship, when she must have foundered. It was therefore a thrilling moment or two until it was discovered that the smoke was really steam, arising from the bilge at the bottom having risen to the heated coal.'[1]

Meanwhile men were working for all our lives to cut through two bulkheads which cut off all communication with the suction of the hand-pumps. One bulkhead was iron, the other wood.

Scott wrote at this time:

'We are not out of the wood, but hope dawns, as indeed it should for me, when I find myself so wonderfully served. Officers and men are singing chanties over their arduous work. Williams is working in sweltering heat behind the boiler to get the door made in the bulkhead. Not a single one has lost his good spirits. A dog was drowned last night, one pony is dead and two others in a bad condition – probably they too will go. Occasionally a heavy sea would bear one of them away, and he was only saved by his chain. Meares with some helpers had constantly to be rescuing these wretched creatures from hanging, and trying to find them better shelter, an almost hopeless task. One poor beast was found hanging when dead; one was washed away with such force that his chain broke and he disappeared overboard; the next wave miraculously washed him on board again and he is fit and well. [I believe the dog was Osman.] The gale has exacted heavy toll, but I feel all will be well if we can only cope with the water.

[1]Wilson's Journal.

Another dog has just been washed overboard – alas! Thank God the gale is abating. The sea is still mountainously high but the ship is not labouring so heavily as she was.'[1]

The highest waves of which I can find any record were 36 feet high. These were observed by Sir James C. Ross in the North Atlantic.[2]

On December 2 the waves were logged, probably by Pennell, who was extremely careful in his measurements, as being 'thirty-five feet high (estimated)'. At one time I saw Scott, standing on the weather rail of the poop, buried to his waist in green sea. The reader can then imagine the condition of things in the waist of the ship, 'over and over again the rail, from the fore-rigging to the main, was covered by a solid sheet of curling water which swept aft and high on the poop.'[3] At another time Bowers and Campbell were standing upon the bridge, and the ship rolled sluggishly over until the lee combings of the main hatch were under the sea. They watched anxiously, and slowly she righted herself, but 'she won't do that often,' said Bowers. As a rule if a ship gets that far over she goes down.

[1] *Scott's Last Expedition*, vol. i, pp. 14–15.
[2] Raper, *Practice of Navigation*, article 547.
[3] *Scott's Last Expedition*, vol. i, p. 13.

ERNEST SHACKLETON

South

A boat journey in search of relief was necessary and must not be delayed. That conclusion was forced upon me. The nearest port where assistance could certainly be secured was Port Stanley, in the Falkland Islands, 540 miles away, but we could scarcely hope to beat up against the prevailing north-westerly wind in a frail and weakened boat with a small sail area. South Georgia was over 800 miles away, but lay in the area of the west winds, and I could count upon finding whalers at any of the whaling-stations on the east coast. A boat party might make the voyage and be back with relief within a month, provided that the sea was clear of ice and the boat survive the great seas. It was not difficult to decide that South Georgia must be the objective, and I proceeded to plan ways and means. The hazards of a boat journey across 800 miles of stormy sub-Antarctic ocean were obvious, but I calculated that at worst the venture would add nothing to the risks of the men left on the island. There would be fewer mouths to feed during the winter and the boat would not require to take more than one month's provisions for six men, for if we did not make South Georgia in that time we were sure to go under. A consideration that had weight with me was that there was no chance at all of any search being made for us on Elephant Island.

The case required to be argued in some detail, since all hands knew that the perils of the proposed journey were extreme. The risk was justified solely by our urgent need of assistance. The ocean south of Cape Horn in the middle of May is known to be the most tempestuous storm-swept area of water in the world. The weather then is unsettled, the skies are dull and overcast, and the gales are almost unceasing. We had to face these conditions in a small and weather-beaten boat, already strained by the work of the months that had passed. Worsley and Wild realised that the attempt must be made, and they both asked to be allowed to accompany me on the voyage. I told Wild at once that he would have to stay behind. I relied upon him to hold the party together while I was away and to make the best of his way to Deception Island with the men in the spring in the event of our failure to bring help. Worsley I would take with me, for I had a very high opinion of his accuracy and quickness as a navigator, and especially in the snapping and working out of positions in difficult circumstances – an opinion that was only enhanced during the actual journey. Four other men would be required, and I decided to call for volunteers, although, as a matter of fact, I pretty well knew which of the people I would select. Crean I proposed to leave on the island as a right-hand man for Wild, but he begged so hard to be allowed to come in the boat that, after consultation with Wild, I promised to take him. I called the men together, explained my plan, and asked for volunteers. Many came forward at once. Some were not fit enough for the work that would have to be done, and others would not have been much use in the boat since they were not seasoned sailors, though the experiences of recent months entitled them to some consideration as seafaring men. McIlroy and Macklin were both anxious to go but realised that their duty lay on the island with the sick men. They suggested that I should take Blackborrow in order that he might have shelter and warmth as quickly as possible, but I had to veto this idea. It would be hard enough for fit men to live in the boat. Indeed, I did not see how a sick man, lying helpless in the bottom of the boat, could possibly survive in the heavy weather we were sure to encounter. I finally selected McNeish, McCarthy, and Vincent in addition to Worsley and Crean. The crew seemed a strong one, and as I looked at the men I felt confidence increasing.

The decision made, I walked through the blizzard with Worsley and Wild to examine the *James Caird*. The 20-ft. boat had never looked big; she appeared to have shrunk in some mysterious way when I viewed her in the light of our new undertaking. She was an ordinary ship's whaler, fairly strong, but showing signs of the strains she had endured since the crushing of the *Endurance*. Where she was holed in leaving the pack was, fortunately, about the water-line and easily patched. Standing beside her, we glanced at the fringe of the storm-swept, tumultuous sea that formed our path. Clearly, our voyage would be a big adventure. I called the carpenter and asked him if he could do anything to make the boat more seaworthy. He first inquired if he was to go with me, and seemed quite pleased when I said 'Yes.' He was over fifty years of age and not altogether fit, but he had a good knowledge of sailing-boats and was very quick. McCarthy said that he could contrive some sort of covering for the *James Caird* if he might use the lids of the cases and the four sledge-runners that we had lashed inside the boat for use in the event of a landing on Graham Land at Wilhelmina Bay. This bay, at one time the goal of our desire, had been left behind in the course of our drift, but we had retained the runners. The carpenter proposed to complete the covering with some of our canvas, and he set about making his plans at once.

Noon had passed and the gale was more severe than ever. We could not proceed with our preparations that day. The tents were suffering in the wind and the sea was rising. We made our way to the snow-slope at the shoreward end of the spit, with the intention of digging a hole in the snow large enough to provide shelter for the party. I had an idea that Wild and his men might camp there during my absence, since it seemed impossible that the tents could hold together for many more days against the attacks of the wind; but an examination of the spot indicated that any hole we could dig probably would be filled quickly by the drift. At dark, about 5 p.m., we all turned in, after a supper consisting of a pannikin of hot milk, one of our precious biscuits, and a cold penguin leg each.

The gale was stronger than ever on the following morning (April 20). No work could be done. Blizzard and snow, snow and blizzard, sudden lulls and fierce returns. During the lulls we could see on the far horizon to the north-east bergs of all shapes and sizes driving along before the gale, and the sinister appearance of the swift-moving masses made us thankful indeed that, instead of battling with the storm amid the ice, we were required only to face the drift from the glaciers and the inland heights. The gusts might throw us off our feet, but at least we fell on solid ground and not on the rocking floes. Two seals came up on the beach that day, one of them within ten yards of my tent. So urgent was our need of food and blubber that I called all hands and organised a line of beaters instead of simply walking up to the seal and hitting it on the nose. We were prepared to fall upon this seal *en masse* if it attempted to escape. The kill was made with a pick-handle, and in a few minutes five days' food and six days' fuel were stowed in a place of safety among the boulders above high-water mark. During this day the cook, who had worked well on the floe and throughout the boat journey, suddenly collapsed. I happened to be at the galley at the moment and saw him fall. I pulled him down the slope to his tent and pushed him into its shelter with orders to his tent-mates to keep him in his sleeping-bag until I allowed him to come out or the doctors said he was fit enough. Then I took out to replace the cook one of the men who had expressed a desire to lie down and die. The task of keeping the galley fire alight was both difficult and strenuous, and it took his thoughts away from the chances of immediate dissolution. In fact, I found him a little later gravely concerned over the drying of a naturally not over-clean pair of socks which were hung up in close proximity to our evening milk. Occupation had brought his thoughts back to the ordinary cares of life.

There was a lull in the bad weather on April 21, and the carpenter started to collect material for the decking of the *James Caird*. He fitted the mast of the *Stancomb Wills* fore and aft inside the *James Caird* as a hogback and thus strengthened the keel with the object of preventing our boat 'hogging' – that is, buckling in heavy seas. He had not sufficient wood to provide a deck, but by using the sledge-runners and box-lids he made a framework extending from the forecastle aft to a well. It was a patched-up affair, but it provided a base for a canvas covering. We had a bolt of canvas frozen stiff, and this material had to be cut and then thawed out over the blubber-stove, foot by foot, in order that it might be sewn into the form of a cover. When it had been nailed and screwed into position it certainly gave an appearance of safety to the boat, though I had an uneasy feeling that it bore a strong likeness to stage scenery, which may look like a granite wall and is in fact nothing better than canvas and lath. As events proved, the covering served its purpose well. We certainly could not have lived through the voyage without it.

Another fierce gale was blowing on April 22, interfering with our preparations for the voyage. The cooker from No. 5 tent came adrift in a gust, and, although it was chased to the water's edge, it disappeared for good. Blackborrow's feet were giving him much pain, and McIlroy and Macklin thought it would be necessary for them to operate soon. They were under the impression then that they had no chloroform, but they found some subsequently in the medicine-chest after we had left. Some cases of stores left on a rock off the spit on the day of our arrival were retrieved during this day. We were setting aside stores for the boat journey and choosing the essential equipment from the scanty stock at our disposal. Two ten-gallon casks had to be filled with water melted down from ice collected at the foot of the glacier. This was a rather slow business. The blubber-stove was kept going all night, and the watchmen emptied the water into casks from the pot in which the ice was melted. A working party started to dig a hole in the snow-slope about forty feet above sea-level with the object of providing a site for a camp. They made fairly good progress at first, but the snow drifted down unceasingly from the inland ice, and in the end the party had to give up the project.

The weather was fine on April 23, and we hurried forward our preparations. It was on this day I decided finally that the crew for the *James Caird* should consist of Worsley, Crean, McNeish, McCarthy, Vincent, and myself. A storm came on about noon, with driving snow and heavy squalls. Occasionally the air would clear for a few minutes, and we could see a line of pack-ice, five miles out, driving across from west to east. This sight increased my anxiety to get away quickly. Winter was advancing, and soon the pack might close completely round the island and stay our departure for days or even for weeks. I did not think that ice would remain around Elephant Island continuously during the winter, since the strong winds and fast currents would keep it in motion. We had noticed ice and bergs going past at the rate of four or five knots. A certain amount of ice was held up about the end of our spit, but the sea was clear where the boat would have to be launched.

Worsley, Wild and I climbed to the summit of the seaward rocks and examined the ice from a better vantage-point than the beach offered. The belt of pack outside appeared to be sufficiently broken for our purposes, and I decided that, unless the conditions forbade it, we would make a start in the *James Caird* on the following morning. Obviously the pack might close at any time. This decision made, I spent the rest of the day looking over the boat, gear, and stores, and discussing plans with Worsley and Wild.

Our last night on the solid ground of Elephant Island was cold and uncomfortable. We turned out at dawn and had breakfast. Then we launched the *Stancomb Wills* and loaded her with stores, gear, and ballast, which would be transferred to the *James Caird* when the heavier boat had been launched. The ballast consisted of bags made from blankets and filled with sand, making a total weight of about 1000 lb. In addition we had gathered a number of boulders and about 250 lb. of ice, which would supplement our two casks of water.

The stores taken in the *James Caird*, which would last six men for one month, were as follows:

30 boxes of matches.
6½ gallons paraffin.
1 tin methylated spirit.
10 boxes of flamers.
1 box of blue lights.
2 Primus stoves with spare parts and prickers.
1 Nansen aluminium cooker.
6 sleeping-bags.
A few spare socks.
A few candles and some blubber-oil in an oil-bag.

Food:
3 cases sledging rations = 300 rations.
2 cases nut food = 200 rations.
2 cases biscuits = 600 biscuits.
1 case lump sugar.
30 packets of Trumilk.
1 tin of Bovril cubes.
1 tin of Cerebos salt.
36 gallons of water.
112 1b. of ice.

Instruments:

Sextant.	Sea-anchor.
Binoculars.	Charts.
Prismatic compass.	Aneroid.

The swell was slight when the *Stancomb Wills* was launched and the boat got under way without any difficulty; but half an hour later, when we were pulling down the *James Caird*, the swell increased suddenly. Apparently the movement of the ice outside had made an opening and allowed the sea to run in without being blanketed by the line of pack. The swell made things difficult. Many of us got wet to the waist while dragging the boat out – a serious matter in that climate. When the *James Caird* was afloat in the surf she nearly capsized among the rocks before we could get her clear, and Vincent and the carpenter, who were on the deck, were thrown into the water. This was really bad luck, for the two men would have small chance of drying their clothes after we had got under way. Hurley, who had the eye of the professional photographer for 'incidents', secured a picture of the upset, and I firmly believe that he would have liked the two unfortunate men to remain in the water until he could get a 'snap' at close quarters; but we hauled them out immediately, regardless of his feelings.

The *James Caird* was soon clear of the breakers. We used all the available ropes as a long painter to prevent her drifting away to the north-east, and then the *Stancomb Wills* came alongside, transferred her load, and went back to the shore for more. As she was being beached this time the sea took her stern and half filled her with water. She had to be turned over and emptied before the return journey could be made. Every member of the crew of the *Stancomb Wills* was wet to the skin. The watercasks were towed behind the *Stancomb Wills* on this second journey, and the swell, which was increasing rapidly, drove the boat on to the rocks, where one of the casks was slightly stove in. This accident proved later to be a serious one, since some sea-water had entered the cask and the contents were now brackish.

By midday the *James Caird* was ready for the voyage. Vincent and the carpenter had secured some dry clothes by exchange with members of the shore party (I heard afterwards that it was a full fortnight before the soaked garments were finally dried), and the boat's crew was standing by waiting for the order to cast off. A moderate westerly breeze was blowing. I went ashore in the *Stancomb Wills* and had a last word with Wild, who was remaining in full command, with directions as to his course of action in the event of our failure to bring relief, but I practically left the whole situation and scope of action and decision to his own judgment, secure in the knowledge that he would act wisely. I told him that I trusted the party to him and said good-bye to the men. Then we pushed off for the last time, and within a few minutes I was aboard the *James Caird*. The crew of the *Stancomb Wills* shook hands with us as the boats bumped together and offered us the last good wishes. Then, setting our jib, we cut the painter and moved away to the north-east. The men who were staying behind made a pathetic little group on the beach, with the grim heights of the island behind them and the sea seething at their feet, but they waved to us and gave three hearty cheers. There was hope in their hearts and they trusted us to bring the help that they needed.

I had all sails set, and the *James Caird* quickly dipped the beach and its line of dark figures. The westerly wind took us rapidly to the line of pack, and as we entered it I stood up with my arm around the mast, directing the steering, so as to avoid the great lumps of ice that were flung about in the heave of the sea. The pack thickened and we were forced to turn almost due east, running before the wind towards a gap I had seen in the morning from the high ground. I could not see the gap now, but we had come out on its bearing and I was prepared to find that it had been influenced by the easterly drift. At four o'clock in the afternoon we found the channel, much narrower than it had seemed in the morning but still navigable. Dropping sail, we rowed through without touching the ice anywhere, and by 5.30 p.m. we were clear of the pack with open water before us. We passed one more piece of ice in the darkness an hour later, but the pack lay behind, and with a fair wind swelling the sails we steered our little craft through the night, our hopes centred on our distant goal. The swell was very heavy now, and when the time came for our first evening meal we found great difficulty in keeping the Primus lamp alight and preventing the hoosh splashing out of the pot. Three men were needed to attend to the cooking, one man holding the lamp and two men guarding the aluminium cooking pot, which had to be lifted clear of the Primus whenever the movement of the boat threatened to cause a disaster. Then the lamp had to be protected from water, for sprays were coming over the bows and our flimsy decking was by no means water-tight. All these operations were conducted in the confined space under the decking, where the men lay or knelt and adjusted themselves as best they could to the angles of our cases and ballast. It was uncomfortable, but we found consolation in the reflection that without the decking we could not have used the cooker at all.

The tale of the next sixteen days is one of supreme strife amid heaving waters. The sub-Antarctic ocean lived up to its evil winter reputation. I decided to run north for at least two days while the wind held and so get into warmer weather before turning to the east and laying a course for South Georgia. We took two-hourly spells at the tiller. The men who were not on watch crawled into the sodden sleeping-bags and tried to forget their troubles for a period; but there was no comfort in the boat. The bags and cases seemed to be alive in the unfailing knack of presenting their most uncomfortable angles to our rest-seeking bodies. A man might imagine for a moment that he had found a position of ease, but always discovered quickly that some unyielding point was impinging on muscle or bone. The first night aboard the boat was one of acute discomfort for us all, and we were heartily glad when the dawn came and we could set about the preparation of a hot breakfast.

This record of the voyage to South Georgia is based upon scanty notes made day by day. The notes dealt usually with the bare facts of distances, positions, and weather, but our memories retained the incidents of the passing days in a period never to be forgotten. By running north for the first two days I hoped to get warmer weather and also to avoid lines of pack that might be extending beyond the main body. We needed all the advantage that we could obtain from the higher latitude for sailing on the great circle, but we had to be cautious regarding possible ice streams. Cramped in our narrow quarters and continually wet by the spray, we suffered severely from cold throughout the journey. We fought the seas and the winds and at the same time had a daily struggle to keep ourselves alive. At times we were in dire peril. Generally we were upheld by the knowledge that we were making progress towards the land where we would be, but there were days and nights when we lay hove to, drifting across the storm-whitened seas and watching with eyes interested rather than apprehensive the uprearing masses of water, flung to and fro by Nature in the pride of her strength. Deep seemed the valleys when we lay between the reeling seas. High were the hills when we perched momentarily on the tops of giant combers. Nearly always there were gales. So small was our boat and so great were the seas that often our sail flapped idly in the calm between the crests of two waves. Then we would climb the next slope and catch the full fury of the gale where the wool-like whiteness of the breaking water surged around us. We had our moments of laughter – rare it is true, but hearty enough. Even when cracked lips and swollen mouths checked the outward and visible signs of amusement we could see a joke of the primitive kind. Man's sense of humour is always most easily stirred by the petty misfortunes of his neighbours, and I shall never forget Worsley's efforts on one occasion to place the hot aluminium stand on top of the Primus stove after it had fallen off in an extra heavy roll. With his frost-bitten fingers he picked it up, dropped it, picked it up again, and toyed with it gingerly as though it were some fragile article of lady's wear. We laughed, or rather gurgled with laughter.

The wind came up strong and worked into a gale from the north-west on the third day out. We stood away to the east. The increasing seas discovered the weaknesses of our decking. The continuous blows shifted the box-lids and sledge-runners so that the canvas sagged down and accumulated water. Then icy trickles, distinct from the driving sprays, poured fore and aft into the boat. The nails that the carpenter had extracted from cases at Elephant Island and used to fasten down the battens were too short to make firm the decking. We did what we could to secure it, but our means were very limited, and the water continued to enter the boat at a dozen points. Much bailing was necessary, and nothing that we could do prevented our gear from becoming sodden. The searching runnels from the canvas were really more unpleasant than the sudden definite douches of the sprays. Lying under the thwarts during watches below, we tried vainly to avoid them. There were no dry places in the boat, and at last we simply covered our heads with our Burberrys and endured the all-pervading water. The bailing was work for the watch. Real rest we had none. The perpetual motion of the boat made repose impossible; we were cold, sore, and anxious. We moved on hands and knees in the semi-darkness of the day under the decking. The darkness was complete by 6 p.m., and not until 7 a.m. of the following day could we see one another under the thwarts. We had a few scraps of candle, and they were preserved carefully in order that we might have light at mealtimes. There was one fairly dry spot in the boat, under the solid original decking at the bows, and we managed to protect some of our biscuit from the salt water; but I do not think any of us got the taste of salt out of our mouths during the voyage.

The difficulty of movement in the boat would have had its humorous side if it had not involved us in so many aches and pains. We had to crawl under the thwarts in order to move along the boat, and our knees suffered considerably. When a watch turned out it was necessary for me to direct each man by name when and where to move, since if all hands had crawled about at the same time the result would have been dire confusion and many bruises. Then there was the trim of the boat to be considered. The order of the watch was four hours on and four hours off, three men to the watch. One man had the tiller-ropes, the second man attended to the sail, and the third bailed for all he was worth. Sometimes when the water in the boat had been reduced to reasonable proportions, our pump could be used. This pump, which Hurley had made from the Flinders bar case of our ship's standard compass, was quite effective, though its capacity was not large. The man who was attending the sail could pump into the big outer cooker, which was lifted and emptied overboard when filled. We had a device by which the water could go direct from the pump into the sea through a hole in the gunwale, but this hole had to be blocked at an early stage of the voyage, since we found that it admitted water when the boat rolled.

While a new watch was shivering in the wind and spray, the men who had been relieved groped hurriedly among the soaked sleeping-bags and tried to steal a little of the warmth created by the last occupants; but it was not always possible for us to find even this comfort when we went off watch. The boulders that we had taken aboard for ballast had to be shifted continually in order to trim the boat and give access to the pump, which became choked with hairs from the moulting sleeping-bags and finneskoe. The four reindeer-skin sleeping-bags shed their hair freely owing to the continuous wetting, and soon became quite bald in appearance. The moving of the boulders was weary and painful work. We came to know every one of the stones by sight and touch, and I have vivid memories of their angular peculiarities even today. They might have been of considerable interest as geological specimens to a scientific man under happier conditions. As ballast they were useful. As weights to be moved about in cramped quarters they were simply appalling. They spared no portion of our poor bodies. Another of our troubles, worth mention here, was the chafing of our legs by our wet clothes, which had not been changed now for seven months. The insides of our thighs were rubbed raw, and the one tube of Hazeline cream in our medicine-chest did not go far in alleviating our pain, which was increased by the bite of the salt water. We thought at the time that we never slept. The fact was that we would doze off uncomfortably, to be aroused quickly by some new ache or another call to effort. My own share of the general unpleasantness was accentuated by a finely developed bout of sciatica. I had become possessor of this originally on the floe several months earlier.

Our meals were regular in spite of the gales. Attention to this point was essential, since the conditions of the voyage made increasing calls upon our vitality. Breakfast, at 8 a.m., consisted of a pannikin of hot hoosh made from Bovril sledging ration, two biscuits, and some lumps of sugar. Lunch came at 1 p.m., and comprised Bovril sledging ration, eaten raw, and a pannikin of hot milk for each man. Tea, at 5 p.m., had the same menu. Then during the night we had a hot drink, generally of milk. The meals were the bright beacons in those cold and stormy days. The glow of warmth and comfort produced by the food and drink made optimists of us all. We had two tins of Virol, which we were keeping for an emergency; but, finding ourselves in need of an oil-lamp to eke out our supply of candles, we emptied one of the tins in the manner that most appealed to us, and fitted it with a wick made by shredding a bit of canvas. When this lamp was filled with oil it gave a certain amount of light, though it was easily blown out, and was of great assistance to us at night. We were fairly well off as regarded fuel, since we had 6½ gallons of petroleum.

A severe south-westerly gale on the fourth day out forced us to heave to. I would have liked to have run before the wind, but the sea was very high and the *James Caird* was in danger of broaching to and swamping. The delay was vexatious, since up to that time we had been making sixty or seventy miles a day, good going with our limited sail area. We hove to under double-reefed mainsail and our little jigger, and waited for the gale to blow itself out. During that afternoon we saw bits of wreckage, the remains probably of some unfortunate vessel that had failed to weather the strong gales south of Cape Horn. The weather conditions did not improve, and on the fifth day out the gale was so fierce that we were compelled to take in the double-reefed mainsail and hoist our small jib instead. We put out a sea-anchor to keep *the James Caird*'s head up to the sea. This anchor consisted of a triangular canvas bag fastened to the end of the painter and allowed to stream out from the bows. The boat was high enough to catch the wind, and, as she drifted to leeward, the drag of the anchor kept her head to windward. Thus our boat took most of the seas more or less end on. Even then the crests of the waves often would curl right over us and we shipped a great deal of water, which necessitated unceasing bailing and pumping. Looking out abeam, we would see a hollow like a tunnel formed as the crest of a big wave toppled over on to the swelling body of water. A thousand times it appeared as though the *James Caird* must be engulfed; but the boat lived. The south-westerly gale had its birthplace above the Antarctic Continent, and its freezing breath lowered the temperature far towards zero. The sprays froze upon the boat and gave bows, sides, and decking a heavy coat of mail. This accumulation of ice reduced the buoyancy of the boat, and to that extent was an added peril; but it possessed a notable advantage from one point of view. The water ceased to drop and trickle from the canvas, and the spray came in solely at the well in the after part of the boat. We could not allow the load of ice to grow beyond a certain point, and in turns we crawled about the decking forward, chipping and picking at it with the available tools.

When daylight came on the morning of the sixth day out we saw and felt that the *James Caird* had lost her resiliency. She was not rising to the oncoming seas. The weight of the ice that had formed in her and upon her during the night was having its effect, and she was becoming more like a log than a boat. The situation called for immediate action. We first broke away the spare oars, which were encased in ice and frozen to the sides of the boat, and threw them overboard. We retained two oars for use when we got inshore. Two of the fur sleeping-bags went over the side; they were thoroughly wet, weighing probably 40 lb. each, and they had frozen stiff during the night. Three men constituted the watch below, and when a man went down it was better to turn into the wet bag just vacated by another man than to thaw out a frozen bag with the heat of his unfortunate body. We now had four bags, three in use and one for emergency use in case a member of the party should break down permanently. The reduction of weight relieved the boat to some extent, and vigorous chipping and scraping did more. We had to be very careful not to put axe or knife through the frozen canvas of the decking as we crawled over it, but gradually we got rid of a lot of ice. The *James Caird* lifted to the endless waves as though she lived again.

About 11 a.m. the boat suddenly fell off into the trough of the sea. The painter had parted and the sea-anchor had gone. This was serious. The *James Caird* went away to leeward, and we had no chance at all of recovering the anchor and our valuable rope, which had been our only means of keeping the boat's head up to the seas without the risk of hoisting sail in a gale. Now we had to set the sail and trust to its holding. While the *James Caird* rolled heavily in the trough, we beat the frozen canvas until the bulk of the ice had cracked off it and then hoisted it. The frozen gear worked protestingly, but after a struggle our little craft came up to the wind again, and we breathed more freely. Skin frost-bites were troubling us, and we had developed large blisters on our fingers and hands. I shall always carry the scar of one of these frost-bites on my left hand, which became badly inflamed after the skin had burst and the cold had bitten deeply.

We held the boat up to the gale during that day, enduring as best we could discomforts that amounted to pain. The boat tossed interminably on the big waves under grey, threatening skies. Our thoughts did not embrace much more than the necessities of the hour. Every surge of the sea was an enemy to be watched and circumvented. We ate our scanty meals, treated our frost-bites, and hoped for the improved conditions that the morrow might bring. Night fell early, and in the lagging hours of darkness we were cheered by a change for the better in the weather. The wind dropped, the snow-squalls became less frequent, and the sea moderated. When the morning of the seventh day dawned there was not much wind. We shook the reef out of the sail and laid our course once more for South Georgia. The sun came out bright and clear, and presently Worsley got a snap for longitude. We hoped that the sky would remain clear until noon, so that we could get the latitude. We had been six days out without an observation, and our dead reckoning naturally was uncertain. The boat must have presented a strange appearance that morning. All hands basked in the sun. We hung our sleeping-bags to the mast and spread our socks and other gear all over the deck. Some of the ice had melted off the *James Caird* in the early morning after the gale began to slacken, and dry patches were appearing in the decking. Porpoises came blowing round the boat, and Cape pigeons wheeled and swooped within a few feet of us. These little black-and-white birds have an air of friendliness that is not possessed by the great circling albatross. They had looked grey against the swaying sea during the storm as they darted about over our heads and uttered their plaintive cries. The albatrosses, of the black or sooty variety, had watched with hard, bright eyes, and seemed to have a quite impersonal interest in our struggle to keep afloat amid the battering seas. In addition to the Cape pigeons an occasional stormy petrel flashed overhead. Then there was a small bird, unknown to me, that appeared always to be in a fussy, bustling state, quite out of keeping with the surroundings. It irritated me. It had practically no tail, and it flitted about vaguely as though in search of the lost member. I used to find myself wishing it would find its tail and have done with the silly fluttering.

We revelled in the warmth of the sun that day. Life was not so bad, after all. We felt we were well on our way. Our gear was drying, and we could have a hot meal in comparative comfort. The swell was still heavy, but it was not breaking and the boat rode easily. At noon Worsley balanced himself on the gunwale and clung with one hand to the stay of the main-mast while he got a snap of the sun. The result was more than encouraging. We had done over 380 miles and were getting on for half-way to South Georgia. It looked as though we were going to get through.

The wind freshened to a good stiff breeze during the afternoon, and the *James Caird* made satisfactory progress. I had not realised until the sunlight came how small our boat really was. There was some influence in the light and warmth, some hint of happier days, that made us revive memories of other voyages, when we had stout decks beneath our feet, unlimited food at our command, and pleasant cabins for our ease. Now we clung to a battered little boat, 'alone,

alone – all, all alone; alone on a wide, wide sea.' So low in the water were we that each succeeding swell cut off our view of the sky-line. We were a tiny speck in the vast vista of the sea – the ocean that is open to all and merciful to none, that threatens even when it seems to yield, and that is pitiless always to weakness. For a moment the consciousness of the forces arrayed against us would be almost overwhelming. Then hope and confidence would rise again as our boat rose to a wave and tossed aside the crest in a sparkling shower like the play of prismatic colours at the foot of a waterfall. My double-barrelled gun and some cartridges had been stowed aboard the boat as an emergency precaution against a shortage of food, but we were not disposed to destroy our little neighbours, the Cape pigeons, even for the sake of fresh meat. We might have shot an albatross, but the wandering king of the ocean aroused in us something of the feeling that inspired, too late, the Ancient Mariner. So the gun remained among the stores and sleeping-bags in the narrow quarters beneath our leaking deck, and the birds followed us unmolested.

The eighth, ninth, and tenth days of the voyage had few features worthy of special note. The wind blew hard during those days, and the strain of navigating the boat was unceasing; but always we made some advance towards our goal. No bergs showed on our horizon, and we knew that we were clear of the ice-fields. Each day brought its little round of troubles, but also compensation in the form of food and growing hope. We felt that we were going to succeed. The odds against us had been great, but we were winning through. We still suffered severely from the cold, for, though the temperature was rising, our vitality was declining owing to shortage of food, exposure, and the necessity of maintaining our cramped positions day and night. I found that it was now absolutely necessary to prepare hot milk for all hands during the night, in order to sustain life till dawn. This meant lighting the Primus lamp in the darkness and involved an increased drain on our small store of matches. It was the rule that one match must serve when the Primus was being lit. We had no lamp for the compass and during the early days of the voyage we would strike a match when the steersman wanted to see the course at night; but later the necessity for strict economy impressed itself upon us, and the practice of striking matches at night was stopped. We had one water-tight tin of matches. I had stowed away in a pocket, in readiness for a sunny day, a lens from one of the telescopes, but this was of no use during the voyage. The sun seldom shone upon us. The glass of the compass got broken one night, and we contrived to mend it with adhesive tape from the medicine-chest. One of the memories that comes to me from those days is of Crean singing at the tiller. He always sang while he was steering, and nobody ever discovered what the song was. It was devoid of tune and as monotonous as the chanting of a Buddhist monk at his prayers; yet somehow it was cheerful. In moments of inspiration Crean would attempt 'The Wearing of the Green'.

On the tenth night Worsley could not straighten his body after his spell at the tiller. He was thoroughly cramped, and we had to drag him beneath the decking and massage him before he could unbend himself and get into a sleeping-bag. A hard north-westerly gale came up on the eleventh day (May 5) and shifted to the south-west in the late afternoon. The sky was overcast and occasional snow-squalls added to the discomfort produced by a tremendous cross-sea – the worst, I thought, that we had experienced. At midnight I was at the tiller and suddenly noticed a line of clear sky between the south and south-west. I called to the other men that the sky was clearing, and then a moment later I realised that what I had seen was not a rift in the clouds but the white crest of an enormous wave. During twenty-six years' experience of the ocean in all its moods I had not encountered a wave so gigantic. It was a mighty upheaval of the ocean, a thing quite apart from the big white-capped seas that had been our tireless enemies for many days. I shouted, 'For God's sake, hold on! It's got us.' Then came a moment of suspense that seemed drawn out into hours. White surged the foam of the breaking sea around us. We felt our boat lifted and flung forward like a cork in breaking surf. We were in a seething chaos of tortured water; but somehow the boat lived through it, half-full of water, sagging to the dead weight and shuddering under the blow. We bailed with the energy of men fighting for life, flinging the water over the sides with every receptacle that came to our hands, and after ten minutes of uncertainty we felt the boat renew her life beneath us. She floated again and ceased to lurch drunkenly as though dazed by the attack of the sea. Earnestly we hoped that never again would we encounter such a wave.

The conditions in the boat, uncomfortable before, had been made worse by the deluge of water. All our gear was thoroughly wet again. Our cooking-stove had been floating about in the bottom of the boat, and portions of our last hoosh seemed to have permeated everything. Not until 3 a.m., when we were all chilled almost to the limit of endurance, did we manage to get the stove alight and make ourselves hot drinks. The carpenter was suffering particularly, but he showed grit and spirit. Vincent had for the past week ceased to be an active member of the crew, and I could not easily account for his collapse. Physically he was one of the strongest men in the boat. He was a young man, he had served on North Sea trawlers, and he should have been able to bear hardships better than McCarthy, who, not so strong, was always happy.

The weather was better on the following day (May 6), and we got a glimpse of the sun. Worsley's observation showed that we were not more than a hundred miles from the north-west corner of South Georgia. Two more days with a favourable wind and we would sight the promised land. I hoped that there would be no delay, for our supply of water was running very low. The hot drink at night was essential, but I decided that the daily allowance of water must be cut down to half a pint per man. The lumps of ice we had taken aboard had gone long ago. We were dependent upon the water we had brought from Elephant Island, and our thirst was increased by the fact that we were now using the brackish water in the breaker that had been slightly stove in in the surf when the boat was being loaded. Some sea-water had entered at that time.

Thirst took possession of us. I dared not permit the allowance of water to be increased since an unfavourable wind might drive us away from the island and lengthen our voyage by many days. Lack of water is always the most severe privation that men can be condemned to endure, and we found, as during our earlier boat voyage, that the salt water in our clothing and the salt spray that lashed our faces made our thirst grow quickly to a burning pain. I had to be very firm in refusing to allow any one to anticipate the morrow's allowance, which I was sometimes begged to do. We did the necessary work dully and hoped for the land. I had altered the course to the east so as to make sure of our striking the island, which would have been impossible to regain if we had run past the northern end. The course was laid on our scrap of chart for a point some thirty miles down the coast. That day and the following day passed for us in a sort of nightmare. Our mouths were dry and our tongues were swollen. The wind was still strong and the heavy sea forced us to navigate carefully, but any thought of our peril from the waves was buried beneath the consciousness of our raging thirst. The bright moments were those when we each received our one mug of hot milk during the long, bitter watches of the night. Things were bad for us in those days, but the end was coming. The morning of May 8 broke thick and stormy, with squalls from the north-west. We searched the waters ahead for a sign of land, and though we could see nothing more than had met our eyes for many days, we were cheered by a sense that the goal was near at hand. About ten o'clock that morning we passed a little bit of kelp, a glad signal of the proximity of land. An hour later we saw two shags sitting on a big mass of kelp, and knew then that we must be within ten or fifteen miles of the shore. These birds are as sure an indication of the proximity of land as a lighthouse is, for they never venture far to sea. We gazed ahead with increasing eagerness, and at 12.30 p.m., through a rift in the clouds, McCarthy caught a glimpse of the black cliffs of South Georgia, just fourteen days after our departure from Elephant Island. It was a glad moment. Thirst-ridden, chilled, and weak as we were, happiness irradiated us. The job was nearly done.

We stood in towards the shore to look for a landing-place, and presently we could see the green tussock-grass on the ledges above the surf-beaten rocks. Ahead of us and to the south, blind rollers showed the presence of uncharted reefs along the coast. Here and there the hungry rocks were close to the surface, and over them the great waves broke, swirling viciously and spouting thirty and forty feet into the air. The rocky coast appeared to descend sheer to the sea. Our need of water and rest was well-nigh desperate, but to have attempted a landing at that time would have been suicidal. Night was drawing near, and the weather indications were not favourable. There was nothing for it but to haul off till the following morning, so we stood away on the starboard tack until we had made what appeared to be a safe offing. Then we hove to in the high westerly swell. The hours passed slowly as we waited the dawn, which would herald, we fondly hoped, the last stage of our journey. Our thirst was a torment and we could scarcely touch our food; the cold seemed to strike right through our weakened bodies. At 5 a.m. the wind shifted to the north-west and quickly increased to one of the worst hurricanes any of us had ever experienced. A great cross-sea was running, and the wind simply shrieked as it tore the tops off the waves and converted the whole seascape into a haze of driving spray. Down into valleys, up to tossing heights, straining until her seams opened, swung our little boat, brave still but labouring heavily. We knew that the wind and set of the sea was driving us ashore, but we could do nothing. The dawn showed us a storm-torn ocean, and the morning passed without bringing us a sight of the land; but at 1 p.m., through a rift in the flying mists, we got a glimpse of the huge crags of the island and realised that our position had become desperate. We were on a dead lee shore, and we could gauge our approach to the unseen cliffs by the roar of the breakers against the sheer walls of rock. I ordered the double-reefed mainsail to be set in the hope that we might claw off, and this attempt increased the strain upon the boat. The *James Caird* was bumping heavily, and the water was pouring in everywhere. Our thirst was forgotten in the realisation of our imminent danger, as we bailed unceasingly, and adjusted our weights from time to time; occasional glimpses showed that the shore was nearer. I knew that Annewkow Island lay to the south of us, but our small and badly marked chart showed uncertain reefs in the passage between the island and the mainland, and I dared not trust it, though as a last resort we could try to lie under the lee of the island. The afternoon wore away as we edged down the coast, with the thunder of the breakers in our ears. The approach of evening found us still some distance from Annewkow Island, and, dimly in the twilight, we could see a snow-capped mountain looming above us. The chance of surviving the night, with the driving gale and the implacable sea forcing us on to the lee shore, seemed small. I think most of us had a feeling that the end was very near. Just after 6 p.m., in the dark, as the boat was in the yeasty backwash from the seas flung from this iron-bound coast, then, just when things looked their worst, they changed for the best. I have marvelled often at the thin line that divides success from failure and the sudden turn that leads from apparently certain disaster to comparative safety. The wind suddenly shifted, and we were free once more to make an offing. Almost as soon as the gale eased, the pin that locked the mast to the thwart fell out. It must have been on the point of doing this throughout the hurricane, and if it had gone nothing could have saved us; the mast would have snapped like a carrot. Our backstays had carried away once before when iced up and were not too strongly fastened now. We were thankful indeed for the mercy that had held that pin in its place throughout the hurricane.

We stood off shore again, tired almost to the point of apathy. Our water had long been finished. The last was about a pint of hairy liquid, which we strained through a bit of gauze from the medicine-chest. The pangs of thirst attacked us with redoubled intensity, and I felt that we must make a landing on the following day at almost any hazard. The night wore on. We were very tired. We longed for day. When at last the dawn came on the morning of May 10 there was practically no wind, but a high cross-sea was running. We made slow progress towards the shore. About 8 a.m. the wind backed to the north-west and threatened another blow. We had sighted in the meantime a big indentation which I thought must be King Haakon Bay, and I decided that we must land there. We set the bows of the boat towards the bay and ran before the freshening gale. Soon we had angry reefs on either side. Great glaciers came down to the sea and offered no landing-place. The sea spouted on the reefs and thundered against the shore. About noon we sighted a line of jagged reef, like blackened teeth, that seemed to bar the entrance to the bay. Inside, comparatively smooth water stretched eight or nine miles to the head of the bay. A gap in the reef appeared, and we made for it. But the fates had another rebuff for us. The wind shifted and blew from the east right out of the bay. We could see the way through the reef, but we could not approach it directly. That afternoon we bore up, tacking five times in the strong wind. The last tack enabled us to get through, and at last we were in the wide mouth of the bay. Dusk was approaching. A small cove, with a boulder-strewn beach guarded by a reef, made a break in the cliffs on the south side of the bay, and we turned in that direction. I stood in the bows directing the steering as we ran through the kelp and made the passage of the reef. The entrance was so narrow that we had to take in the oars, and the swell was piling itself right over the reef into the cove; but in a minute or two we were inside, and in the gathering darkness the *James Caird* ran in on a swell and touched the beach. I sprang ashore with the short painter and held on when the boat went out with the backward surge. When the *James Caird* came in again three of the men got ashore, and they held the painter while I climbed some rocks with another line. A slip on the wet rocks twenty feet up nearly closed my part of the story just at the moment when we were achieving safety. A jagged piece of rock held me and at the same time bruised me sorely. However, I made fast the line, and in a few minutes we were all safe on the beach, with the boat floating in the surging water just off the shore. We heard a gurgling sound that was sweet music in our ears, and, peering around, found a stream of fresh water almost at our feet. A moment later we were down on our knees drinking the pure ice-cold water in long draughts that put new life into us. It was a splendid moment.

JUDITH BEVERIDGE

Whale

Grennan makes some incisions into the intestinal wall –
an awful smell comes out – we swear, turn our heads away.
The hiss of the gas escaping sounds as though a whole

convoy of trucks were suddenly pumping down on their
pneumatic brakes. It washed ashore six days ago. Every
one said that a build-up of gases might make it explode.

We rope the carcass ready to tow it to where we know
there's an old whale fall. Davey says the beast looks almost
as big as the Hindenburg blimp. We move out, bumping

over the breakers. Soon the carcass starts to cannonade
like a back-firing lorry; then maggots, white as sea spray,
stream out. The stiff easterly keeps blowing them into

our eyes and mouths along with the terrible stink. Davey
and I stagger, turning nauseous under the sky's wide, emptying
plate. We can't say what's worse: being covered in bits

of exploded whale, or these writhing maggots. Grennan
keeps steering, making for the horizon. Finally we dump it,
amazed at how long it takes for bubbles to stop rising.

KEM NUNN

The Dogs of Winter

Once outside the boy swung the boat north, taking them past the headland Fletcher had climbed the day before, allowing a first look at the bay and the distant point. The sea was deeper here and the Zodiac rode easily on the ground swell, smooth green water rushing beneath the bow as overhead the sky continued to clear with the sunlight pooling upon the water.

Chunks of driftwood bobbed in the light, together with beds of kelp, some bearing various crustaceans – bright spots of pink or orange which had been loosed from their moorings by the power of the swell, and on at least two occasions, sea lions raised sleek dark heads to look at them as they passed, and Fletcher was reminded of what he had seen on the landing – the scars wrapping Harmon's body like the seams of a baseball. He had not heard about the man being bitten, but clearly this is what had occurred. And this was as good a place as any – the very heart of what surfers called the Red Triangle. The fish came to the mouths of the rivers to spawn. The sea lions came to eat the fish. The great white came to eat the sea lions, and being of poor sight, were wont to get the occasional surfer as well, though generally these were spit back out when the mistake had been so noted and thereby leaving some, like Drew Harmon, to tell the tale.

Such lines of thought were, of course, counter-productive, and Fletcher set about preparing his equipment the better to drive them from his mind. He fooled with his light meter and gauged the angle of the sun. There were bumps on the horizon but as yet no sets. He heard the boy say something behind him and turned to see that he was pointing toward the shore. Fletcher followed the boy's finger with his eye. What he saw were three dark figures on boards angling toward them from the point, and as he followed their line of attack, he saw the first set beginning to build on the reef.

He had no doubt he and the boy were still well off the shoulder, but there was no way he was going to quell the beating of his heart as the set began to build, no way to escape that adrenaline rush you got from seeing a truly big wave from ground zero. It never looked quite real to him. And his first reaction was always the same and had been for twenty years; he could never quite believe he was putting himself in its way.

Fletcher had been around long enough to know the drill. The surfers would hang back a little on the inside, wait out the set, then sprint for the outside. Once outside, the trick was in knowing where to wait. The advantage enjoyed on this particular morning by Jones and Martin was that they were surfing with someone who had already charted the break and could show them their line-ups. It would be up to Fletcher and the boy to find their own.

It was to this end that Fletcher now set himself, and he watched as the first wave of the set passed over the reef and began to break. At close quarters, it was an unnerving spectacle, and yet a thing to behold, full of terror and fluid beauty. The amount of water involved was such that it was like watching a piece of the earth become liquid, as if in some cataclysm, or at the hour of creation. The wave rose first with great mass, like a hill, but this hill was made of liquid, in constant flux, and even as you watched it, it would change its form, turning itself to a long dark wall as the face went vertical and then beyond vertical as the crest began to feather and finally to pitch forward, to strike the water far out in front of the face – thus creating the vaunted green room of surfing myth – the place to be if you were to be there at all, on a board, at the eye storm, encompassed by the sound and the fury, bone dry in a place no one had ever been, or would be again, because when the wave was gone the place was gone too and would exist only in memory, or perhaps, if the right person was there, in the right place, with the right equipment, it would exist on film – a little piece of eternity to hang on the wall.

There were five waves in the set. Fletcher reckoned their size at twenty-five feet. A pair broke perfectly and could have been ridden. Three others broke in sections that were too long to make – sections of two, maybe three hundred yards – and when a section like that would go off, Fletcher was aware of the little beads of water that had gathered along the sides of the Zodiac jumping with the impact – suggesting no doubt some principal of physics he had, in his present position, no desire to contemplate at length.

When the set had passed, he and the boy looked at each other. For a moment, Fletcher felt at a loss, as if in need of a translator, at which point, it occurred to him that this, of course, was ridiculous.

'We're going to have to get closer,' Fletcher said. He made an effort to sound casual about it, as if this was business as usual.

The boy nodded. He turned in a more northerly direction, running almost parallel to the coast. He did not, however, give the boat much throttle, suggesting, Fletcher decided, that he understood what was needed but was in no great hurry to get there. In fact, Fletcher might have said the same for himself and he made no move to speed things up. But he had seen his shot and it had given him something with which to cut his fear. The hunt had begun. It was something he had not experienced in a long time and it was good, he thought, to have it back. It was good to be here.

The sea was smooth in the wake of the set, and Fletcher was aware of the three surfers angling toward the line up from the north even as he and the boy angled in from the south. He watched as the three men stroked to some invisible point of Drew's choosing then stopped and swung themselves up to straddle their boards. He held up a hand for the boy to cut back on his throttle and the boy did so. He and the boy were now even with the surfers, separated from them by a hundred yards of ocean.

It was his general plan to keep them in his sights, to hold south and west, then move in with the next set. He would set up on the south side of the peak, fifty yards from the impact zone. He would shoot with the sun at his back, lighting the wave at its highest and most dramatic point, then ride up and over a still unbroken shoulder as the wave passed by. The boy would have to gun it to get them over the shoulder, but it was nothing he had not already done on the bars and Fletcher did not anticipate a problem. Afterward it would only remain to negotiate the mouth of the river at the session's end.

'This is good,' Fletcher told the boy. He lowered his hand. 'We'll wait right here.'

The boy nodded, cut back on his throttle and raised the engine, allowing them to drift. Fletcher could find no suitable line up to the north, but to the south, there was a large island at the mouth of the river, and it seemed to him that if they held to its westernmost tip, they would be about right.

Turning toward the bay, he found them at the south end of a long strip of sand, and he picked that point where the sand ended in rock as his second line-up. He said as much to the boy. 'Here and here,' he said, and he showed the boy the tip of the island and the end of the sand. 'We stay between these two points. Next set, we'll move in a little closer. You just head right toward them.' He waved at the surfers. 'I'll be taking the pictures.' He held up the camera for the boy to see, as if this would make him understand. 'But as soon as I say go, you give it the gas, turn us out to sea, and get us over the shoulder.' He paused a moment. 'Just like you did on the inside. That was good, how you got us over those bars.'

The boy smiled at him when he said that and Fletcher thought maybe this would work after all. The boy remained silent and Fletcher went back to fiddling with his stuff. He worked now with the lens, sighting through it, adjusting his focus. Through the lens, he could see the surfers very clearly, seated on their boards, a few yards apart. He could see Robbie Jones and Sonny Martin talking to one another. He could see Drew Harmon slightly apart, arms folded on his chest, eyes on the horizon.

He checked his watch and his line-ups. They had moved too far north along the sand and he said as much to the boy. The boy tilted his engine so that it was back in the water and made a little circle, bringing them back on line. Fletcher nodded. He looked at his watch once more, checked his light meter, and sighted through his lens.

It was always odd, he thought, these little moments of calm on a big day. He felt the sun warming the shoulders of his wet suit. He put down the camera and reached over the side of the boat. He splashed his face and hands, allowing the water to roll down his back, as he had begun to sweat in the thick suit. He looked at the boy. The boy was checking the line-ups, holding to them. He gets it, Fletcher thought. He's going to do all right. At which point, Fletcher saw him look out to sea, and he saw his expression change.

Fletcher followed the boy's gaze. He was just in time to see Drew Harmon flatten himself on the deck of his board and begin to windmill toward the horizon. Jones and Martin followed suit and Fletcher felt his heart jump. It did not seem to him that they should have to paddle with such intent less they had been too far inside, and he looked toward the horizon himself and saw that, in fact, things had taken a turn.

A wave had begun to build and its peak had indeed shifted and it was further outside than he had anticipated. He looked at the boy, telling him to go, but the boy had tilted the engine out of the water and was now having difficulty in getting it back in. Fletcher moved to the stern to see if he could help and when the boy looked at him, he could see that the boy was scared.

The engine was hung up on something and when they got it into the water, it stalled. Fletcher watched as the boy tugged at the starter line. It was a short line and a one-man job and it was the boy's engine. Fletcher sat back on the rubber seat, striving to master his fear. They were still, he reasoned, far enough south to be well off the peak. Even with a stalled engine, they might ride out a wave or two. Surely it would start in the end. It was the thought with which he sought to console himself as the first gust of an offshore wind kicked down upon them from the mouth of the river.

Fletcher looked at the boy tugging furiously on the line. The wind was cold and dry, laced with the scent of pine. It suggested high pressure and clear skies. It was all he could have asked for. It was also pushing the rubber boat in a northerly direction and he knew that if this continued, if they did not get under power soon, they would be in deep shit, the blessing become the malediction. He looked toward the beach, as if he might have some say in the matter, and it was while he was looking that he heard someone yell. It was a kind of war whoop echoing across the water. Fletcher knew it to be the voice of one of the surfers, and he knew only too well what it meant and he turned to the sea. What he saw there was a wave as large as any he had yet been forced to deal with in thirty years on the water.

The thing seemed to stretch from beyond the point to the mouth of the river – one endless dark wall which, when it hit the reef, would no doubt blot out the sky. Already it was beginning to draw from the depths, adding exponentially to itself even as Fletcher watched it, and it was here that he heard the second war whoop and that he saw the surfers. There were two of them and he believed it to be Jones and Harmon.

Harmon was scratching for the lip, trying to make it over the top. Robbie Jones was behind him, but even as Fletcher watched, he saw the kid swing the board around and begin to stroke toward the river mouth, paddling for all he was worth in an effort to overcome the water flowing back up the face to meet him. To Fletcher's astonishment, the kid was going for it. He was trying to catch the wave.

It was an impossible moment. Fletcher had no doubt the kid was undergunned, that he was already too late into the wave. There was a good ten feet of water above his head, another twenty below him, with the wave just about to jack as he got to his feet. Instinctively, Fletcher went for his camera.

The wave face was dark with the amount of water it contained but with the sunlight splitting the last thin bands of cloud, lighting up the yellow rails of the Brewer gun and the red stripes of the wet suit in such a way that, as Fletcher brought the scene into focus, he could actually see the colors reflected on the face of the wave, and suddenly he was drawing a bead on Robbie Jones. Even as the boy jerked on the starter line. Even as the wave bore down upon them, because in some dark corner of his head, Fletcher knew very clearly what he was looking at. The only shot ever taken from the water at Heart Attacks. In epic conditions. A rider up. The thing he had come for. He saw his shot and he pulled the trigger.

It was really quite perfectly done. Robbie Jones was on his feet, crouched slightly with the board bucking beneath him but driving cleanly down the face of the wave. Fletcher held him in frame, leading him slightly, firing away, aware now too of yet one more sound – something apart from the thunder of the wave, the clicking of the camera. He was aware of the absurd sputtering of the Zodiac's outboard as it kicked to life and yet even as he put out his hand in anticipation of its acceleration, the foolish thing leaped beneath him.

Fletcher's hand, extended backwards, caught at the craft's rail. The rail in question, however, was wide and fat, made of synthetic rubber, slick with spray, and it afforded no purchase. Nor was there time for a second grab. Just like that and Fletcher was going over, still clutching the old orange housing he'd once risked his life to save, ass first into the icy Pacific.

Fletcher's first impulse, upon sinking beneath the surface, was to believe this was not happening. He had been thirty years a waterman. Only a kook of the first order would get caught looking, would fall out of the boat half a mile from shore, his camera still in his hands. The thought, however comforting, was a fleeting one and quickly erased by the sub-fifty-degree water which filled his head and he knew that he had indeed taken the fall, that he'd done it in the path of one of the biggest waves he had ever had to deal with, in shark-infested waters, hundreds of yards from a beach upon which no palms swayed in tropic light.

The second thought that came to him was that he would survive. He was no stranger to the open ocean. He had rolled waves for thirty years and he would roll this one. At which point, he rose to the surface and looked once more into the teeth of the dark monster already beginning to feather before a luminous sky and he was not so sure. The urge to panic rose in his chest. He felt it as a physical presence, a dull pain which sapped the strength from his arms. Still, he knew what to do, and he knew that knowing it was the key. Or rather, he wanted to believe that knowing it would prove the key. While one voice outlined a course of action, another cursed him for his ineptitude, calling loudly for panic and surrender, as if, by drowning, he could simply call the whole thing off. Fletcher gathered himself before the approaching wall of water. He took two or three deep breaths, held the last one, bent himself in the middle and kicked out with his legs, swimming for the bottom.

He held the old orange housing before him, kicking hard, hoping to get deep enough so that the wave would not suck him back to the surface. The water grew colder with each kick, and darker. He could see no more than a few inches in front of his face when he felt it come. It came first as a movement in the ocean around him. And then the explosion, prenaturally low. A thing felt more than heard. A function of blood. A promise of what was to come. And yet when the turbulence hit, it did so out of all proportion even to what he had expected. The camera was jerked from his grasp. His wet suit filled as the frigid water flushed down his back. His first thought was that the damn thing had burst a seam, that it would carry him down, as surely as if he had been encased in concrete.

There was no swimming against it now. The shock waves buffeted him as though he were some badly outclassed fighter pinned to the ropes, his sense of direction lost to the sea. Again, he sought to cling to a single thought, that it was a question of time – twenty seconds, max, and that though these twenty seconds might seem an eternity, he had been here before. This too would pass. At which point he began to flail, clutching at handfuls of water, clawing in what he could only hope was the proper direction, forgetting everything save the desire to breathe.

And in time it came. The release. The blackness gone to gray, the sudden explosion of sunlight and sound as his head broke through the layers of foamy water, ringing as if from a blow. His first thought, however, was for the next wave, and instinctively he turned toward the reef where indeed the second wave of the set had already begun to build and he saw the boy for the last time.

The Zodiac had managed somehow to ride over the top of the first wave, for Fletcher saw it quite clearly, not fifty yards ahead of him. He could see the boy sitting in what appeared to be some attitude of calm at the stern of his craft in his blue pants and red flannel shirt, his red ball-cap. The Zodiac, however, showed no indication of having regained power. It was simply sitting on the face of the wave, about midway up, with the crest already beginning to feather above it.

One could see what was to come. The thing was inevitable and Fletcher turned from it. Once again he dove. And once again he felt it come, the distant thunder, the pressure in the ears. This time, however, he got under it. Maybe it was the thrashing he had taken on the first wave, making for an inspired effort. Maybe the second wave was simply smaller. He allowed himself a momentary respite and for a moment he simply drifted, allowing his air to take him up. But he was deep and once more the need for oxygen pushed him, kicking and clawing for the surface, dreading what he would find there, for he did not believe he could roll a third wave. Nor did he know if the turbulence was pushing him toward the shore or only holding him in place. When he surfaced, however, the sea was awash once more in light, and there was nothing on it save the movement of the swell itself, which might be seen as an undulation of the horizon, as if some procession of unnamed beasts roamed there, skulking in the blue light.

He felt the cold now, as he had not felt it before. It was a gnawing thing. It worked upon the bone. He fought once more the almost overwhelming urge to panic, to simply thrash wildly, but set out instead with a measured stroke, offering himself instructions on hand position and follow-through, keeping his head down, breathing, at least for the present, on every other stroke, in some attempt to escape the impact zone before the next set could find him there. He had gone some distance when he became aware of a voice booming at him from across the water. What he saw was Drew Harmon. The man was stroking toward him on the deck of his big wooden board, a bright blue O'Neil hot lid in place over his outsized head, his nose and beard dripping water, his eyes full of demented light.

The man stroked up to him, pulled himself up to straddle the board, reached around behind his back, and, like the magician in a magic show, produced a single Churchill swim fin which he proceeded to pass to Fletcher. Fletcher received it with fingers from which any feeling had long ago departed.

'Swim for the point on the north side of the river mouth,' Harmon shouted at him. He waved at some rocks Fletcher was able to see only when the ground swell had raised him to sufficient height.

'Don't let yourself get pulled past those rocks,' Drew said. 'You get caught in the rips at the mouth, you're going back out to sea.'

And so saying, the man was gone. He flattened back out on his board and paddled off toward a distant line-up. Fletcher, for his part, had not said a word. The thought occurred to him that he was perhaps, at this particular point, incapable of it. Eventually he tried. He managed a kind of grotesque call. He found his voice was hoarse with the cold and weak, and he did not believe that the man had heard him and did not himself have any clear idea of what he was trying to say. Perhaps he was trying to tell him to look for the boy. Perhaps it was for himself that he called. But the man was gone, and Fletcher was alone once more with his blue hands and his ice-cream headache and his Churchill swim fin, which, eventually, he managed to work on over his bootie and with it in place he began to swim. He swam for the rocks and the beach he could not exactly see.

TIM WINTON

Breath

A monster storm showed up before autumn even arrived. On the forecast maps it looked like a tumour on the sea between us and the southern iceshelf. The moment he saw it Sando began planning our attempt on the Nautilus. On the Saturday and Sunday before the front arrived the swell in its path hadn't yet gathered momentum. We'd have to wait for the passage of the storm and catch the swell in its wake. Which meant I'd have to wag school if I wanted to make the trip.

Before the wind had even stirred the trees I knew I wasn't ready for the Nautilus. On the night the storm descended I lay in bed feeling the roof quake, wondering how I could plausibly avoid the whole endeavour. For two days black squalls ripped in from the sea and rain strafed the roads and paddocks and forest. On the morning of the third day, while it was still full dark and spookily still, I woke to a rumble that caused the house stumps to vibrate. If you didn't know any better you'd have thought a convoy of tanks was advancing up our drive and into the forest behind us. It was a low, grinding noise, a menacing pulse that didn't let up for a moment. I got out of bed feeling queasy. I packed a towel and wetsuit into my school bag, ate a couple of cold sausages from the fridge and waited for the dawn.

I got to the bus stop outside the butcher's about a half-hour early, figuring that if Sando didn't come then I'd just go ahead and take the bus to school. This morning school was an attractive option. But a few moments later, Loonie showed up blowing steamy breath on his hands, and before we'd even begun to speak the VW with its trailer and dinghy pulled in.

It was quite a drive west through the forest and then out along fishing tracks to the lonely little beach inshore of the island. All the way over Sando and Loonie psyched themselves up, each feeding off the other's nervous energy, while I sat pressed to the window, silent and afraid.

For any soul with a taste for excitement the mere business of launching Sando's dinghy should have been thrill enough for one day. The cove was a maelstrom with waves breaking end to end across it and the shorebreak heaved down with such force it sent broken kelp and shell-slurry into the air. We hauled the boat bow-out, timed our launch between waves and got the motor going, but we almost came to grief as a rogue set rumbled into the bay. By that stage there was nowhere for us to go but out, so we headed straight at those looming broken lines of foam with the throttle wide open in the hope they'd green up again before we reached them. We grabbed any handhold we could find. I felt the wind rip at my hair. And somehow we made it. As we slammed up each in turn we were airborne and the prop bawled before we landed again with a shattering thump. Loonie hooted like a rodeo rider; he'd have flapped a hat had there been one available. We found safe water, but it wasn't a good start to my day at the Nautilus. I rode the rest of the way rattled and sweating in my wetsuit. The granite island and its clump of seals were awash. The sea beyond was black and agitated.

We pulled up near the break during a lull and stood off in deep water to landward just to wait and watch before anchoring. There wasn't much to see at first except a scum of spent foam on the surface.

Ocean and air seemed hyper-oxygenated; everything fizzed and spritzed as if long after the passage of previous waves there was energy yet to be dissipated. The land behind us was partly obscured by the island and a low, cold vapour the morning sun failed to penetrate. Nothing shone. The sea looked bottomless.

Only when the first new wave arrived did I see what really lay before us. It came in at an angle, just a hard ridge of swell, but within a few seconds, as it found shallow water, it became so engorged as to triple in volume. And there at its feet lay the great hump of rock that gave the place its name. The mass of water foundered a moment, distorting as it hit the submerged obstacle. The wave reared as though climbing the obstruction and then sagged drastically at each end before the yawning lip pitched forward with a sound that made me want to shit.

Fifteen foot, said Loonie.

Yeah, Sando replied. And it's breakin in three feet of water.

In fact there were times when the wave broke over no water at all. Every set brought a smoker that sucked everything before it as it bore down, dragging so much water off the rock as it gathered itself that when it finally keeled over to break the granite dome sat free and clear before it. At these moments the trough of the wave actually sank below sea level. It was a sight I had never imagined, the most dangerous wave I'd ever seen.

We watched a couple of sets and then anchored up at a distance before Sando dived in and led us out. All three boards were Brewers – long, heavy Hawaiian-style guns. They were the same equipment we used at Old Smoky and Sando kept saying how good and solid they felt. He kept up the usual inspirational patter, but I was sullen with fright. Every time he tried to make eye contact I looked away, paddling without conviction until he drew ahead with Loonie at his elbow going stroke for stroke.

They sat up together outside the boil while I hung well back in deep water. Behind us the dinghy yanked at its rope, disappearing between swells. Sets came and went but everything passed by unridden. The waves were big but even at half the size I thought they'd be too sudden, way too steep, and the shallow rock beneath made them unthinkable. True, it was an awesome sight but the whole deal only broke for fifty yards or so; it was hardly worth the risk. I watched Sando and Loonie out there, right in the zone, letting wave after wave go by as if they'd come to the same conclusion despite themselves.

Then a wide one swung through and Sando went for it.

I saw the distant flash of his teeth as he fought to get up sufficient speed. A moment later it was vertical and so was he. As he got to his feet it was obvious the board was too long for the contour of the wave; he was perilously slow to turn. The wave hurled itself inside out. Sando staggered a moment, almost falling out of the face altogether. But he kept his feet and cranked the Brewer around with a strength I knew was beyond me. The fin bit. He surged forward as the wave began to lurch and dilate, reef fuming and gurgling below. The lip pitched over him. He was gone a moment, like a bone in the thing's throat. And then a squall of spume belched him free and it was over. He skidded out into the deep, dead water ahead of me and let the board flutter away.

I dug my way across, retrieved the Brewer and steered it back to where he lay with his knees up and his head back.

Jesus, he murmured. Oh Jesus.

I sat beside him, holding the big board between us. He slowly got his breath back but he was wild-eyed.

When you go, he said, go wide and early.

Don't think so, I muttered.

He took his board, checked the fin and got on.

You get half a second, that's all; it's brutal.

I shook my head.

C'mon, Pikelet. You know what's what.

That's why I'm stayin right here.

I didn't bring you here to watch, did I?

I said nothing.

It'll put some fizz in your jizz.

I felt plenty scared but not panicked; this time I knew what I was doing.

Shit, he said. I thought I brought surfers with me. Men above the ordinary.

I shrugged.

Pikelet, mate. We came to play.

He was grinning as he said it but I felt a sort of menace from him then. I didn't give a damn. My mind was made up. He wheeled around in disgust and I watched him paddle back out to where Loonie scratched uncertainly between looming peaks.

When Sando sat up beside him Loonie straightened a little, as if fortified by his presence, and only a few moments later he took the place on. But the wave he set himself for was a shocker. It was wedge-shaped and rearing – butt-ugly – even before he got going. As he leapt to his feet you could see what was about to happen. Yet the next few awful seconds earned Loonie honour in defeat. The wave stood, hesitated, and then foundered with Loonie right at the crest. He'd assumed his desperate crouch, pointed the board to the sanctuary of the channel, but he was going nowhere but down. The wave subsided beneath him, sucked him with it. Great overpiling gouts of whitewater leapt off the reef and the most I could see of Loonie was a threshing arm. Half his board fluttered thirty feet in the air. For a horrible moment the granite dome of the reef was completely bare. Then all that broken water mobbed across the rock, driving Loonie before it, boiling off into the deep ahead of me while I sat there, rigid.

The air was hissing, the sea bubbled underfoot, and I knew Loonie was down there somewhere in the white slick having the shit kicked out of him, but I didn't move until I heard Sando's furious yell.

It was whiteout down there. The water was mad with current. It was like diving blind into a crowd, and I groped, hauled off at angles until I saw the bluish contours of the seabed below. I dived again and got nowhere. I hit the surface, saw Sando – still yards off – hauling himself my way, and then I heard Loonie's gasp and turned to see his upraised arm. He was twenty yards behind me, even closer to the boat than I was.

When I got there I swept him up onto my board and listened to him puke and breathe and puke some more. The back was out of his wetsuit and there was skin off his shoulders. His nose bled, his legs trembled, but by the time Sando reached us he was laughing.

FIONA CAPP

That Oceanic Feeling

Entranced by that mythical line where sea meets sky, Tennyson's Ulysses regarded all experience as 'an arch wherethrough gleams that untravelled world whose margin fades for ever and for ever when I move'. I once saw that archway in a dream. It loomed up out of the water at Portsea, like the famous Torii gateway in the sea off Miyajima Island in Japan, except that my dream arch was an imposing Roman structure made of stone. I was just about to go to Europe for six months. In the dream, I knew that Europe and all the promise of life yet to be lived lay on the other side of that portal. What did it matter that Portsea faces the Antarctic?

Surfers were forever hovering on the verge of departure, of launching into the unknown. All that horizon-gazing while waiting for waves gave life a particular orientation. If I had an image of the future, it was of an ocean horizon – the fine thread upon which all our hopes hang. In certain conditions when you were out surfing, clouds massing along the horizon or the quality of light could make islands appear in the distance. Atlantis beckoned. What stopped the inquisitive surfer from paddling out in pursuit of these chimera was the endless promise of perfection embodied in the approaching waves. Your destination came to you, each wave with its own story of distant lands.

One 35 degree day at Jan Juc, I was gazing out to sea when I noticed something strange happening on the horizon, a trick of light bouncing between sky and sea. It was low tide and the water was a translucent turquoise. White cabbage butterflies and other small insects were darting over the waves.

'Take a look at the horizon,' I said to a surfer with whom I had fallen into conversation. I had never seen anything like it but did not want to mention what it looked like for fear of sounding melodramatic.

'Looks like a tidal wave, doesn't it?' he replied.

Exactly what I was thinking: it looked as if a giant wave was steaming towards us in a great wall of foam.

Most of our waves in southern Australia are generated by cyclones in the high latitudes of the Indian and Southern oceans close to Antarctica. Coasts in the southern hemisphere are, as Rachel Carson so hauntingly put it, 'washed by waves that have come from lonely, unvisited parts of the ocean, seldom crossed by vessels, off the normal routes of the air lines'. Even a surfer in California can find herself riding waves born in the west-wind belt of the screaming Sixties that circle Antarctica. Oceanographers can spot the signature of these Southern Ocean waves in their frequency, rapidity and the direction from which they come.

Our waves come from the deep south but our cultural ties bind us to the north. My abiding sense that Europe lay on the other side of the horizon had always complicated my feelings about surfing. When younger, I felt that my two touchstones – surfing culture and European culture – belonged to two separate, incompatible universes. Modern surfing, as opposed to the ancient Hawaiian art, was a New World, not an Old World, phenomenon. Although surfing was now popular in France, Spain, England and other parts of Europe – I once saw a surfer in the Baltic Sea – it was the *idea* of surfing that helped me close the gap between my two spheres. I had written my first novel in the belief that surfing could symbolise something much larger than the act itself – that it could transcend culture and place. But it was not until *Night Surfing* was published in France and Britain that I knew with certainty that surfing as a metaphor for life could indeed resonate as powerfully for someone in Paris or London as it could for an Australian.

I was talking to Jack about all this one day, when his eyes lit up at the phrase the 'idea of surfing'. He showed me a book of wave photography to which he had contributed the text. His essay described a surfing epiphany he had experienced during a trip to Florence in 1996. He was manager of the Surfworld Museum at the time and had been asked to assist with the production of surfing graphics and film footage to complement a surfwear pavilion at an Italian trade fashion show. He had always longed to visit Europe and now, at fifty-two, his chance had come, courtesy of surfing. So great was his joy at finally finding himself there that he would wake in the early hours of the morning in his hotel bed in Florence and find himself laughing out loud.

Jack was brought up a Catholic but after a year spent in a boarding school run by the Sisters of Mercy where beatings and screamed abuse were part of the daily routine, he learned to regard the church with deep wariness. His gradual rejection of Catholicism, however, did not mean the end of his spiritual quest. Soon after his arrival in Florence, he wandered jet-lagged into the Basilica of Santa Croce, a Franciscan church dating from the second half of the thirteenth century. The iconography of his childhood, he was surprised to find, was still deep within him. 'Surrounded by 600 year old frescoes and the tombs of Renaissance figures such as Michelangelo and Machiavelli,' he later wrote, 'I found myself in a small alcove to the lower left of the main altar. A cluster of candles burned before a statue of the Virgin. Who had lit each one of these, I wondered, and for what hopes did those flames flicker? From the corner of my right eye a movement caused me to half turn. There, in the gloom and grey of the vault that reached above the main altar, a single shaft of light had hit the gold embossed crucifix suspended perhaps four metres above the ground. In that instant, for the first time in my life, I understood with great clarity the basis of my own religious impulse, its relationship to the great sweep of human endeavour, and my long and fruitless search for peace, with its metaphor of the perfect wave.'

Later, he writes about sitting on a stone wall outside Fiesole, talking with some friends about surfing. 'It was Giannino who used the term "the *idea* of surfing" in describing how the activity itself blurred at the edges and became something far bigger, into which all people could reach for whatever it was they sought . . . In the upper reaches of the Arno Valley, somewhere out there in the night, a perfect wave stood up, feathered, then peeled off flawlessly along the edges of my mind.'

NAM LE

The Boat

The storm came on quickly. The crosswind surged in, filtering through the apertures in the rotten wood, sounding like a chorus of low moans. The boat began to rock. Hugging a beam at the top of the hatch, Mai looked out and her breath stopped: the boat had heeled so steeply that all she saw was an enormous wall of black-green water bearing down; she shut her eyes, opened them again – now the gunwale had crested the water – the ocean completely vanished – and it was as though they were soaring through the air, the sky around them dark and inky and shifting.

A body collided into hers, slammed her against the side of the hatch door. The boat righted and she slipped again, skidding in jets of water down the companionway. The hatch banged shut. Other bodies – she was on top of them – thighs and ribs and arms and heads – jammed this way and that with each groaning tilt, writhing toward space as though impelling the boat to heave to, back into the wind. The rocking got worse. Light was failing fast now and inside the hold it had become uncannily dim.

Inches away from Mai's face, a cross-legged man tipped forward, coughed once into his hands, then keeled back onto his elbows. His face was expressionless. When the smell arrived she realised he had vomited. In the swaying half-dark, people pitched forward and back, one by one, adding to the slosh of saltwater and urine in the bilge. People threw up in plastic bags, which they then passed on, hand to hand, until the parcel reached someone next to a scupper.

'Here.'

Mai pinched the bag, tried to squeeze it out through the draining slit, but her fingers lost their hold as the boat bucked. The thin yellow juice sprayed into her lap.

On the steps below her, an infant started crying: short choking bursts.

Instantly she looked for Truong – there he was, knees drawn up to his chin, face as smooth and impassive as that of a ceramic toy soldier. Their eyes met. Nothing she could do. He was wedged between an older couple at the bottom of the steps. Where was Quyen? She shook off the automatic anxiety.

Finally the storm arrived in force. The remaining light drained out of the hold. Wind screamed through the cracks. She felt the panicked limbs, people clawing for direction, sudden slaps of ice-cold water, the banging and shapeless shouts from the deck above. The whole world reeled. Everywhere the stink of vomit. Her stomach forced up, squashed through her throat. So this was what it was like, she thought, the moment before death.

She closed her eyes, swallowed compulsively; tried to close out the crawling blackness, the howl of the wind. She tried to recall her father's stories – storms at sea, waves ten, fifteen metres high! – but they rang shallow against what she'd just seen: those dense roaring slabs of water, sky churning overhead like a puddle being mucked with a stick. She was crammed in by a boatload of human bodies, thinking of her father and becoming overwhelmed, slowly, with loneliness. As much loneliness as fear. Concentrate, she told herself. And she did – forcing herself to concentrate, if not – if she was unable to – on the thought of her family, then on the contact of flesh pressed against her on every side, the human warmth, feeling every square inch of skin against her body and through it the shared consciousness of – what? Death? Fear? Surrender? She stayed in that human cocoon, heaving and rolling, concentrating, until it was over.

She opened her eyes. A procession of people stepping over her, measuredly, as though hypnotised, up the companionway and onto the deck. She got up and followed them.

The night sky was starless. Only moonlight illuminated everything, emanating from a moon low and yellow and pocked, larger than she had ever seen it before. Its surface appeared to her as clear and as close as the ridges of a mountain from a valley. Pearly light bathed the stunned and salt-specked faces of the hundred people on deck, all of whom had expected to die but were instead granted this eerie reprieve.

Nobody talked. Night, empty of sound, held every soul in thrall – the retching, the complaint of babies, the nervous breathings, now all muted. The world seemed alien, somehow beyond the reach of Mai's mind – to be beneath the giant moon, and have nothing but space, and silence, all around.

A fog rolled over the water.

Mai looked sternward and saw Quyen slumped, arms outstretched, collapsed to one knee. Her head lolled against her left shoulder. Her forearms were bleeding from rope burn – she must have been stranded on deck when the storm came in; someone had strapped her, spread-eagled, to a low horizontal spar, and saved her life.

Mai searched for Truong.

From below deck there now came a humming of prayer. Then someone gasped – Mai swung to find a face, then several, turning pale, hands to mouths beneath stupefied eyes.

'Do you hear?'

'What is it?'

'Be quiet! Be quiet!' an urgent voice commanded. 'Listen.'

But when the noise on the boat ceased, there still came from every direction the sound of people whispering, hundreds of people, thousands, the musical fall and rise of their native tongue. Barely intelligible. Sometimes right next to Mai's ear and she would whip around – but there would be nothing except the close, grey fog.

In a whisper, 'It's nothing – the wind, that's all.'

'Who's there?' someone demanded loudly, unsteadily, from the prow.

No answer, just the lapping of low murmurs.

On the foredeck, a man turned to his companion.

'Here?'

The second man nodded. Beneath the moonlight Mai recognised him. It was Anh Phuoc, the leader of the boat. He was, Quyen had told her, one of those mythic figures who'd already made his escape and yet returned, again and again, to help others.

He nodded and looked out into the haze.

And now she realised where they were – where they must be. Everyone had heard about these places. They had ventured into the fields of the dead, those plots of ocean where thousands had capsized with their scows and drowned. They stared into the fog. All drawn into a shared imagination, each in some space of unthinking as though they had leapt overboard, some madness possessing them, puncturing the glassy surface of the water and then plunged into black syrup, coming up into breath but panicked, disoriented, flailing in a viscid space without reference or light or sound.

'Try to sleep.'

It was Quyen; she had untangled herself from her station and crawled forward. Mai turned to her, then looked away. There was a sort of death in her face.

'I saw Truong, down . . .' Mai began, then saw that he had appeared silently behind his mother. He stood close by Quyen without touching her. For a moment Mai was seized with a desire to take the boy up and press him hard against her chest, to keep him – his stillness, self-containment, whatever it was about him – close to her. But she, too, was contained, and didn't move. She began to smell incense from the hold. People praying to their ancestors. It lightened her head. A dim thought struggled, stabilised, in her mind – maybe the voices on the water were those of their ancestors. Maybe, she thought, they were answering their prayers. What did they know? What were they so desperate to communicate?

'It's over now.'

She let herself pretend Quyen was speaking to her and not to Truong.

'The storm's over, Child. Try to sleep.'

Mai submitted, and when she closed her eyes, knowing they were both beside her, she found the hum of the phantom voices almost lulling – almost like the wash, when she dozed off, of a monsoon starting, or a wedding, dim-sounding on distant midday streets. A sea wind bearing men's voices up from the wharf. At times she thought she almost recognised a voice. When her eyes opened a second later it was morning: the moon had disappeared and the cloud streaks were already blue-bruised against a sky the colour of skin.

The first five days they'd travelled on flat seas. It had been hot, and Mai had faced the choice of being on deck and burnt by the sun or being below in the oven-heated hold. In the beginning people swam in the ocean, trailing ropes off the slow-moving junk, but afterward the salt on their bodies cooked their skin like crispy pork.

She spent as much time as she could bear out of the hold, which simmered the excrement of a hundred people. Their boat was especially crowded, Quyen had explained, because it carried two human loads: another boat organised by the same guide had at the last minute been confiscated by the Communists.

Each family kept mostly to itself. Mai was alone. She stayed close to Quyen and Quyen's six-year-old son, Truong. He was a skinny child with an unusually bony frame and a head too big for his body. His eyes, black and preternaturally calm, were too big for his head. He spoke in a watery voice – rarely – and, as far as Mai could tell, never smiled. He was like an old man crushed into the rude shape of a boy. It was strange, she thought, that such a child could have issued from Quyen – warm and mischievous Quyen.

When Mai first met him they'd been gliding – silently, under cover of night – through a port full of enemies. Even then his demeanour had been improbably blank. The war had that to answer for too, she'd thought – the stone-hard face of a child barely six years old. Only when the boat shifted and his body leaned into hers had she felt, astonishingly, his heartbeat through his trunk – an electric flurry racing through the concavities of his back, stomach and chest. His body furious with life. He was engaged in some inward working out, she realised, and in that instant she'd grasped that nothing – nothing – was more important than her trying to see whatever it was he was seeing behind his dark, flat eyes.

Two nights later, as Mai had been trying to sleep on deck, the song began. The faint voice drifted out of the hold with a familiar undertow. It was an old Vietnamese folk song:

> I never thought to be a soldier's wife,
> You were not born to foreign lands preside;
> Why do the streams and hills our love divide?
> Why are we destined for this faithless life?

In the shade of the hibiscus hedge her mother had once sung the same words to her during the years her father was away at war. The hibiscus flowers outside their kitchen in Phu Vinh, which bloomed only for a single day. And though dusk came, her mother would keep singing the soldier's wife's lament, her long black hair falling over Mai's face soft as a mosquito net, and Mai would trace the darkening red of the flowers through that curtain of hair.

Mai followed the song into the hold. She stopped at the bottom of the companionway steps; in the darkness she could just make out Quyen's form, lying on her side in front of Truong as though shielding him. Her voice was thin, attenuated in some way, stripped of vibrato. It didn't slide up to notes the way traditional singers' did. Mai stood on the dark steps and listened:

> The path of wind and rain is yours to take,
> While mine does mourn an empty room and bed;
> We reach to touch each other, but instead . . .

Her mother, who had waited each time her husband went to sea, again when he left to fight the Communists, and then – five years later – when he left once more, to report for re-education camp. That was supposed to have been the last time. He was supposed to have been gone for ten days – the prescribed sentence for low-ranked soldiers. Mai remembered: on the eleventh day the streets were swept, washed, festooned with lanterns – women in their best and brightest outfits. The war had been lost, their husbands and fathers were coming home. Mai and Loc wore clothes their mother had borrowed. All through the afternoon they'd waited, through the night, too, the lanterns growing more and more dazzling, the congee and suckling pig cold, congealed. The next morning Mai's mother sent for word but received none. What could she – could any of them – do?

Overcome with feeling, Mai wanted to ask Quyen to stop singing – not to stop singing. Never to stop. How could she explain it all? Afterward, she had seen her mother caught on that cruel grade of time, growing old, aging more in months than she had in years – and yet she had given no comfort to her. She had been a daughter selfish with her own loss. From that day on, she never again heard her mother sing.

Squatting down, Mai dried her eyes with her sleeves. The song continued. With a shock, Mai realised Quyen's mouth was not moving. She was asleep. The singing cut off as Truong lifted, turned his head, staring at Mai with large obsidian eyes. Stunned, she said nothing. She looked back at his pale face, the slight, girlish curve beneath his nose to his lips. The intentness of his gaze. Then, slowly, she felt whatever turmoil broke and banked inside her becoming still. Watching her the whole time, Truong opened his mouth and took a deep breath:

> You took my love southeast before I asked
> Whereto you went, and when you should return;
> Oh warring soul! through bitter years you learned
> To treat your sacred life like leaves of grass.

Quyen stirred. Her eyelids still closed, she murmured, 'Yes, you miss your father too. Don't you, my prince?'

He stopped singing. Shadows sifted in the darkness.

Here was how it began: her mother brought her through the dim kitchen into the yard. Her father had been released, three months prior, from re-education camp, and immediately admitted into the hospital in Vinh Long. He had gone blind. The doctors were baffled because they could identify no physical abnormality, no root cause. His re-education had blinded him. Mai, in the meantime, continued trundling every day from corner to corner, selling cut tobacco to supplement their family income. Her father's sickness was not unlike the war: something always happening elsewhere while she was forced on with her daily routine.

That day had been a slow one and she'd come home early. In the yard, beneath branches of mastic and white storax flowers, next to the deciduated hibiscus hedge, her mother had hooked her fingers under her waistband and handed her a damp roll of money. The ink faded from the sweat of counting and recounting.

'Child can spend it however Child likes but try to keep, *nha*?'

Knowing her mother's usual frugality, Mai struggled to respond, but her mother said nothing more, wiping her hands stiffly on her pyjama pants and turning back into the house.

Two days later she told Mai to go visit her father at the hospital.

'Child is a good child,' he told her after a long silence, his eyes fixed on some invisible locus in the air. He'd barely reacted when she came in and greeted him – it was only her second visit since he'd returned from re-education camp. What had they done to him there? She remembered him being gaunt three months ago, when he'd first returned, but now his whole face was sunken – as though its foundation had finally disintegrated, leaving his features to their slow inward collapse. His eyes extruded from their deep-set sockets like black stones.

'How is Ba?'

'Ba is unwell,' he said, rubbing his stubbled chin. He spoke to her as if to a servant. He didn't even look in her direction.

Mai hesitated. 'Can Ba see?'

He didn't seem blind to her. She'd always imagined blindness to be a blacking out – but what if it wasn't? What if he *could* see – his eyes seemed outwardly unchanged – but had now chosen not to? What if his eyes were already looking elsewhere?

She said, 'Ba will get better.'

'Child is a big girl now. How old is Child now?'

'Sixteen.'

'Heavens,' he cried. Then jokingly, 'So Child has a boyfriend, *ha*?'

Mai blushed and her father's hand searched for her head, patted it. Instinctively she twisted her cheek up into his rough palm. She'd come with so much to say – so much to ask – but he might as well have been deaf as blind. He laughed humourlessly. 'At sixteen, Ba had to look after Ba's whole family.'

Mai didn't reply. She felt insolent looking at his face when he didn't look back.

'Look after your mother,' he said.

Look at me, she wanted to say. She considered moving into his fixed line of sight but didn't dare. Just once, she thought. Just look at me once, Ba, and I'll do anything you say.

'And obey her, *nha*?'

'Yes, Ba.'

He gave a single nod, then smiled, but it was nothing more than a flexing of his lips.

'Listen to your mother. Promise, *nha*?'

'Yes, Ba.'

'Child.' His voice lowered conspiratorially and, her breath quickening, Mai stooped down closer to him. He was going to talk to her. Once, that had been her whole life. He smelt like rusted pipes. 'Stop it,' he whispered. She held her breath, watching his eyes. They were still locked in midair. 'Stop crying, Child.'

She held herself still as he patted her head again.

'Good girl,' he said.

The next day her mother put her on a bus to Rach Gia. It was a five-hour trip, she was told. Here was a plastic bag for motion sickness. In the market she would be picked up by an uncle she had never met. 'Give this to him,' her mother said, and pressed a fold of paper, torn from an exercise book, into her confused hands. Just before she got on the bus, her little brother, Loc, tugged at her shirt and asked if she minded if he used her bicycle.

'Use your own bicycle.'

She boarded. Watched the two of them through the scuffed, stained window. Then, on the street, her mother raised one hand from her thigh in a hesitant motion, as though halfway hailing a cyclo.

'Ma?'

Mai pushed through the scree of indifferent bodies and rushed out to her mother. She stood there, breathing hard, sensing the larger finality in their parting. Her mother asked if she still had the money. Yes. Remember not to let anybody see it. Yes. Her mother smiled abstractedly, then brought her hand onto Mai's head and eased down, combing hair between her fingers.

'Child,' she said softly, 'remember, *nha*? Put your hat on when Child gets off.'

Mai stammered: 'Child hasn't said goodbye to Ba.'

Her mother's hand followed the contours of her skull down into the inlet of her neck, a single motion. 'Don't worry,' she said. 'Ma will say. For Child.'

As the bus pulled out, a residue of memory surfaced in Mai's mind. Seeing her father off the first time – seven years ago, when he left for the war – her mother had clung fast to his elbow, her body turned completely into his, her face creased as though it were having trouble holding together a coherent emotion. But the second time – five years later, by the end of the war – her face had completely smoothed itself over. It had learned how to be expressionless.

Mai looked out of the back window – searching for her mother's face – but the street, like a wound, had closed over the space where it had been.

After hearing him sing, Mai caught herself, time and time again, searching for Truong. She was most at ease sitting in the shade of the hatch door, facing the prow, watching him with the other children. The only structure on the foredeck was the pilothouse and the children played in a small clearing behind it – a concession of territory from the adults teeming all around. Many of the children were twice Truong's age. He played with them laconically, indifferently, often leaving a game halfway through when he was bored, inevitably pulling a small group along – eager for him to dictate a new game.

Unlike the others he didn't constantly look around to find his family. He lived in a space of his own absorption. Quyen, too, seemed content to let him be. Hemmed in always by dozens of other sweaty, salt-gritted bodies, Mai watched him, stealing solace, marvelling at how he could be in the sun all day and remain so pale.

It seemed impossible she'd known him only a few days.

According to Quyen, Truong's father – her husband – had already made his escape. She told Mai that he had arrived safely in Pulau Bidong, one of the larger Malaysian refugee camps, eight months ago. He was waiting for them there.

Why hadn't they travelled together?

'We are going to America,' Quyen continued, passing over Mai's question. 'My husband has already rejected one offer from Canada. He says he has made friends in the Red Crescent.'

'Red Crescent?'

'Do you have any family there?'

After a while Quyen, misreading Mai's silence, continued, 'You are probably going to Australia, no? Many people are going there now.'

'No. I don't know.'

'You don't know?' She pursed her lips in mock decision: 'Then Mai will come with us.'

'*Thoi,*' started Mai uncertainly.

'You must come. That one likes you,' Quyen said, gesturing at Truong. 'He talks about you all the time.'

Mai flushed with pleasure, not fully understanding why – as she knew Quyen was lying. 'He is very good,' she said. 'Very patient.'

'Yes,' Quyen replied. She reflected for a moment. 'Like his father.'

'And who has ever heard of a young boy who can sing like that? It's a miracle. He will make you rich one day.'

'*Thoi,* don't joke.'

She looked at her friend, surprised. 'I am not joking.'

Together they turned toward him. He stood skinny and erect, his clothes hanging from his limbs as though from a denuded tree's branches. His hands directing the ragtag crew to throw their sandals into a pile. Mai wondered briefly if it made Quyen proud – seeing all those children scrambling to obey her son. The game was one her brother used to play. Relaxing her mind, Mai could almost fool herself into thinking he was there, little Loc, springing away as the designated dragon swung around to protect his treasure hoard. He was about the same age as Truong. Her thoughts started to drift back to her last meeting with her father, at the hospital, when Quyen interrupted:

'That one was an accident.'

Mai immediately blushed, said nothing.

'He slid out in the middle of the war.'

How could she joke about such a thing? Mai still remembered her father's photo on the altar those five years, the incense and prayer, the hurt daily refreshing in her mother.

'You must miss him,' said Mai, 'Truong's father.'

Quyen nodded.

'When were you married?'

'Nineteen seventy-two,' Quyen answered, 'in the middle of

everything.' For a moment her expression emptied out, making her seem younger. 'I was your age then.'

'Maybe more accidents will happen,' Mai said, swallowing quickly through her words, 'when you see him again. When we reach land.'

Quyen snorted, then started laughing. Her face had recomposed itself now – was again knowing, shrewd, self-aware. She was pretty when she laughed. 'Maybe,' she said. She prodded Mai. 'And what about you?'

But the mention of land – coming even from her own mouth – cancelled out any joke for Mai. She had been trying not to think about it. From every quarter everyone now discussed, obsessively, their situation: they were on a broken-down junk, stranded in the Eastern Sea – here, or maybe here – an easy target for pirates – everyone knew about the pirates, had heard stories of boats being robbed and then rammed, of women being taken, used, dumped. On top of that they were starving, some of them beginning to get sick. No one, however, gave voice to the main fear: that they might not make it.

Mai pushed the dread down. Desperate to change the subject, she said the first thing that came to mind. 'Wasn't it dangerous to escape,' she asked, 'with Truong so young?'

Her laughter subsiding, Quyen settled into a smile. 'It was because of him,' she said at last, 'that I decided to escape.' The smile hardened on her mouth.

They both turned toward him again. It had been three days. Watching him – letting in the thought of another day, and after that, another – Mai realised that Quyen's determination, as much as she tried to take part in it, felt increasingly superficial to her. She studied the boy's face. Above his awkward body it remained as stony and impassive as ever.

In Rach Gia, in the milling market, Mai had been met by a man with a skewed look who talked to a spot behind her shoulder. He called her name by the coriander-selling place. She was waiting for him, her hat on, next to a grease stand, petrols and oils and lubricants spread out like lunch condiments.

'Mai,' she heard, 'Mai, *ha*?' and, still sick from the lurching bus trip – it had been her first ride in an automobile – she was swept up by this man who hugged her, turning her this way and that.

'Child has the letter?' he grunted into her ear.

She was confused. He said it again, thrust her out at arm's length and glared straight at her for the first time. She tried hard not to cry.

'Heavens,' he said, hastily letting go of her and stepping back. His face spread in an open, unnatural smile before he walked away. All at once Mai remembered her mother's instructions. The folded paper. She ran after him and pressed it into his hand. He read it, furtively, refolded it into a tiny square, and then he was Uncle again.

The first hiding place was behind a house by the river. Uncle told her to climb to the top of a plank bed and stay there, don't go anywhere. She lay with the corrugated aluminium roof just a few thumbs above her head, and in the middle of the day the heat was unbearable. The wooden boards beneath her became darkened and tender with her sweat.

A few days later Uncle came to get her – it was after the worst of the afternoon heat – and made her memorise a name and address in Rach Gia in case anyone asked her questions. She felt light-headed standing up.

'When Child reaches land,' he told her, 'write to Child's mother. She will say what to do next.' She nodded dumbly. It was the first and final confirmation of her life's new plan: she was leaving on a boat. He looked at her and sighed. 'She said nothing for Child's own protection.' He gave her another abbreviated hug. 'Does Child understand?' He wasn't, in all likelihood, her real uncle – she knew that now – but still, when he left, she felt in her stomach a deep-seated fluster. It was the last she saw of him.

The second hiding place was a boat anchored beneath a bridge on the Loc Thang river. Mai stayed down below deck for days and days, with sixty people maybe, among cargo sacks of sweet potatoes. No one talked; every sound in the dark was rat-made. She caught herself whimpering and covered her mouth. Once in a while the owner brought a few kilos of rice and they cooked it with potatoes over low kerosene flames and ate, salting their bit, chewing quietly. People coughed into their sleeves to muffle the sound. Parents fed their babies sleeping pills.

One night the owner appeared with another man who came in and tapped her on the shoulder. He tapped five other people as well. They all followed him out of the boat into the hot dark strange openness. A rower waited nearby and after some hesitation and muted dissent they climbed into his canoe, sitting one behind the other, Mai in the middle. The new man – the guide – instructed the rower to cross to the other side of the river. But he didn't, he kept on paddling downstream for what seemed to Mai like hours and hours. At one point she found herself falling asleep. She woke to the sound of wood tapping hollowly against wood. They were pushing into the midst of a dark cluster of houseboats. The rower stopped, secured a lanyard to one of the boats and leapt aboard. He lit a small lantern and began passing large drums reeking of diesel into the canoe. Moments later they moored against the riverbank. The rower crept onshore with a hoe and exhumed something long and grey from beneath a coconut grove.

'Detachable sail,' someone whispered.

Mai turned around. The speaker was a young woman. She

sounded as though she might have been pointing out bad produce at a market stall.

'It's a detachable sail,' the woman repeated.

Mai began asking her what that was when the rower turned, silencing them both with a glare. A moment later Mai felt a cupped hand against her ear.

'My name is Chi Quyen.' The woman used the word *Chi,* for 'older sister'. She reclined, smiling grimly but not unkindly, then leaned forward again. 'Chi too is by herself.'

Mai nodded. Shyly, she lifted a finger and crossed her lips.

For a long time they glided soundlessly, close to shore, and then they entered a thick bed of reeds. They stopped. The rower turned around, shook his heavy head and made the sign for no talking. It was dark. He struck a match and lit an incense stick and planted it in the front tip of the canoe. After a while Mai became confused. No one else seemed to be praying. When the stick burned down the guide asked the rower, in a low voice, to light another one. At least an hour passed. Occasionally Mai made out the rower's profile, hard and sombre. She took the dark smell of sandalwood into her body.

The canoe swayed. 'Maybe they're waiting,' a new voice whispered gruffly. 'Move out of the reeds so they can see the signal.'

'Keep your head down!' the rower spat.

At that moment Mai realised the incense stick – its dim glow, its smoke, perhaps – was their signal.

Someone else said, 'They won't wait.'

'Move out of the reeds,' the man repeated.

Mai felt a hot breath in her ear: 'If they come, follow Chi, *nha*? Jump out and swim into the reeds. You can swim, no?'

'If who comes?'

'Fuck your mother, I said keep your head down!'

Someone behind her hissed and the canoe rocked wildly from

side to side. The rower whirled around. Then, through the reeds, a light like a car beam flashed on and off. Fumbling, the rower lit a new incense stick, planted it at the canoe tip and paddled, swiftly and silently, back out. They saw it ahead, barely visible in the weird, weakly thrown light from the banks. An old fishing trawler, smaller than she'd imagined – maybe fifteen metres long – sitting low in the water. It inched forward with a diesel growl. A square pilothouse rose up from the foredeck, a large derrick-crane straddling its back deck, and the boat's midsection congested with short masts and cable rigs. Two big eyes painted on the bow. The canoe drew alongside and three men leaned over the gunwale above them and pulled them up, wrist by wrist. Everyone was aboard within a minute. Before being ushered down the hatch, Mai looked back and saw the canoe, abandoned in the boat's wake, rocking on the dark river.

Inside the hold, the stench was incredible, almost eye-watering. The smell of urine and human waste, sweat and vomit. The black space full of people, bodies upon bodies, eyes and eyes and eyes and if she'd thought the first boat was crowded, here she could hardly breathe, let alone move. Later she counted at least two hundred people, squashed into a space meant for fifteen. No place to sit, nor even put a foot down; she found a crossbeam near the hatchway and hooked her arm over it. Luckily it was next to a scupper where the air came through.

Quyen settled on the step below her, whispering to a young boy. She caught Mai's eye and smiled firmly.

The boat continued its creeping pace. People padded the engines with their clothes to reduce the noise.

'Quiet,' an angry voice shushed downward. 'We're near the gate.'

But no one had been speaking. Through the scupper Mai peered into the night: their boat was gliding into a busy port. Pressed hard beneath her was the body of the boy Quyen had been talking to.

'Natural gate a hundred metres long,' she heard suddenly. The water carried the low sound clearly. Then she realised the voice came from above deck, so subdued the person might have been talking to himself. 'About ten metres wide. On the rising tide.'

Then another voice under the wind: 'Viet Cong . . . manned with two M30s —'

'Automatic, no?'

'Machine guns.'

'What did Phuoc say about the permit?'

In the darkness, thought Mai, to feel against you the urgent flutter of a child's heart. The hopped-up fragility of it.

A tense sigh: 'Even with the permit.'

'Leave at night and they shoot. They shoot anything.'

The speakers paused for a short while. Then a voice said, 'We'll find out soon enough.'

She settled forward against the young boy, not wanting to hear any more. Trying to block it all out: the voices, the smell. It was unnerving to think of all those other bodies in the darkness. Black shapes in the blackness, merging like shadows on the surface of oil. She crouched there, in the silence, beneath the hatchway. Spying on the bay through the scupper. Gradually, inevitably, the dark thoughts came. Here, in the dead of night, contorted inside the black underbelly of a junk – she was being drawn out into an endless waste. What did she know about the sea? She was the daughter of a fisherman and yet it terrified her. She watched as Quyen reached back and with a surprisingly practised gesture pressed her palm against the boy's forehead. From above, watching the set of his grim face, Mai thought of her father. Their last meeting. His blindness. He'd taught her not to blame the war but how could she not? – all the power of his own sight seemed still intent on it.

Through the crack of the scupper the land lights, like mere tricks of her eyes, were extinguished one by one. Someone cut the engines.

She pulled the young boy's body closer to her; it squirmed like a restless animal's.

'Truong,' a voice whispered sharply from beneath them.

She peered down. It was Quyen.

'Don't be a nuisance, Child.' Quyen looked up at Mai, then said ruefully, 'This is my little brat. Truong.'

'Yours?' Mai frowned. 'But — '

From the deepest part of the hold, several voices shushed them. In the silence that followed, even the tidal backwash seemed loud against the hull. Then a grind of something against the boat. Mai had never heard a sound so sudden and hideous.

'What is it?'

'A mine? I heard they put mines —'

The metal shrieked each minute movement of the boat.

'Heavens!'

'But boats pass here, must pass here every day —'

Fiercely: 'Quiet!'

The sound sheared off – leaving behind a deep, capacious silence. Mai stiffening at every creak of the boat, every dash of water against its rotten sidewood. Then, without warning, the call and fade of a faraway voice. She crushed her cheek against the crown of the young boy's head and for the first time felt him respond – both of his small fists clamping her forearm. She shut her eyes and trained herself to his frenzied heartbeat, as though its pulse – its fine-knitting rhythm – carried the only possible thread of their escape. Long minutes passed. The boat glided on, pointed headfirst into the swell. Finally the fierce voice coughed:

'We're safe for now.'

Murmurs rose up. The hatch was lifted. Under the sudden starlight Mai could see the whole of the boy's face, arching up to meet the fresh air.

'Child,' said Quyen, 'greet Chi. Properly.'

He looked up at Mai – his eyes black and clear and unblinking. '*Chao* Chi,' he said in his reed-thin voice.

All around them people's faces were untensing, bodies and voices stirring in restless relief. But Mai, clutching this strange young boy, found herself shivering in the warm night, relief only a sharp and unexpected condensation in her eyes.

Once the storm passed, six days out, everything changed.

Fishermen on the boat agreed that this storm had come on faster than any they'd ever experienced. It destroyed the caulking and much of the planking on the hull. The inboard was flooded, and soon afterward, both engines cut out completely.

What food had been left was spoiled. Water was short. Anh Phuoc, whose authority was never questioned, took charge of rationing the remaining supply, doling it out first to children, then the infirm, then everyone else. It amounted to a couple of wet mouthfuls a day.

The heat was unbearable. Before long the first body was cast overboard. Already a handful of people had been lost during the storm, but this was the first casualty witnessed by the entire boat. To the terrible drawn-out note of a woman's keening, the bundle was tossed, a meek splash, into the water.

Like everyone else, Mai looked away.

After the storm it seemed to Mai that a film had been stripped from the world. Everything became more intense – the sun hotter, the light more vivid, the sea darker, every word a discordant affront to the new silence. The storm had forced people into their privacies: the presence of others now assailed each person's solitude in facing up to the experience of it. Children turned introverted, playing as though conducting conversations with themselves.

Even time took on a false depth: the six days before the storm stretched out, merged with memory, until it seemed as though everything that had ever happened had happened on the boat.

A man burned his clothes to let up smoke. He was quickly set upon, the fire smothered – the longer they drifted, the more fearful they became of pirates. That night another bundle was thrown overboard. Minutes later they heard a thrashing in the water. It was too dark to see anything, yet, still, everyone averted their gaze.

Thirst set in. Some people trapped their own urine. Some, desperate for drinkable water, even allowed themselves the quick amnesia and prayed for another storm. It was fantastic to be surrounded by so much water and yet be dehydrated. Mai soon realised she wouldn't make it. The day following the storm she imitated some of the other youth, hauling up a bucket tied to the bowline. Under the noon sun the seawater was the colour of amethyst.

She drank it. It was all right at first. It was bliss. Then her throat started scalding and she wanted to claw it out.

'You stupid girl,' Quyen reproached her, demonstrating how to use her fingers to induce vomiting. She hugged her fiercely. 'Heavens, you can't wait? We're almost there.'

But what did Quyen know? Mai had heard – how could she possibly have not? – that other boats had successfully made the crossing in two days. She tried to sleep, to slide beneath the raw scour of pain in her throat. They'd been out seven days. How much longer? Her father was persistent in her thoughts now – all those weeks, even months, he'd spent on this same sea, in trawlers much like this one. He'd been here before her.

That afternoon, when she awoke, her muscles felt as though they had turned to liquid. She could feel her heart beating slurpily. She followed the weakening palpitations, counterpointing them to the creak and strain of the boat, the occasional luff of the sail. The sun brilliant but without heat. She was even thirstier than before.

'I'm not going to make it,' she said. Saying it touched the panic, brought it alive.

'Don't speak,' said Quyen. 'Go back to sleep.'

Mai struggled into a half-upright position. She made out a small group of children next to the bulwark, then pressed her imagination to find him again, little Loc, turning with a snarl as he growled, 'Dragon!' She smiled, bit back tears. Behind him, her old schoolfriend Huong was selling beef noodles in front of the damp, stink-shaded fish market. Straight through the market she followed her daily route, picking up speed, past fabric stalls and coffee yards, the dusty soccer field where sons of fishermen and truck drivers broke off from the game to buy cigarettes, and then to the wharf, her main place of business, among the taut hard bodies crating boxes, the smell of fish sauce, the rattling talk of men and the gleaming blue backs of silver fish, ice pallets, copper weighing scales bright in the sun, the bustle of docking and undocking, loading and unloading —

A bare-chested man turned around and looked directly at her.

'Ba?'

It filled her with joy to see him like that again: young and strong, his eyes clear and dead straight. He looked like he did in the altar photograph. It was her father before the war, before re-education, hospitalisation. Back when to be seen by him was to be hoisted onto his shoulders, gripped by the ankles. His hands tough, saltish with the smell of wet rope. She moved toward him, she was smiling, but he was stern.

'Child promised,' he said.

During his long absences at sea she had lived incompletely, waiting for him to come back so they could tell to each other each moment of their time apart. He spoiled her, her mother said. Her mother was right and yet it changed nothing: still he went away and still, each time, Mai waited.

Her sudden, fervent anger startled her.

'Why send Child away? Child obeyed Ba.' Her mind sparked off the words in terrific directions. 'Child could have waited for Ba to get better.' They had promised each other. He had left for ten days and returned, strange and newly blind, after two years. A thought connected with another: 'It was Ba who left Child.'

He stood there, tar-faced, empty-eyed, looking straight at her. She lifted her hands to her mouth, unable to believe what she had just said. The words still searing the length of her throat.

'Child is sorry,' she whispered. 'Ba and Ma sacrificed everything for Child. Child knows. Child is stupid.'

He would leap off the boat and swing her into the crook of his arm, up onto his shoulders. Her mother fretting her hands dry on her silken pants, smiling nervously. I can't get it off me, he would say. His hands quivering on either side of Mai's ribcage – It's stuck, I can't get this little beetle off me!

She missed him with an ache that was worse, even, than the thirst had been. All she'd ever known to want was his return. So she would enjoy the gift of his returning, and not be stupid.

'Child is sorry.'

He didn't respond.

'Child is sorry, Ba.'

'Mai.'

He was shaking her. She said again, 'Child is sorry,' then she felt fingers groping around in her mouth, a polluting smell and then her eyes refocused and she realised it was not her father she saw but Truong, standing gaunt over her.

'Thank heavens,' came Quyen's murmur.

Looking at him she finally understood, with a deep internal tremor, what it was that had drawn her to the boy all this time. It was not, as she had first assumed, his age – his awkward build. Nothing at all to do with Loc. It was his face. The expression on his face was the same expression she had seen on her father's face, every day, since he'd returned from re-education. It was a face dead of surprise.

She gasped as the pain flooded back into her body. She was awake again, cold.

'Mai's fever is gone,' Quyen said. She smiled at Mai, a smile of bright industry – such a smile as Mai had never hoped to see again. Unexpectedly she was reminded of her mother, and, to her even greater surprise, she found herself breaking into tears.

'Good,' whispered Quyen. 'That's good.'

Mai wiped her eyes, her mouth, with the hem of her shirt. 'I'm thirsty,' she said. She looked around for Truong but he seemed to have slipped away.

'You should be. You slept almost two days.'

It was evening. She stood up, Quyen helping her. Her legs giving at first. Slowly she climbed up the hatch. On deck she shielded her eyes against the sunset. An incandescent red sky veered into the dark ocean. Rows and rows of the same sun-blotched, peeling faces looked out at nothing.

'Everyone's up here,' Quyen whispered, 'because down there are all the sick people.'

'Sick people?'

Mai checked the deck, then searched it again with growing unease. He'd been standing over her. Keeping her voice even, she asked, 'Where is Truong?'

'Truong? I don't know.'

'But I saw him – when I woke up.'

Quyen considered her carefully. 'He was very worried about you, you know.'

He wasn't in the clearing with the other children. Mai shuffled into the morass of arms and legs, heading for the pilothouse. Nobody made way for her. At that moment Truong emerged from the companionway. She almost cried out aloud when she saw him – gone was the pale, delicate-faced boy she'd remembered: now his lips were bloated, the skin of his cheeks brown, chapped in the pattern of bruised glass. An awful new wateriness in his gaze. He stood there warily as though summoned for punishment. Mai mustered her voice:

'Is Child well?'

'Yes. Are you better?'

'Truong, speak properly!' scolded Quyen.

'How is Chi Mai?'

'Well. Better.' She leaned toward him, probing the viscosity of his eyes. His face's swollenness gave it a sleepy aspect.

'Ma said Chi Mai was very sick.'

'Chi is better now.'

'Tan and An were more sick than Chi,' he said. 'But Ma says they were lucky.'

Mai smiled at Quyen; she hadn't heard him talk so much before. His voice came out scratchy but steady. He stood before them in a waiting stance: legs together, hands by his sides.

'Chi is glad for them.'

'They died,' he said. When Mai didn't respond he went on: 'I saw the shark. All the uncles tried to catch it with that –' he pointed to a cable hanging off the derrick-crane – 'but it was too fast.'

'Truong!'

His eyes flicked to his mother. Then he said, 'Fourteen people died while Chi Mai was sleeping.'

'Child!'

He balled up his hands by his sides, then opened them again. 'Chi Mai isn't sick any more, *ha*?'

'That's right,' Mai and Quyen said together.

It was difficult to reconcile him with his frail, wasting body.

Seeing him, Mai's own body felt its full exhaustion. 'Now . . . let's see . . .' She lifted one hand until it hovered between them, palm down. 'Child wants to play slaps?'

His black eyes stared at her with something akin to pity.

'Pretend this is the shark,' she exclaimed. Quyen glanced up at her. Immediately – horrified, shocked by herself – Mai pulled back her hand. 'Chi is just joking.'

Later that evening, a young teenage girl with chicken legs wandered over to the gunwale and in a motion like a bow that didn't stop, toppled gracefully over the side.

'Wait!' someone cried.

'Let her be,' another person said. 'If she wants to, let her be.'

'Heavens, someone save her. Someone!' The first man stumbled to his feet, wild-eyed.

'You do it. Go on. Jump.'

He stood like a scarecrow, frozen. Everyone watched him. He walked to the side and looked down at the shiny, dusk-reflecting water.

'I can't see her,' he said.

'She must not have any family,' Quyen whispered to Mai.

'She has the right idea,' another low voice said. 'Is there any better way to go?'

'*Thoi,*' Anh Phuoc said, coming over. '*Thoi,* that's enough.'

Re-education camp. For two years those two words had framed the entirety of her imaginative life. Her father, of course, hadn't talked about it when he returned – nor her mother. Now, for the first time, someone talked to her about it. Anh Phuoc had fought in the same regiment as her father – had been sentenced to a camp in the same district. No, he hadn't known him. By the time the Communists took Ban Me Thuot in March 1975, the Americans were long gone and the Southern regiments in tatters – soldiers deserting, taking cover as civilians, fleeing into the jungles. Escape on every man's mind. Soon they all learned there was no escaping the Communists: not in the country they now controlled. They were skilled, he said, at turning north against south, village against village. He fell quiet.

Mai waited. She watched him remembering. Nine days had passed and now she noticed how severely he had aged: his eyes gone saggy, his skin mottled with dark sunspots.

'In the camps,' he said, 'they do what they do best. They take a man – and then they turn him against things.'

From the back deck a middle-aged woman started wading in their direction through the sprawl of bodies. She held the port gunwale with both hands for balance.

'Husbands against wives,' he went on. 'Children against parents. Your only chance is to denounce everyone, and everything, they tell you to.'

The woman reached them. She made her complaint in a hoarse voice. She was owed water. She had tendered hers to another child who had collapsed, she said, and pointed aft. Anh Phuoc held Mai's eyes for a second, then followed the woman.

Her father wouldn't have denounced her – she was sure of that. Not in his own heart. But again she understood how necessary it was to stay on the surface of things. Because beneath the surface was either dread or delirium. As more and more bundles were thrown overboard she taught herself not to look – not to think of the bundles as human – she resisted the impulse to identify which families had been depleted. She seized distraction from the immediate things: the weather, the next swallow of water, the ever-forward draw of time.

'Mai!'

It was Anh Phuoc. She stood up, hauled herself on weak legs along the gunwale, toward the rear of the boat. Past the hatch she suddenly saw Truong – propped up against the rusty mast of the derrick-crane, his chin drooping onto his chest, arms bony and limp by his sides.

Mai leapt forward, swiping her elbows and knees from side to side to clear space. The surrounding people watched listlessly.

'Water!'

No one reacted. She looked around and spotted an army flask – grabbed it, swivelled the cap open, held it to his mouth. A thin trickle ran over his rubbery lips before the flask was snatched away. She looked up and saw a man's face, twisted in hate the moment he struck her, his knuckles hard as a bottle against her cheek. She fell over and covered Truong's body.

'She stole water.'

'I'll pay it back,' said Anh Phuoc roughly.

Truong started coughing. Mai sat back, her cheek burning, and mumbled apology in the direction of the man. He was picking the flask up from the ground. People glanced over, disturbed by the waste. There had been a minor outcry the previous evening when a woman – an actress, people said – had used the last of her ration to wash her face.

Truong squinted up at Mai. Everything about him, the dark sore of his face, his disproportioned, skeletal limbs, seemed to be ceding its sense of solidity. She touched his blistered cheek with her fingers – was reminded of the sting on her own cheek from the man's blow.

'Ma,' he wheezed.

'It's all right,' she said. 'Ma is coming. Chi is here.'

'Where's Quyen?' asked Anh Phuoc. He stood up quickly and walked off.

Truong said, 'Child wanted to count the people.'

He coughed again, the air scraping through his throat. Watching him, a helpless feeling welled up within Mai and started to coalesce at the front of her skull. 'Child,' she whispered.

Quyen arrived. She seemed to be moving within a slower state, her face drawn, hair tangled. She saw Truong and bent down to him. 'Look,' she murmured, 'you hurt yourself.'

'He fainted,' said Mai.

'Why didn't Child stay with Ma?'

'I don't like it down there,' he said.

'Oh, Mai,' Quyen exclaimed, turning to her. 'Are you all right?'

'He shouldn't be in the sun. He needs more water.'

'It's too dark to count down there,' Truong said. He brought up his arms, dangled them loosely over his knees. An old man's pose. Quyen squatted down and enfolded him, clinching him between her elbows, raking one hand through his hair and cupping his forehead with the other.

'I was so tired,' said Quyen. 'Thank you.'

'He needs more water.'

'Does Child know?' She was speaking to Truong. 'Does Child know how lucky he is? To have Chi Mai look after him?'

Anh Phuoc leaned down close to both of them. 'Come with me,' he muttered. They followed him forward to the pilothouse, everyone watching as they passed. Once inside he closed the door. Carefully, he measured out a capful of water from a plastic carton and administered it into Truong's mouth.

The sight – even the smell – of the water roused an appalling ache in Mai's stomach, but she said nothing.

'Good boy,' said Anh Phuoc.

Quyen's eyes followed the carton. 'Is that all there is?'

Holding the tiller with one hand, he reached down and opened the cupboard beneath it. Three plastic white cartons.

'That's all,' he said, 'unless it rains.'

'How long will it last?'

'Another day. Two at the most.'

Her temple still aching, Mai looked out of the pilothouse windows. From up here she could see the full length and breadth of the boat: every inch of it clogged with rags and black-tufted heads and sunburned flesh. Up here would be the best place to count people. She wrenched her eyes away from the water carton and looked out instead at the sky. Not a cloud in sight. But the sky was full of deceit – it looked the same everywhere. She looked at the horizon, long and pale and eye-level all around them. Whatever direction she looked, it fell away into more water.

The tenth day dawned. Engines dead, the boat drifted on. Grey shadows strafing the water behind it. The detachable sail hoisted onto a short mast's yard and men taking turns, croaking directions to each other as they tried to steer the boat, as best they could, to the south.

Mai watched Truong with renewed intensity. Since Mai's recovery Quyen had kept to herself, remaining huddled, during the day as well as night, underneath the companionway stairs where they all slept. That morning Mai had found her sitting in the slatted light, staring vacantly into the dark hold. Squeezed between two old women.

'How is Truong?' Quyen asked her quietly.

Mai said: 'I keep telling him to come down.'

'He doesn't like it down here.'

Mai nodded, not knowing what to say.

Quyen dropped her chin and closed her eyes. Mai looked her over. She didn't look sick.

'Is Chi all right?'

Quyen nodded almost impatiently. One of the women beside her spat into her hands. When Quyen looked up her face was distant, drawn in unsparing lines.

'Look after him, *nha*? Please.'

Above deck, each hour stretched out its hot minutes. Mai lay on her back under the derrick-crane, her head against someone's shin, limbs interwoven with her neighbours'. Truong wedged beside her. The crane cast a shadow that inched up their bodies. She threw her sleeve over her face to ward off the sweltering sun. At one point a wind blew in and the boat began to sway, lightly, in the water. She was riding her father's shoulders. Her mother watching them happily. Whenever he was home he brought with him some quality that filled her mother so there was enough left, sometimes, for her to be happy.

Truong started singing. Softly – to himself – so softly she wouldn't have heard him if her ear hadn't been inches from his mouth. She gradually shifted her arm down so she could hear better. He sang the ballad from the third night. It seemed an age ago. She listened, hardly daring to breathe, watching the now-darkening sky knitting together the rigs and cables of the crane above them as though they were the branches of trees.

When he finished, the silence that surged in afterward was unbearable. Mai reached across her body and gently took hold of his arm.

'Who taught Child how to sing like that?'

He didn't answer.

The next morning, back below deck, she woke up to find a puddle of vomit next to his curled-up, sleeping body. It gleamed grey in the early light of dawn.

'The child has the sickness,' a voice said without a second thought. It was one of the old women who had camped with them beneath the companionway stairs. The hatch was open and light flowed in like a mist, dimly illumining the three other bodies entangled in their nook. The deeper recess of the hold remained black.

'No, he doesn't,' said Mai.

'Poor child. He is not the last. Such a pity.'

'Be quiet!' Mai covered her mouth, abashed, but no one reproached her. Several bodies stirred on the other side of the stairs.

Barely awake, Quyen rolled over to her son and propped herself up on an elbow. She brushed his cheek with her knuckles. For a second, in the half-light, Mai thought she saw an expression of horror move across her friend's face.

'Child is sorry,' Mai murmured to the old woman.

Truong's eyes were glazed when he opened them. He looked like a burnt ghost. He leaned over, away from his mother, and dry-retched. There was nothing left in him to expel. Another of their neighbours, a man who smelled of stale tobacco, averted his legs casually.

'What it can do to you,' the old woman said, her gums stained crimson from chewing betel leaves, 'the ocean.'

'Does Child's stomach hurt?' asked Mai.

'Yes.'

'What it can steal from you and never give back. My husband, both my daughters.'

'It's just a stomach ache,' said Quyen, then looked up as though daring the old woman – or anyone – to disagree. A gang of eyes, unmoving, inexpressive, watched them from the shadows.

That evening, Anh Phuoc ladled out the last rations of water. He shuffled wearily through the boat, repeating the same account to anyone who stopped him, intoning his interlocutors' names as though that were the only consolation left him to offer them. Weak moans and thick silences trailed him.

When Mai poured her ration into Truong's cup, Quyen frowned, and then flinched away. 'Thank you,' she said at last. For the first time she used the word for 'younger sister'.

'It's nothing. I already took a sip.'

'Poor child,' repeated the old woman, shaking her head.

Truong took some water in, then coughed some of it out. People looked over. In the dusk light his face was pallid and shiny.

He opened his mouth. 'Ma,' he said.

'I'm here,' said Quyen.

'Ma.'

Quyen bit her lips, wiped the sweat from Truong's brow with a corner of her shirt. Finally his eyes focused and he seemed to look straight at Mai.

'It's so hot,' he said.

'*Thoi,*' said Quyen, dabbing above his eyes, around his hairline.

'I want to go up.'

'Sleep, my beloved. My little prince. Sleep.'

Mai wanted desperately to say something to him – something useful, or comforting – but no words came. She got up to close the hatch door.

The old woman took out a betel leaf and inserted it into the slit of her toothless mouth.

His sickness followed the usual course. Muscle soreness and nausea in the early stages. That evening his blisters began to rise, some of them bleeding pus. He became too weak to swallow water.

In the middle of the night, Mai woke to find Truong half draped over her stomach. His weight on her so light as to be almost imperceptible, as though his body were already nothing more than bones and air. 'Everything will be fine,' she whispered into the darkness, her thoughts still interlaced with dream, scattered remotely across space and grey sea. Back home she'd slept on the same mat as Loc. Her mother by the opposite wall. She reached down and touched Truong's brow.

He stirred awake.

'Is Child all right?'

'I want to go up.'

The skin on his face was hot and moist. Mai lifted her eyes and noticed Quyen, mashed in the shadow of the companionway steps, staring at both of them.

'Take him,' she said dully.

Mai found a spot for them by the pilothouse, surrounded by sleeping families. When dawn came, Truong's head slid with a slight thud onto the planking. Half asleep, Mai sought his shoulder, shook it. His body gave no response. She sat up and shook him again. His clothes stiff with dried sweat. Nothing.

'Truong,' whispered Mai, feeling the worry build within her. She poked his cheek. It was still warm – thank heavens! – it was still warm. She checked his forehead: hotter than it had been last night. He was boiling up. His breath shallow and short. With agonising effort she cradled his slight, inert body and bore him up the stairs into the pilothouse.

Anh Phuoc was slumped underneath the tiller, sleeping. Three infants were laid out side by side on the floor, swaddled in rags.

He woke up. 'What is it?' He saw Truong in her arms. 'Where's Quyen?'

She laid him down. Then she turned to find Quyen.

'Wait.' Anh Phuoc got up, surveyed the boat through the windows, then retrieved a flask from behind the bank of gauges. He unscrewed the cap and poured a tiny trickle of water into a cup. 'This was for them,' he said, gesturing at the motionless babies. 'How they've lasted twelve days I don't know.' He screwed the flask cap back on and then, with tremendous care, handed her the cup. 'But they won't make it either.' He paused. 'Let me find Quyen.'

Truong wouldn't wake up. Mai dipped one finger into the cup, traced it along the inner line of his lips. Once it dried she dipped her finger again, ran it across his lips again. She did this over and over.

One time she thought she saw his throat twitch. His face – the burnt, blistered skin, its spots and scabs – the deeper she looked, the more his features dissociated from one another until what she looked into, as she tended him, was not a face, but a brown and blasted landscape. Like a slow fire it drew the air from her lungs.

Commotion on deck. Someone shouting. She jolted awake, checked Truong – he was still unconscious, his fever holding. A weird tension suffusing the air. Another death? Mai opened the pilothouse door and asked a nearby woman what was happening.

'They saw whales,' the woman said.

'Whales?'

'And then land birds.'

It was as though she were sick again, her heart shocked out of its usual rhythm. 'Land? They saw land?'

The woman shrugged.

All at once Quyen burst out of the hold, her hair dishevelled and her eyes watery and red. She spotted Mai.

'Here!' Mai called out excitedly. 'Chi Quyen, here!' She stood on tiptoes and scanned all the horizon she could see. Nothing. She looked again. 'Someone said they saw land,' she announced aloud. Realising people were scowling at her, she turned toward Quyen. Too late she caught a new, rough aspect in her eyes. Quyen strode up into Mai's face.

'Where's my son?'

She pushed into the pilothouse. Mai stumbled back, tripping over the doorsill.

Inside, Quyen saw Truong and rushed toward him, lowering her head to his. She emitted a throaty cry and twisted around to face Mai.

'Stay away,' she declared. 'You've done enough!' Her voice was strained, on the verge of shrillness.

'Chi,' gasped Mai.

'I've changed my mind,' Quyen went on, the pitch of her words wavering. Her expression was wild, now – cunning. 'He's my son! Not yours – mine!'

'*Thoi,*' a man's voice interjected.

Mai spun and saw Anh Phuoc in the doorway.

'What's the matter?'

Quyen glared at him. He waited for her to speak. Finally, her tone gone sullen, she said, 'She took my son.'

He sighed. 'Mai was looking after him.'

Quyen stared at him, incredulous, then started laughing. She clamped both hands over her mouth. Then, as though in embarrassment, she dipped her head, nuzzling Truong's chest like an animal. Mai watched it all. The thick dense knot back behind her temple. Quyen's body shuddered in tight bursts awhile, then slowly, hitchingly, it began to calm. It seemed for a moment as though Quyen might never look up again. When she did, her face was utterly blanched of expression.

'Mai wouldn't hurt Truong,' said Anh Phuoc tiredly. 'She loves him.'

Quyen threw him a spent smile. 'I know.' But she didn't look at Mai. Instead, she turned and again bent over the unconscious shape of her son. That was when she began to cry – silently at first, inside her body, but then, breath by breath, letting out her wail until the whole boat could hear.

He was her shame and yet she loved him. What did that make her? She had conceived him when she was young, and passed him off to her aunt in Da Lat to raise, and then she had gotten married. With the war and all its disturbances, she had never gone back to visit him. Worse, she had never told her husband.

'He would leave me,' she told Mai. 'He will.'

But she couldn't abandon her only son – not to the Communists – not if she could find a way out of the country. Even if he didn't want to leave, and even if he didn't know her. Her aunt had balked and Quyen had been forced to abduct him. She'd been wrong to have him – she knew that – but she'd been even more wrong to give him away. Surely, she thought, she was right to take him with her. But when she saw him weakening – then falling sick – she realised that perhaps he was being punished for her shame. Whether he lived or died – perhaps it wasn't for her to decide.

She begged Mai to forgive her.

Mai didn't say anything.

'He doesn't love his own mother,' said Quyen.

'That's not true.'

Quyen leaned down and unstuck his hair from his forehead, and parted it. They'd moved him back down into the hold, under the companionway stairs, for shade.

Quyen sniffed. 'It's fair. What kind of mother watches that happen to her only son – and does nothing?'

'You were sick.'

Quyen turned to her with a strange, shy expression, then lowered her gaze.

'I knew you would take care of him,' she said.

'Of course.'

'No.' She looked down at her son's fevered face. 'Forgive me. It was more than that. My thoughts were mad.' She gave out a noise like a hollow chuckle. 'I thought of asking you . . .' she said. 'I was going to ask you to take him in – to pretend he was your son.' She shook her head in wonderment. 'He likes you so much. Yes. I thought – just until I could tell my husband the truth.'

Mai remained quiet, her mind turbulent.

Quyen sniffed again. '*Thoi*,' she declared. 'Enough!' Caressing her forearm – still scored with rope marks from the storm six days ago – she smiled into the air. 'It's my fault.'

'Chi.'

'Whatever happens to him.'

Mai stared down, unsteadily, at the marred, exposed field of Truong's face.

'You don't have to answer,' Quyen continued in her bright voice. 'Whatever happens, I deserve it.'

He entered into the worst of it that afternoon, moving fitfully into and out of sleep. His breath short, irregular. Their neighbours kindly made some space for him to lie down. When some children came to visit, Quyen rebuffed them without even looking. Mai sat silently opposite them, next to the old betel-gummed woman, transfixed by her friend's intensity.

Then, at the end of the afternoon – after five long hours – Truong's small body suddenly unclenched and his breath eased. The lines on his forehead cleared. It seemed, unbelievably, that he had prevailed.

'It's over,' Mai said joyfully. 'Chi, the fever has broken.'

Quyen cradled him in her lap, rocking him lightly. 'Yes, yes, yes, yes,' she sighed, 'Sleep, my beloved.'

His clothes were soaked with sweat. For a fleeting moment, as Mai saw his face unfastened from its distress, the fantasy crossed her mind that he was dead. She shook it off. Quyen's hair fell over her son's face. They both appeared to her strangely now, as if at an increasing remove, as if she were trying to hold them in view through the stained, swaying window of a bus.

Truong hiccuped, opened his eyes and rasped, 'Ma has some water?' With an almost inaudible moan Quyen hunched over and showered his brow with kisses. Outside, the evening was falling, the last of the light sallow on his skin. After a while Truong gathered his breath again.

'Ma will sing to Child?'

'Sing for the poor child,' said the old woman.

Quyen nodded. She started singing: a Southern lullaby Mai hadn't heard for years, her voice more tender than Mai had imagined it could be.

Truong shook his head weakly. 'No – not that one.' He made an effort to swallow. 'My favourite song.'

'Your favourite song,' repeated Quyen. She bit her lip, frowning, then swung around mutely, strickenly, to Mai.

Mai reached out to stroke Truong's hair. She said, 'But Child must sleep, *nha*?' She waited for him to completely shut his eyes. Quyen found her hand and held it. Mai cleared her throat, then, surprised to find her voice even lower, hoarser than Quyen's, she started singing:

> I am the vigil moon that sheds you light
> My soul abides within the Thousand Peaks;
> Where drunk with wine and Long-Tuyen sword you seek
> And slaughter all the leopards of the night.
>
> And in the steps of Gioi Tu, seize Lau-Lan
> And quash the Man-Khe rivers into one.
> You wear the scarlet shadow of the sun:
> And yet your steed is whiter than my palm . . .

Abruptly her voice broke off, then she swallowed, picked up the thread of melody again, and sang it through, her voice as hard as Quyen's face was tender, her voice resolute and unwavering, sang it through to the very end.

The old woman nodded to herself.

The next morning – the morning of their thirteenth day – a couple of the fishermen sighted land. A swell of excitement, like a weak current, ran through the boat. People looked at one another as though for the first time.

'We made it,' someone quietly announced, returning from deck. He paused on the companionway, his head silhouetted against the sunlight. In the glare, Mai couldn't make out his face. He said, 'We're safe now.' The words deep in his throat.

Quyen and Truong were underneath the stairs. Mai had left them to themselves during the night. Now, with those others strong enough, Mai followed the man above deck. Outside, the dawn sun steeped through her as though her body were made of paper. Dizziness overwhelmed her when she saw the half-empty deck – had they been so depleted? She thought, with an odd pang, of Truong, his incessant counting. Then she saw the prow, teeming with people, all peering ahead, attitudes stalled in their necks and shoulders. She made her way forward, then spotted, far ahead, the tiny breakers on the reefs, and behind those, the white sand like a bared smile. Birds hanging in mid-air over the water.

During the night she had come to her decision. Her thoughts starting always with Truong and ending always with her father, upright in his hospital bed, staring at some invisible situation in front of him. A street with its lights turned off. She came into morning feeling a bone-deep ache through her body. The boat would land – they would all land – Mai would write to her family, and wait for them, and then she would look after Truong as if he were her own child. The decision dissolved within her, rose up with the force of joy. She would tell Quyen. She would look after him, completely, unconditionally, and try not to think about the moment when Quyen might ask her to stop.

Nearly weightless in her body, Mai descended the companionway.

When she reached the bottom she spun and searched behind the stairs. There they were. The hold awash with low talk.

'Chi Quyen.'

She was about to call out again when she sensed something amiss. Quyen's back – folded over Truong's sleeping form – it was too stiff. The posture too awkward.

Mai moved closer. 'Chi?' she asked.

Quyen's crouched torso expanded, took in air. Without turning around she said, 'What will I do now?' Her voice brute, flat.

Mai squatted down. Her heart tripping faster and faster, up into her throat.

Quyen said, 'He didn't.'

She said, 'All night. He wouldn't wake up.'

She was wrong, thought Mai. What did she know, thought Mai. When she'd left last night, Truong had been recovering. He'd been fine. He'd been asking Mai, over and over, to sing to him. What could have happened?

Quyen shifted to one side. He was bundled up in a blanket. The bundle tapered at one end – where his legs must have been. Mai could see no part of him. How could this be the end of it? She wrung the heels of her hands into her eyes, as if the fault lay with them. Then she felt Quyen's face, cool with shock, next to her own, rough and wet and cool against her knuckles, speaking into her ear. At first she recoiled from Quyen's touch. What was she saying? She was asking Mai for help. She was asking Mai to help her carry him. It was time, she said. Time, which had distended every moment on the boat – until there had seemed to be no shape to it – seemed now to snap violently shut, crushing all things into this one task. They were standing – when had they gotten up? – then they were kneeling, facing each other over the length of him. Quyen circumspect in her movements, as though loath to take up any more space than her son now needed. She seemed not to see anything she looked at. Together, the two of them brought the bundle aft, through the shifting, silent crowd, past the derrick-crane, where a group of the strongest men waited. There, the wind turned a corner of the blanket over and revealed the small head, the ash beauty of his face, the new dark slickness of his skin. With a shudder Quyen fell to it and pressed and rubbed her lips against his cheek.

Anh Phuoc, standing with three other men, waited for Quyen to finish before touching her shoulder.

He said, to no one in particular, 'We'll make land soon.'

As though this were an order, Mai took Quyen's arm and led her the full span of the boat to the prow. Again, the crowd parted for them. They stood together in silence, the spray moistening their faces as they looked forward, focusing all their sight and thought on that blurry peninsula ahead, that impossible place, so that they would not be forced to behold the men at the back of the boat peeling the blanket off, swinging the small body once, twice, three times before letting go, tossing him as far behind the boat as possible so he would be out of sight when the sharks attacked.

WILLIAM LANGEWIESCHE

The Outlaw Sea

Since we live on land, and are usually beyond sight of the sea, it is easy to forget that our world is an ocean world, and to ignore what in practice that means. Some shores have been tamed, however temporarily, but beyond the horizon lies a place that refuses to submit. It is the wave maker, an anarchic expanse, the open ocean of the high seas. Under its many names, and with variations in color and mood, this single ocean spreads across three-fourths of the globe. Geographically, it is not the exception to our planet, but by far its greatest defining feature. By political and social measures it is important too – not merely as a wilderness that has always existed or as a reminder of the world as it was before, but also quite possibly as a harbinger of a larger chaos to come. That is neither a lament nor a cheap forecast of doom, but more simply an observation of modern life in a place that is rarely seen. At a time when every last patch of land is claimed by one government or another, and when citizenship is treated as an absolute condition of human existence, the ocean is a realm that remains radically free.

Expressing that freedom are more than forty thousand large merchant ships that wander the world with little or no regulation, plying the open ocean among uncountable numbers of smaller coastal craft and carrying nearly the full weight of international trade – almost all the raw materials and finished products on which our land lives are built. The ships are steel behemoths, slow and enormously efficient, and magnificent if only for their mass and functionality. They are crewed from pools of the poor – several million sailors of varying quality, largely now from southern Asia, who bid down for the jobs in a global market and are mixed together without reference to such petty conventions as language and nationality. The sailors do not enjoy the benefit of long stays in exotic ports, as sailors did until recently, but rather they live afloat for twelve months at a stretch, enduring a maritime limbo in the ships' fluorescent-lit quarters, making brief stops to load and unload, and rarely going ashore. They are employed by independent Third World 'manning agents,' who in turn are paid for the labor they provide by furtive offshore management companies that in many cases work for even more elusive owners – people whose identities are hidden behind the legal structures of corporations so ghostly and unencumbered that they exist only on paper, or maybe as a brass plate on some faraway foreign door. The purpose of such arrangements is not to make philosophical points about the rule of law, but to limit responsibility, maximize profits, and allow for total freedom of action in a highly competitive world. The ships themselves are expressions of this system as it has evolved. They are possibly the most independent objects on earth, many of them without allegiances of any kind, frequently changing their identity and assuming whatever nationality – or 'flag' – allows them to proceed as they please.

This is the starting point of understanding the freedom of the sea. No one pretends that a ship must come from the home port painted on its stern, or that it has ever been anywhere near. Panama is the largest maritime nation on earth, followed by bloody Liberia, which hardly exists. No coastline is required either. There are ships that hail from La Paz, in landlocked Bolivia. There are ships that hail from the Mongolian desert. Moreover, the registries themselves are rarely based in the countries whose names they carry: Panama is considered to be an old-fashioned 'flag' because its consulates handle the paperwork and collect the registration fees, but 'Liberia' is run by a company in Virginia, 'Cambodia' by another in South Korea, and the proud and independent 'Bahamas' by a group in the City of London.

The system in its modern form, generally known as 'flags of convenience,' began in the early days of World War II as an American invention sanctioned by the United States government to circumvent its own neutrality laws. The idea was to allow American-owned ships to be re-flagged as Panamanian and used to deliver materials to Britain without concern that their action (or loss) would drag the United States unintentionally into war. Afterward, of course, the United States did join the war – only to emerge several years later with the largest ship registry in the world. By then the purely economic benefits of the Panamanian arrangement had become clear: it would allow the industry to escape the high costs of hiring American crews, to reduce the burdens imposed by stringent regulation, to limit the financial consequences of the occasional foundering or loss of a ship. And so an exodus occurred. For the same reasons, a group of American oil companies subsequently created the Liberian registry (based at first in New York) for their tankers, as a 'development' or international aid project. Again the scheme was sanctioned by the U.S. government, this time by idealists at the Department of State. For several decades these two quasi-colonial registries, which attracted shipowners from around the world, maintained reasonably high technical standards, perhaps because behind the scenes they were still subject to some control by the 'gentlemen's club' of traditional maritime powers – principally Europe and the United States. In the 1980s, however, a slew of other countries woke up to the potential for revenues and began to create their own registries to compete for business. The result was a sudden expansion in flags of convenience, and a corresponding loss of control. This happened in the context of an increasingly strong internationalist democratic ideal, by which all countries were formally considered to be equal. The trend accelerated in the 1990s, and paradoxically in direct reaction to a United Nations effort to impose order by demanding a 'genuine link' between a ship and its flag – a vague requirement that, typically, was subverted by the righteous 'compliance' of everyone involved.

These developments were seemingly as organic as they were calculated or man-made. For the shipowners, they amounted to a profound liberation. By shopping globally, they found that they could choose the laws that were applied to them, rather than haplessly submitting to the jurisdictions of their native countries. The advantages were so great that even the most conservative and well-established shipowners, who were perhaps not naturally inclined to abandon the confines of the nation-state, found that they had no choice but to do so. What's more, because of the registration fees the shipowners could offer to cash-strapped governments and corrupt officials, the various flags competed for business, and the deals kept getting better.

The resulting arrangement, though deeply subversive, has an undeniably elegant design. It constitutes an exact reversal of sovereignty's intent and a perfect mockery of national conceits. It is free enterprise at its freest, a logic taken to extremes. And it is by no means always a bad thing. I've been told, for example, that the cost of transporting tea to England has fallen a hundredfold since the days of sail, and even more in recent years. There are similar efficiencies across the board. But the efficiencies are accompanied by global problems too, including the playing of the poor against the poor and the persistence of huge fleets of dangerous ships, the pollution they cause, the implicit disposability of their crews, and the parallel growth of two particularly resilient pathogens that exist now on the ocean – the first being a modern strain of piracy, and the second its politicized cousin, the maritime form of the new, stateless terrorism. The patterns are strong in part because they fit so well with the long-standing realities of the sea – the ocean's easy disregard for human constructs, its size, the strength of its storms, and the privacy provided by its horizons. Certainly the old maritime traditions of freedom are involved, but something new is happening too. It is not by chance that the more sophisticated pirate groups and terrorists seem to mimic the methods and operational techniques of the shipowners. Their morals and motivations are different, of course, but all have learned to work without the need for a home base and, more significantly, to escape the forces of order not by running away, but by complying with the laws and regulations in order to move about freely and to hide in plain sight.

The result has been to place the oceans increasingly beyond governmental control. To maritime and security officials in administrative capitals like London and Washington, D.C., steeped in their own traditions of national power, these developments have come in recent years as a surprise. For public consumption, the officials still talk bravely about the impact of new regulations and the promise of technology, but in private many admit that it is chaos, not control, that is on the rise. They have learned what future historians may be able to see even more clearly, that our world is an ocean world, and it is wild.

The *Kristal* was therefore a typical casualty of modern times. It was an all-purpose tanker, 560 feet long, that had been built in Italy in 1974 and for more than a quarter century had restlessly wandered the world, riding the downward spiral of the maritime market under a progression of names, owners and nationalities. By the winter of 2001, at the advanced age of twenty-seven, it was flying the flag of Malta – a registry of convenience, with a typically shoddy track record and a reputation for allowing owners to operate their vessels nearly as they pleased. The ship belonged to an obscure but law-abiding Italian family whose members had an understandable penchant for privacy.

They owned it through the device of a Maltese holding company that existed only on paper, as a mailing address in the capital, Valletta, the home port painted on the ship's stern. There is no evidence that the *Kristal* ever stopped there, though it did sometimes sail through the Mediterranean and so must occasionally have passed by. It was crewed by manning agents primarily in Karachi, Pakistan, but also in Spain and Croatia. Its business and maintenance were managed through layers of other companies, variously of Switzerland and Monaco.

Though the *Kristal* was well painted and regularly passed inspections, it was at least five years beyond the ideal retirement age, and had grown decrepit and difficult to maintain. Its owners kept it sailing anyway, apparently with the intention of squeezing a final few years of profitability from the hull before selling it to other operators still lower on the food chain or, if none could be found, directly to a shipbreaker for the scrap value of its steel. They were unable to attract business from the major oil companies, most of which now try to apply stringent standards to the tankers they charter and generally shy away from vessels past the age of twenty because of the risk of breakup, and the expense and negative publicity caused by spills. Nonetheless, there were plenty of other, less exacting customers available, as long as the *Kristal* could transport their cargoes at a low enough price. Indeed, the *Kristal* was constantly busy. Throughout the previous year, it had engaged in a regular globe-circling trade, carrying molasses from India to western Europe, kerosene from Latvia to Argentina, and soy oil from Argentina around Cape Horn to India again. The molasses was a sign of the *Kristal*'s decline: it is the product left over from refined sugar, a low-value cargo carried on the cheap by ships that are typically one step removed from the grave. There is little risk to the principals involved – the customers and shipping companies – because the hulls and cargoes are insured, and in the event of an accident and a spill, molasses disperses easily and disappears without a trace. It is no small matter in choosing a ship that the same is generally true of Third World crews.

The *Kristal*'s customer in February 2001 was a subsidiary of the big British sugar company Tate & Lyle, which had contracted with the ship's owners to bring a full, heavy load of twenty-eight thousand tons of molasses from two ports on the west coast of India to an unspecified European destination that would be decided en route on the basis of the market. The crew consisted of thirty-five men of various nationalities, mostly Pakistani – about ten men more than usual for a ship of this type, because they would need to carry out repairs while under way. Such repairs are standard in the industry, and have the double advantage of allowing ships to avoid both costly layups and the prying eyes of inspectors. Aboard the *Kristal* most of the repairs consisted of chipping with hammers and chisels at heavy rust that had spread like a cancer under the paint, across the main deck and through the hull. There have been reports, difficult to substantiate but entirely plausible, that considerable and illicit welding was also being performed, and that as a result, one of the cargo tanks could not be used – a restriction that may have caused the crew to load the molasses improperly, placing severe strains on the ship's structure. It seems unlikely that this could have occurred without at least the tacit approval of the ship's management company. Be that as it may, the crew certainly knew about the *Kristal*'s precarious condition and were glad for their jobs nonetheless.

The captain was a forty-three-year-old Croatian named Allen Marin – one of many such officers from former Communist states, who are known to be competent and able to live on low salaries and who now constitute something of a global officer caste at service on increasing numbers of ships. Marin was an affable character, and he was well liked by his subordinates aboard, though some of them thought that he seemed strangely uninterested in the technical aspects of running the ship. It was noticed, for instance, that during the important final loading of the molasses in India, he and the chief mate, another Croatian, went ashore overnight, leaving supervision of the work to a junior officer. No one objected, of course. The *Kristal* was a molasses ship, but a fairly happy one. The attitude was to let the captain have his fun.

On February 4, 2001, the *Kristal* set out across the Indian Ocean on a route that would take it through the Suez Canal and the Strait of Gibraltar. The days passed in monotonous succession, broken by the routine of alternating six-hour watches, the anticipation of work and of rest. During their time off, the men ate and slept, and they relaxed by playing Ping-Pong or watching films in the messrooms. There were three such messrooms: one for the officers and cadets, who were both Croatian and Pakistani; one for the skilled sailors, who included three Spaniards; and one for the remainder of the crew, the ordinary hands, all of whom were Pakistani. The Pakistanis were fed a native cuisine of spicy foods that conformed to Islamic restrictions. They liked to watch Indian films, full of weddings and dancing. The others preferred the slicker Hollywood fare, with sleek women and guns. The two groups knew each other well, and they mixed, but they were not necessarily pals. They called the superstructure where they lived the 'iron house,' because it was made of metal and hemmed them in. It stood aft on the hull and rose five levels above the main deck to the bridge. It was not uncomfortable, but after a while it seemed small. The crew's conversations there were almost exclusively about the ship, because after many months together it provided all that was left to be said.

The Indian Ocean was calm. Word came that the destination would be Amsterdam. There was a period of concern partway to the Red Sea, when a portion of the main deck suddenly bulged upward, breaking some welds. Captain Marin reported the problem to the management company and received a private reply, presumably to carry on. Only one crewman expressed grave concern. He was one of the three Spaniards, a bearish, bearded forty-one-year-old pumpman named Juan Carlos Infante Casas, who despite his enormous physical strength had a reputation as a worrier. Infante Casas's duties included operating the valves and cargo pumps and sounding the tanks from overhead on the deck. Like the other Spaniards, both of whom were mechanics, he came from Galicia, along La Costa del Morte, Spain's western Atlantic shore. He had gone to sea out of restlessness as a young man, and had never married, and still lived with his mother, to whom he was close. The sailor's life was not the adventure he had hoped it would be, but he had stayed with it for lack of choice, and for nearly two decades he endured the steady loss of income and security experienced first by Europeans and Americans and then by all the successive rearguards of whatever nationality on the increasingly anarchic seas. After six months aboard, he was looking forward to leaving the ship just a few days ahead, at a scheduled fuelling stop and partial crew change in Gibraltar. In messroom conversation he said that he knew the *Kristal* too well to trust it on the winter Atlantic. The other Spaniards felt more equable, though they too were scheduled to leave at Gibraltar. The older of them was a lean, greying man, nearly sixty, named José Manuel Castineiras, who said that he neither regretted nor enjoyed his life at sea but considered it his destiny. It was easier for him than for his friend Infante Casas, therefore, when, after the *Kristal* passed through the Suez Canal, word came that the Gibraltar stop had been eliminated: the ship would fuel instead at Ceuta, on the Moroccan side of the strait, and the crew change would be delayed until Amsterdam. That too was destiny.

The passage through the Mediterranean was uneventful. To keep to schedule, Captain Marin maintained the full engine speed of 88 rpm, driving the heavy ship westward at 11 knots through six-foot waves that were typically steep for that sea. The hull shuddered sometimes, but the ship barely pitched, and it rolled side to side by only 5 degrees – not enough even to spill coffee. Spray wet the forward deck. The crew chipped rust. Life in the iron house continued normally.

The *Kristal* arrived at Ceuta on February 24. A storm was forecast for the Atlantic ahead, along the Portuguese and Spanish coasts, and gale warnings were in effect farther to the north. Marin ordered four hundred tons of bunker fuel, enough for another twelve days. While the ship took on the fuel, Juan Carlos Infante Casas went ashore and called his mother, who doted on him. When she answered the phone, he said, '*Hola Espańal*,' which is what he always said. He told her that he was calling from Ceuta and that his return to Galicia had been delayed.

His mother said, 'We have arranged your room.'

Infante Casas said, 'Please don't make special arrangements.'

She said, 'Why? What do you mean?'

He said, 'The ship is in very bad condition.' He told her its name, which until then she had not known. He asked about the weather in Galicia. She reported that it was very nice.

But her view was limited, as land views are, by the orderly little neighbourhood that surrounded her and the sky immediately overhead. At most, she might have seen on television a simplified prediction that tomorrow the sun would hide behind clouds. While fueling in Ceuta, Captain Marin had access to more sophisticated forecasts, as well as to reports of troubles ahead: ships were reporting heavy seas off the coasts of Portugal and Spain and to the north toward England across the notorious Bay of Biscay. In earlier, more orderly times he might have been expected to go gently on his ageing ship, and to wait in port until the weather had passed. But on the free-market sea, where profit margins are slim, delays of even a few hours can seem unacceptably costly, and a captain who develops a reputation for timidity will soon find that someone has taken his place. As soon as the fueling was finished, therefore, Captain Marin ordered the ship to get under way, and in the last hours before midnight of February 24 he sent the *Kristal* sailing fast past the Rock of Gibraltar and on into the Atlantic night.

At once the ride grew rough. The waves at first were about twelve feet high – black swells more felt than seen, across which the *Kristal* pitched and rolled and occasionally bashed. Such conditions were not in themselves immediately worrisome: the local winds remained light, and in technical terms the sea state seemed to be only about Force 5, on a scale of 12. Nonetheless, the swells were evidence of a significant disturbance ahead, the barometer was falling, and it was clear that worse was to come. With a schedule to keep, Captain Marin maintained full engine speed. The weather's resistance slowed the ship by about 2 knots as it fought northwestward to round the Cabo de São Vicente, on the Portuguese coast.

At 2:00 a.m. a twenty-five-year-old Pakistani deck cadet named Naeem Uddin joined the officers on the bridge to begin his regular six-hour watch. Uddin was a tall, docile man who had grown up in northern Pakistan, on the border with China, as the son of a security guard. Under the mistaken impression that the merchant marine would provide some of the discipline and pride of a naval career, he had trained to become a deck officer at an academy in Karachi. After three years aboard working ships, he knew better now, but with debts piled up behind him and only three months of required sea time remaining before he would qualify for his first license, he felt committed to the life. To make the best of it, he provided the discipline for himself, working hard without complaint and never commenting on the wisdom of his superiors. He was not, however, without judgment.

When Uddin came onto the bridge, the second officer, whose watch it was, told him that the autopilot was being overwhelmed and that he should take the helm and steer. This was one of Uddin's standard duties, and he performed it well, carefully holding the headings as commanded, but not without wondering, as the hours went by, whether there wasn't some better way to handle the coming storm than busting straight through.

Two years before, while serving aboard an Iranian freighter, Uddin had witnessed an argument between a chief mate and captain about just such a question. They were on a trip from India to the Persian Gulf, with a typhoon ahead. The chief mate had recommended taking a circuitous route along the coast to avoid the worst of the storm, but the captain, worried about the additional eight hours that such a tactic would cost, had said, 'Whatever happens, we're going straight.' As a result, they were caught by winds and waves so great that the ship's engine was overpowered, and with rpm set for 'full ahead,' they found themselves dead in the water, sometimes even moving backward. Since the rudder was ineffective at less than about 5 knots, they lost the ability to steer, and for twenty-four hours they drifted helplessly until the storm abated. It was merely by luck that they had sufficient sea room and did not end up wrecked against a lee shore. This time, as Uddin knew, the sea room was more limited: the land was out of sight, but should the *Kristal* turn or drift to the right, it would soon encounter breakers on the Iberian shores.

Steadily over the next two days the weather grew worse. The *Kristal* struggled northward in the open ocean off the Portuguese coast, headed for a point abeam Spain's Cape Finisterre, where it would be able to turn slightly eastward and take a straight line across the outer Bay of Biscay for the famously stormy French island of Ouessant, which stands at the southern entrance to the English Channel. By now the sea was so rough that all work on the deck had stopped. Sleep was difficult, movie watching nauseating, Ping-Pong a dangerous sport. Ships coming from the north warned of still rougher stuff ahead. Captain Marin seemed detached as usual. He maintained full forward speed.

By late afternoon on February 26, on the second day after passing Gibraltar, the *Kristal* was offshore of Galicia, in the vicinity of Cape Finisterre, and conditions by any standard were severe. The sea state by now was at least Force 9. To the men on the bridge, the ocean seemed to be coming apart. The wind howled out of the north, above a constant roar of crashing water. Sheets of heavy spray rose to smash against the bridge's wings. The waves were steep and breaking, as high as thirty feet. They regularly buried the bow and sometimes swept across the entire deck, engulfing the ship to the superstructure and filling the aft passageways faster than they could drain.

The view from a ship of such conditions is in some ways a privileged one – a rare display of the ocean's power that may seem exhilarating even to a crew fighting for survival. There is for instance a home video of a violent storm in the Mediterranean off Spain that was shot in December 2000 from the bridge of a Greek-owned gasoline tanker named the *Castor*, in which waves are seen burying the deck, and in the background the Polish crew can be heard laughing and whooping with delight. That crew proved foolish, because the *Castor* then cracked severely and, threatening at any moment to sink or explode, embarked on what became an epic voyage under tow as a 'leper ship' that for six weeks was refused entry by every port of refuge. Nonetheless, there is no denying the abstract beauty of a heavy storm at sea.

But the crew of the *Kristal* knew their ship too well to indulge in abstraction. At 2:00 a.m., when Naeem Uddin entered the bridge for his watch, he found not only the scheduled second officer on duty there but the chief mate and the third officer as well, both of whom had stayed on past the end of their normal watches. To Uddin they looked afraid, as was he. The blackness of the night was streaked with the white of breaking waves. The ship was rolling and pitching violently. Through the spinning clear-screens on the bridge's windows he could see the familiar bow light moving wildly as the ship plunged into the oncoming waves. It was just possible to make out the masses of water boiling across the deck. When Uddin took the helm, he found the ship difficult to steer. It was hogging over the crests, surfing down the watery slopes, sagging and staggering through the troughs, and slewing left and right by 20 degrees. There was no chance of keeping the *Kristal* exactly on heading. Uddin fought back with large rudder movements, trying to average the swings.

It was not Uddin's first experience with the *Kristal* in a storm. The previous winter, he had steered through similar waves while bound for Ireland with another load of molasses, and he had watched the ship rolling to 25 degrees, uncomfortably close to its capsize threshold. On that occasion, however, the captain had been another man, a Spaniard, who had ordered Uddin to turn the *Kristal* into the waves and slow it to its minimum manoeuvring speed, easing the ship's motion at the cost of a delay. This time, in contrast, Captain Marin was below in his cabin, probably asleep. He had stood duty on the bridge for much of the time since leaving Ceuta, and undoubtedly he needed to rest. Meanwhile, the helm's instrumentation showed that the engine power was still set at the full 88 rpm – as if before retiring, the captain had given an order to maintain the maximum possible speed at all costs.

Uddin was very aware of his low rank as a cadet, and he continued to steer without comment, but he sensed that this ride was more dangerous than any he had known before. Though the *Kristal* was rolling less steeply than it had during the previous year's storm, initially to only about 15 degrees, it was shaking, slamming, and pitching severely through the waves. From the changes in vibration it seemed that the propeller at times was either cavitating or coming partially out of the water. In combination the motions were complex. The ship would roll three or four times, bury its bow, and, while struggling upward, oscillate more rapidly. Uddin realized that enormous strains were being placed on a weak and rusty hull.

The officers on the bridge realized it too. They were openly anxious about the ship's fate, and they repeatedly asked one another, 'What's going to happen? What should we do?' The second officer radioed to another vessel that was out that night, running downwind from the north. He asked about the ocean ahead. The answer came back that it was very, very rough, with really high seas – meaning worse even than these. The chief mate and the second officer discussed running for shelter on the Spanish or Portuguese coast, but perhaps because of the dangers of such lee shores, they did not pursue the idea. On several occasions when the slamming grew most intense, the chief mate asked Uddin to alter the course off-wind, 10 or 20 degrees to the west, but each time, the rolling grew so severe that he hastily remanded the order and had the ship brought back to face the storm. The best compromise seemed to lie among compass headings slightly to the left of the weather, by which the ship took the waves on the starboard bow. It wasn't much of a solution. The *Kristal* continued to pitch and slam, and at the extremes rolled past 25 degrees, making it impossible for the officers to stand without holding on, and causing loose objects in the bridge to slide, topple, and crash. Significantly, the chief mate did not rouse the captain or take it upon himself to break the schedule and reduce the engine speed.

Uddin's watch was shortened that night in recognition of the fight he had put up to maintain control. He went below before dawn, had a cold meal, and retreated to his cabin to rest. His cabin was on the galley deck, three levels below the bridge and two above the main deck, at the front of the superstructure, overlooking the bow. It had a bunk, a cabinet with drawers, and a forward-facing porthole covered by a curtain. Uddin undressed, put on nightclothes, and lay in his bunk as usual with his head toward the bow and his feet toward the stern. By wedging himself against the bunk's preventer board, he managed to sleep.

Uddin woke to a series of severe jerks accompanied by the crash of crockery in the galley and the shouts of men. It was 12:30 in the afternoon. He felt groggy and disoriented, and he assumed that the weather had turned worse – that the ship must simply have rolled more heavily than before. But then he noticed that his feet were higher than his head, and indeed that the entire Cabin sloped down steeply, toward the bow. The engine had stopped; there were no vibrations. Uddin heard the splashing of water. He scrambled out of the bunk, went to the porthole, and drew aside the curtain. The scene outside sent a shock of terror through him: the *Kristal*'s hull had broken nearly in two from below and had folded down at the midpoint into a V that was awash and flexing, hanging together merely by the skin of the deck. The ship was dead in the water. Storm waves surged through the breach.

Though in retrospect there had been plenty of advance warning of the *Kristal*'s vulnerability to such a disaster, it was a confusing moment aboard the ship. As the break occurred, the men on duty up on the bridge saw a cloud of steam rising from rupturing pipes on the main deck. Some of the crew later said that despite the conditions at the time, the Spanish pumpman, Juan Carlos Infante Casas, was moving along a catwalk directly above the break and that he ran for his life upslope toward the superstructure, barely keeping ahead of the advancing water as the hull angled downward into the sea. His friend, the lean, graying mechanic, José Manuel Castineiras, was in the messroom, finishing a lunch of chicken and soup, when he felt a shock and heard the ship rupture. It made a sharp, heavy crack like a cannon shot, after which within seconds the deck pitched down. Castineiras rushed outside onto a passageway and clambered hand over hand up a stairway to the deck above, where he and other crewmen broke out life jackets and put them on. The general alarm rang. Without further encouragement they began to gather at the assigned muster stations and prepared to abandon ship.

Later it was pointed out that even mortally wounded tankers tend to float for a while, that molasses does not burn, and that there was really no need for the crew to hurry. The old adage was mentioned that sailors should only step *up* into lifeboats – meaning as a desperate last resort when the hull sinks away below their feet. In reality, however, things are rarely so clear. This crew was not sitting in some office thinking back on an event; they were enduring the chaos of the ocean in present time, clustered on the tilted deck of a broken hull in a ferocious winter storm, facing the prospect of imminent death. The air temperature was 39 degrees, and the water was only slightly warmer. The *Kristal* was equipped with an inflatable life raft in a canister and two open lifeboats on davits, one on each side. Had the stern suddenly sunk, as it seemed about to do, it would have tangled the lines, dragged the lifeboats down, and possibly taken the raft as well. It is regrettable that many of the men panicked, and that in their haste some of those now outside on the deck had not dressed beyond the shorts and T-shirts that they happened to have on. But it is also understandable that they urgently lowered the lifeboats to the level of their deck and began to scramble in.

Captain Marin and the chief mate had been dining in the officers' messroom when the accident occurred, and they quickly climbed to the bridge. As they rushed in, they heard the third engineer, on duty in the subsurface confines of the engine room, repeatedly calling on the intercom for clarification. The engineer was a Serb. All he knew for certain was that the engine had shut down automatically, that the overload alarm had sounded, and that his attempt to restart the equipment had failed. His first concern was the loss of control – and indeed with the bow rearing out of the water and acting as a sail, the broken hull had slewed to the west and was rolling dangerously. Now the fire alarms were going off (perhaps triggered by the steam), and he could make no sense of this at all. Captain Marin was nearly as confused, which may explain why he later reported that he had a fire in the engine room, when in reality he did not. He knew in any case that the ship was lost, and he ordered the third engineer to abandon his efforts immediately and evacuate the engine space. Then he got on the radio and broadcast the first emergency call.

In his cabin, braced against the sloping floor, Naeem Uddin donned a heavy sweater over his nightclothes, put on fresh coveralls and shoes, and strapped himself tightly into a life jacket. Oddly, he heard no alarms. Indeed, he was impressed by the silence of the dying ship – a quiet interrupted only by the crash of falling objects and the rhythmic banging of his cabin door, which had popped open and was swinging with the *Kristal*'s rolls. He staggered down the deserted hallway, shouldered through a doorway to the outside, and emerged into the roar of the storm.

By then maybe ten minutes had passed since the catastrophe had struck. Uddin was the last man to arrive at the port-side muster station, which was on the leeward, or downwind, side of the ship. By the time he got there, the port lifeboat was hanging just outboard of the deck, and twenty-two of the crew had already climbed in. The Croatian chief engineer was there, as were all three Spaniards and a large number of ordinary Pakistani sailors. Uddin saw two higher-ranking Pakistanis standing on the deck beside the boat – including the third officer who had been on the bridge the night before and who now held a crew list and was checking off names. That third officer was Kenneth Romal, a Karachi native not quite twenty-eight. He was listening to shouted instructions and reassurances from Captain Marin, who stood two levels higher, on the port bridge wing. The men in the lifeboat were silent, and obviously very afraid. Uddin climbed into the bow of the lifeboat, which because of the ship's angle was pointed slightly down. He noticed that the water below was thick with molasses and that the spilled cargo was calming the waves.

Third Officer Romal had a handheld radio with which he could communicate with Captain Marin. Romal climbed in and sat in the stern of the lifeboat, at the helm. After another ten minutes the captain gave the order to abandon ship; and the chief mate, standing one deck higher, used a brake arrangement to ease the port lifeboat into the waves. Twenty-five men were aboard. They unshackled the boat from the cables, started the small diesel engine, and drifted clear of the ship. From the bridge wing the captain radioed for them to try to hold position close by in case of difficulties as, next, the chief mate lowered the starboard lifeboat, with an additional eight men aboard. The starboard lifeboat was on the ship's upwind side, bearing the full brunt of the storm, and as the captain had expected, it had trouble in the heavy seas: the sailors could not disengage the boat from one of the davit wires, which with every high wave now bowed like a giant snare, threatening to slip under the boat and capsize it. Uddin saw the captain and first mate rushing about the bridge, anxiously monitoring the struggle below. Eventually the starboard lifeboat was freed, but then it began to drift forward along the broken hull toward the breach – a rage of breaking seas that would surely kill the men if they were swept in. Uddin and the others in the port lifeboat watched in frustration, unable to intervene. Somehow the starboard lifeboat moved clear. Carrying its crew of eight frightened sailors, it faded rapidly into the storm and disappeared from view.

Captain Marin and the chief mate were alone now on the ship. Some in the port lifeboat argued for returning to try to pick them up, but the captain insisted by radio that the ship was about to sink, and they should steer clear. It was an extraordinarily brave gesture – a private display of honour by a man who must have known that his own captains ashore, those shadowy companies whose schedules he had so dutifully served, would never have taken such risks for him. For the moment, at least, this did not seem to matter. He and the first mate checked through the entire superstructure for possible stragglers, even going below the waterline to search the engine room. When they were sure that no one had been left behind, they returned to the deck and managed to deploy the inflatable life raft despite the ferocious winds. They then jumped down into it and floated free.

Aboard the port lifeboat, the crew had already lost sight of the *Kristal*. The suddenness with which they found themselves alone was almost as frightening as the facts that faced them. They had no idea of their destination, or even if their plight was known. Third Officer Romal was steering skilfully, but they were twenty-five sailors without survival suits on a bitterly cold ocean, moving among mountainous seas in a small open boat, battered by wind and spray. They veered between roaring breakers, any of which could have rolled or swamped their fragile craft. Sitting at the bow with the masses of water hissing and rearing overhead, Uddin tried and failed to estimate the size of the waves. They were later said to be thirty feet high – a clinical measure that cannot convey the sea's dimensions as perceived by the *Kristal*'s terrified castaways, and does not account for the likelihood that several times an hour under such conditions, waves would come along that were nearly twice that high. Romal kept cautioning the men to stay calm, but with limited success. Uddin and several others remained functional. But one man was seasick and vomiting, and many of the others were seized by a dangerously unreasoned animalistic craving to survive. Romal was able to maintain control for awhile, but the storm pressed in relentlessly and did not allow his men the space to collect their minds.

It is useful to remember the reductive effect that fear has on thoughts and reactions: whether during shipwrecks or in other disasters, a sort of tunnel vision may set in and narrow people's views. There is another story that comes to mind. It took place three years before the *Kristal*'s demise, on a similarly old and rusty ship named the *Flare*, which had set off from Rotterdam on a stormy winter crossing to Montreal. The *Flare* was a dry-bulk carrier, flagged in Cyprus, carrying a typically multinational crew of twenty-five men. The voyage was extremely rough, with waves exceeding fifty feet. For two weeks the *Flare* slammed and whipped, flexing so wildly that, according to one survivor, the tops of the deck cranes appeared at times to touch. Late one night as it approached the Canadian coast, the *Flare* broke cleanly in two. No one at first was lost. The entire crew clustered on the stern section, which listed to the side and began to sink. Strangely, the engine continued to turn, slowly driving the hulk on an erratic course through the night. The crew managed to launch one lifeboat, but it broke away before anyone could climb aboard. The men were terrified and panicky, and for good reason, since ultimately twenty-one of them died. But before the end, on the sinking stern, there was a moment of savage joy when a ship floating in the opposite direction suddenly loomed out of the darkness ahead, as if it were coming to rescue them. The frightened men cheered – only then to their horror to see the name *Flare* painted in block letters on the side. It was of course their own detached bow section, and it passed them by.

Something similar happened to the crew in the *Kristal*'s port lifeboat. They never saw the *Kristal* again, but at the top of a wave Uddin spotted another ship coming toward them and gave a shout, and the men responded euphorically, with only one idea in mind – the immediate salvation to be found on that ship's deck. Apparently they gave no thought to the difficulty of climbing a ship's sides in such seas, and they never discussed the possibility of waiting for helicopters and rescue divers, which, if they were not already on the way, could quickly now have been brought into play.

Third Officer Romal seemed to be as single-minded as the others: he headed the lifeboat at full speed for the ship while the excited men fired off rocket flares. On the handheld radio Romal began broadcasting, 'Mayday! Mayday! This is *Kristal Lifeboat Number One*, with twenty-five men on board!' He repeated this several times until the ship answered.

It took twenty minutes to close the distance.

The rescuer was a Panamanian-flagged gas carrier named the *Tarquin Dell*, with a Filipino captain and crew. Gas carriers are high-sided vessels, and this one was particularly so, with a main deck about thirty feet above the waterline because the ship was riding empty. For the same reason, the *Tarquin Dell* was difficult to manoeuvre in the storm and was prone to heavy rolling. The captain managed to turn it beam-on to the waves in an attempt to shelter the waters on the downwind side. The lifeboat circled around the stern and tucked in close against the ship's hull. Far above, on the *Tarquin Dell*'s deck, sailors in life jackets stood along the railing, holding on against 30-degree rolls and attempting to lower light 'heaving lines' to which the lifeboats bow and stern lines might be secured. This proved almost impossible to accomplish. The *Tarquin Dell*'s sailors were trying to tie one-handed knots on a rolling deck, and the few lines that they did secure broke. They kept at it. The situation in the lifeboat was nearly uncontrollable. Many of the crewmen were crying and shouting, and the boat was brutally slamming against the ship's unyielding hull. At the lifeboat's bow, Uddin broke a boat hook while fending off and then continued to work with the stub. The storm waves seemed to be undiminished by the *Tarquin Dell*'s mass. One of the waves rose so high that it came within a foot of simply depositing the lifeboat on the ship's deck. But then the boat dropped away, and the final line broke. The boat started drifting rapidly forward, and the *Tarquin Dell*'s crew ran along with it, until one of them heaved a heavy line into the centre of the crowd below.

At that point things went very wrong. Rather than securing the line to prevent further drift, a cluster of desperate men grabbed it, each higher than the last, until very quickly they were standing, half hanging from the line, and unbalancing the lifeboat. Again Third Officer Romal shouted desperately, 'Sit down! Calm down!' But then a large wave broke over the boat, and the boat swamped and tilted, and all but one man, who was sitting toward the stern, were washed into the Atlantic Ocean.

Despite their life jackets, most of the men seem to have gone deep. The gray-haired Spaniard José Manuel Castineiras felt the tangle of flailing legs and arms as he fought his way back to the surface. When he emerged, he saw that the lifeboat was flooded and floating low in the water, and that the man who had not washed out was still sitting upright in place, but whether because he had suffered a heart attack or a blow to the head, he was dead. He was a Pakistani 'galley boy,' fifty-some years old, and the first of the *Kristal*'s crewmen to die. Castineiras swam to the lifeboat and somehow crawled in. Several others followed. The *Tarquin Dell* had pivoted in the wind and drifted some distance away, and it was no longer providing protection from the storm. Rolling heavily, it began a series of difficult manoeuvres to set itself up for another try.

When Naeem Uddin was washed out of the lifeboat, he heard the screams of others before the ocean closed over his head. Time then slowed for him. He felt himself sinking as his life jacket started sliding up his chest. Astoundingly, like other life jackets provided by the *Kristal*, it was not equipped with a crotch strap. He grabbed the life jacket with both hands before it escaped over his shoulders, and he rode it to the surface. When his head emerged from the water, he found himself in a wilderness of waves, with neither the ship nor the lifeboat in sight. Three other crewmen floated within view, including the burly, black-bearded pumpman, Infante Casas.

The ocean was shockingly cold. Eventually, from the top of a large wave, Uddin spotted the swamped lifeboat and in the background the *Tarquin Dell*. Holding his life jacket to his chest with one hand, he began to swim. He had on his nightclothes, his heavy sweater, his coveralls, and his shoes. He used an improvised side-stroke, and because of the life jacket, he stayed mostly on his back. This meant that when occasionally he caught sight of the lifeboat, it appeared upside down and above his head. He navigated by those sightings. He swam for fifteen minutes or more, past crewmen floating helplessly. At times he cried. He knew he was going to die, and he wondered what it would be like, whether he would feel pain. The cold water had hurt him at first, but it no longer did. He had visions of his mother, his father, his sister, his brothers. He recited a Muslim prayer in preparation for the end. He said, 'There is no god but God, and Muhammad is the last prophet of God.' When he got to the lifeboat, he noticed that it was riding nose high in the water because there were five men in it, including Castineiras and the dead galley boy, and they were all sitting at the stern. Uddin tried to crawl aboard, and he was surprised to find that he lacked the strength. He hung on to a rope until he found a way to drape himself over the lifeboats gunwale and roll in.

Later, another Pakistani sailor arrived, and Uddin helped him aboard. They sat toward the bow for balance. The waves were relentless. There were seven people in the lifeboat – which meant that eighteen remained in the water, though most of them were no longer in view. The *Tarquin Dell* was back in action, providing its limited lee shelter, now with an innovation: a heavy rope strung in a loop from bow to stern, to which the men in the water could cling, and the lifeboat could be attached. But the situation was grim. Uddin and Castineiras both saw dead bodies floating nearby. For the men still alive, the *Tarquin Dell*'s sailors threw life rings into the water and dangled ropes and a rope ladder over the side. After a while Infante Casas swam up, holding his life jacket under one arm. Castineiras thought that Infante Casas got into the lifeboat, and Uddin thought that he did not. What is certain is that Infante Casas grabbed a dangling rope with his powerful arm, that a wave washed over him, and then he was gone.

It seemed obvious by now that others were dying too. The lifeboat was being repeatedly washed over by the waves, and was threatening at any moment to capsize again. Uddin realized he could not afford to be passive – that he would have to fight to survive. He slid out of the lifeboat, swam along the *Tarquin Dell*, and despite being slammed repeatedly against the hull, somehow caught the rope ladder and began to climb. The climbing was slow. Uddin's leg was badly cut, and it was warm with blood. The ladder swung violently as the *Tarquin Dell* rolled, and waves continued to clutch at him, sapping what little of his strength remained. About halfway up the side of the ship, bruised and battered and still vulnerable to the ocean's surface, he simply could not move anymore. He did not pray or think of his family then; his mind was empty. He hung on. Vaguely, he felt someone grab his collar from above. It was a strong grip, and he fainted.

When he regained consciousness a few minutes later, he was lying in a small room along with three other survivors – the only *Kristal* crewmen who ever actually found the sought-for safety of the *Tarquin*

Dell's deck. All of them were blue with cold. Someone gave Uddin a bowl of soup. Someone bandaged his leg. The captain of the *Tarquin Dell* came and said that a helicopter was on the way. Soon afterward, Uddin heard the whacking of its blades.

The helicopter was a bulbous Sikorsky, with a rescue diver and a winch. It had come from La Coruña, an old port city on the Galician coast, a half-hour flight away. The pilot was a local star, a man universally known by his first name, Evaristo. He went after five men still loose and alive in the water, winching them up in a double harness two at a time. Then he swung over to the flooded lifeboat and picked up the survivors there too. Castineiras was the second to last to leave. By then another helicopter was coming onto the scene. Evaristo flew his load of hypothermic survivors to the hospital in La Coruña. By the time they got there, one of them had gone into cardiac arrest; he was rushed into the emergency room and revived. Evaristo headed immediately back to sea to find the dead and search for the missing. The second helicopter meanwhile had easily retrieved Captain Marin and the *Kristal*'s chief mate from their life raft and had rescued the eight men in the starboard lifeboat too. Uddin and the three others aboard the *Tarquin Dell* were the last to be plucked from the scene.

In business terms, the damage control began within hours. As the crew recovered in La Coruña, at the hospital and a hotel, the *Kristal*'s managers sent a representative to the city and employed guards to keep unauthorized visitors away – meaning mainly the press. They hired a crisis-management public relations firm in London, issued a terse statement of regret, and endured a few days of national coverage in Spain before the news of the disaster faded away. The survivors were rapidly repatriated to the far points of the globe and were paid their salaries to the end of their contracts, as the contracts required. The families of the dead were offered lump sums by the *Kristal*'s insurance company, in London. To receive this money, the next of kin first had to sign 'quitclaims' promising not to pursue further action. Against the advice of the international seafarers' union, almost all of them signed. The amounts were kept private, and they involved commitments to silence, but it is known that most were small, that the payouts varied according to nationality, and that the Spanish got the most because before they signed the quitclaim, they made a little fuss.

The *Kristal* broke entirely in two and floated for several days until first the bow section sank, and then the stern. As for the human tally, eleven men had died, almost a third of the *Kristal*'s crew. Only four of the bodies were ever retrieved. Among those who were never found was Infante Casas. It seemed poignant and strange that the Spaniard had sailed the world for years, only to die here, off his own Galician shores. When Castineiras left the hospital in La Coruña, it took him less than an hour to make the trip home. But perhaps the saddest loss of all was that of the one sailor who by measure of his performance should have survived – the young, levelheaded Third Officer Romal. After he was swept from the lifeboat, he was never seen again.

The *Kristal* was by no means the only ship to sink that year. Dozens of other large ships were lost (hundreds, if one includes small merchant vessels), as they had been every year before, and have been every year since. During the 1990s alone, several thousand sailors died, worldwide.

The frustration is that a large body of regulations exists to keep such maritime disasters from happening. Most of the regulations are generated by a specialized United Nations agency called the International Maritime Organization (IMO), which is based in London and since 1958 has issued a plethora of technical standards for the maintenance and operation of large ships at sea. The IMO is a typically idealistic construct for bringing order to the world – a democratic assembly of 162 member nations, all of them determinedly equal, who work with the assistance of a technical staff and the consultations of accredited nongovernmental groups to establish regulatory packages known as 'conventions,' which the individual member states are free to adopt (or not) in their sovereign maritime laws. The enforcement of those laws is a separate question, and it is spotty, because the arrangement allows the IMO no enforcement powers of its own. Most of the individual states have neither the expertise nor the inclination to enforce their own official standards, and they rely instead on independent technical organizations known as 'classification societies,' which are not hired by the states, but rather selected and paid for by each ship's owner. Considerations of conflict of interest are not allowed to intrude. Nonetheless, the IMO has been influential and indeed has become the universal reference for life at sea. Thumbing through the international conventions, hefting the books of regulations, or browsing the logbooks and certificates required to be carried aboard a ship, one might easily conclude that thanks to good government in London, the situation is completely under control.

In other words, the ocean looks tight in print, much as many increasingly ungovernable nations still do by formal description. The problem, as some insiders will admit in private, is that the entire structure is something of a fantasy floating free of the realities at sea. Worse, from the point of view of increasingly disillusioned regulators, the documents that demonstrate compliance are used as a façade behind which groups or companies can do whatever they please. This does not mean that the paperwork is counterfeit – which is one of the more unimaginative concerns of U.S. officials with whom I've spoken. Far from it; the documents are as authentic as can be. The *Kristal* illustrates the point. It was designed, built and maintained to full IMO standards, and it operated under the well-known maritime authority of a modern democratic state. It passed both scheduled and spot inspections on a regular basis. It was supervised and approved by the Italian classification society known as RINA, one of the world's top five. Its IMO safety procedures were approved by Det Norske Veritas, the Norwegian classification society considered to be the best in the business. Its crew knew of course that the *Kristal* was an unsafe ship, rusting away beneath its paint, harried by its owners, and was probably, as a result, handled recklessly. But when occasionally the ship was boarded by official inspectors whose role in principle should have been to intervene, the crew treated them with the well-practised wariness of the Third World poor toward the police – a superficial acceptance underlain by fundamental disregard. When I finally tracked down Naeem Uddin in London, he told me he had never even considered the possibility that I suggested to him – that in theory the inspectors were functioning on the crew's behalf. Two years after the sinking, he remained visibly traumatized by the experience. But he was pragmatic too. Indeed, he'd had a job precisely because the *Kristal*'s inspections had meant hardly anything at all.

DEBORAH CRAMER

Climate and Atlantic

Thirty-five hundred years before the birth of Christ, an ancient civilization flourished in the fertile valley between the Tigris and Euphrates Rivers. Its people were successful traders, productive farmers, and highly skilled artisans. They built magnificent cities and temples, created the world's first script and carved it into stone and clay tablets, and harnessed the river floodwaters to irrigate fields of barley and wheat. A little more than one thousand years later, beset by internal strife, threatened with invasion by desert nomads, and weakened by a failing agricultural economy, these once robust Mesopotamian cities collapsed.

Sumerian agricultural practices help explain the sudden decline of one of the world's earliest civilizations. Tree cutting, dam building, and river diversion destroyed soil fertility. Fields, silted and salted by extensive irrigation systems, could no longer produce rich harvests of wheat. A long and severe drought added a final stress, forcing people to abandon their cities and farms. When the drought began, rains high in the Taurus Mountains of eastern Turkey slackened. Less water drained into tributaries feeding the Tigris and Euphrates. These rivers, whose floodwaters nourished the fertile crescent, the cradle of civilization, subsided. Windblown sand fell instead of rain, and an ancient civilization came to a close. Falling lake levels in Turkey, at the headwaters of the Tigris and Euphrates, and in the Dead Sea implicate drought as a possible cause for the demise of this ancient civilization.

The seafloor in the Gulf of Oman, downwind of the rivers, offers further evidence of aridity at the time the cities were abandoned. Dust, stirred up dry winds, blew from the Tigris–Euphrates Valley out to sea, where it settled on the bottom. High levels of eolian dust in the Gulf of Oman hold the memory of an ancient time that left a sparse record of its end.

Water continues to be a precious commodity in the Middle East. The Tigris and Euphrates still rise and fall in tune with rainfall in mountain highlands, rising in years when rain is abundant, falling in times of drought. In the dry periods, the waters of the Euphrates River subside as much as 40 percent. A scarcity of water has provoked political tensions. Turkey, controlling the headwaters, has turned off the water supply to downstream Iraq and Syria. Iraq and Syria, arguing about the effect of a Syrian dam on river flow, have called in their armies to help settle the dispute.

At the southern edge of the sand seas of the Sahara lies a vast semi-arid plain, the Sahel. On this steppe reaching across Africa from Atlantic to the Red Sea, clumps of short grasses and clusters of thorny acacia and mimosa rise from the bare sand. Some rain falls here, more than in the Sahara, but less than in the grassy savanna and lush forest further south. In the Sahel, this place of conversion where barren desert gives way to green savanna, nomadic herdsmen roam, seeking water for their cattle, goats, sheep, and camels. At the plain's southern edge, where the growing season is longer, sedentary farmers cultivate crops of sorghum and millet. Whether nomads or farmers, dwellers of the Sahel await the rains, but they are unreliable. The entire year's water, 4 to 8 inches (10 to 20 centimeters) in the north and 20 to 24 inches (50 to 60 centimeters) in the south, can fall in a few torrential downpours. The heavy rain washes away seeds and crops and rushes toward the sea. Even after a storm, the soil, only lightly touched by the rain, often remains dry.

Sahel, translated from Arabic, means fringe or shore. Its boundary, like the inconstant line between land and sea, ebbs and flows with the coming and going of rain, advancing and retreating from season to season, year to year. When rain is plentiful, the Sahel recedes. When rain is scarce, it advances, drying out the savanna, cutting shorter the already short growing season, leaving the herds thirsty. Satellite images of the Sahel show its edges keeping pace with the rain, growing and shrinking as much as 190 miles (300 kilometers) over the years, as the rains depart and return. Beginning in 1968, drought has devastated the Sahel, withering crops and killing animals. Thousands of people have died of starvation, and millions more have abandoned their sand-drenched villages for the poverty of urban shantytowns. Rainfall is half what it was in earlier years.

Life in the Sahel was not always so tenuous. In the years between 1200 and 1600, the wealthy city of Timbuktu stood at the crossroads of trans-Saharan trade. Prosperous merchants gathered in this city on the bend of the Niger River to trade salt, copper, dates, and figs from the north for gold, ivory, and slaves from the south. Farmers tilled soils fertilized by the Niger's abundant floodwaters and exported wheat throughout West Africa. Muslim scholars came to study at the city's great mosques. Today, the fabled city of Timbuktu is but a shadow of its former glory. Shifting sands encroach upon the decaying mosques, and the Niger's subsiding waters have cut off access into the town. A scant 5 to 9 inches (12 to 22 centimeters) of rain falls each year. Nearby, sand swallows a village built around an oasis that has dried up.

To the east, Lake Chad is disappearing. Once fed by swift-flowing rivers, once spreading over 10,000 square miles (26,000 square kilometers) the largest lake of the Sahel has shrunk to two ponds half its former size. Fishing villages built at its edge now sit stranded in dry sand and grass, miles away from the receding shore. Outside the villages, the land is bare. Great herds of elephants, zebra and giraffe once flourished in the Sahel, when rivers were full and grasslands were lush and green. Today, fossilized bones and empty gravel tracks mark the riverbeds where large animals once gathered.

What is the difference between feast and famine in the Tigris–Euphrates Valley, between life and death in the Sahel? In the Sahel and in the Middle East, nonsustainable agricultural practices – excessive woodcutting, overgrazing, and misuse of water – may further degrade land stressed by drought, but drought – and its relief – are caused by rain. And rain is the gift of the sea; its scarcity or abundance is determined hundreds of miles away, in currents circling Atlantic, in warm water rushing north with the Gulf Stream and cold deep water sinking to the bottom of the Labrador Sea. In the life-giving rains that once drenched the Sahel, in the floodwaters of the Tigris and Euphrates Rivers, even in the bitter chill of a London winter, the voice of the sea sings. The signal from the sea is muted, subtle, carrying the memory of wind long since dissipated. Though hard to read and to understand, its influence is immense, shaping and sustaining earth's climate. Atlantic currents surge or thin, water temperature and wind rise or fall, and faraway lands feel the effects. During the Mesopotamian drought, North Atlantic cooled. When her tropical waters warm, desiccation comes to the Sahel. Changes in climate have led to the poverty and wealth of nations and the rise and fall of civilizations, and climate answers the call of distant, enigmatic seas.

The sea stirs, ever so slightly, and the winds and rains reply. A minute sea change resounds and fills the atmosphere. The Labrador Sea chills, barely perceptibly, between 1 and 2 degrees Fahrenheit (2 and 3 degrees Celsius), and the water disappears into the depths, unleashing heat into the air. The westerlies strengthen, infusing northern Europe with heat and moisture from the sea. Abundant snows thicken the glaciers of Scandinavia, while to the south, Alpine ice shrivels under dry skies and the edges of the Mediterranean and the Middle East thirst for rain. Such has been the climate in Europe for the past twenty-five years; such was the climate when the Mesopotamian civilization collapsed.

This gentle song of the sea, dryly known as the North Atlantic Oscillation, swells and subsides, rises and falls, and the climate adjusts. During the 1950s and 1960s, the waters of the Labrador Sea warmed slightly. The seawater floated – too warm to sink and release its heat to the wind. Europe, its radiator turned down, cooled. The westerlies shifted south, causing drought in Scandinavia but bringing needed rain to Morocco, Spain, and the Tigris–Euphrates headwaters. Atlantic still spawned deep, cold currents, but they originated farther east, in the Greenland Sea, and easterly winds conveyed their heat to Labrador.

In the marginal lands of the Sahel, the sea brings forth life, announces death. There, throughout most of the year, a dry, dusty wind, the Harmattan, part of the northeast trades, sweeps across the Sahara and through the Sahel. From the south come the monsoons, moisture-laden winds blowing off the Atlantic. Where they meet, the warm, moist air rises, thunderheads fill the sky, and it rains. This point of convergence, near the equator, migrates with the sun, moving south in the winter and north in the summer, bringing rain when the sun is highest. When the rains come, lines of storms known as Lignes des Grains cross the Sahel, bearing the awaited water.

Moving north across the Sahel, between lush savanna and desert, annual rainfall declines precipitously, decreasing approximately one quarter of an inch (six millimeters) each mile (1.6 kilometers). Should the moist winds tarry or fade, depriving the Sahel of even these sparse rains, drought soon follows. For the last thirty years, monsoon rains have forsaken the Sahel, lingering 150 miles (240 kilometers) to the south, out over the sea. Some scientists attribute their absence to the destruction of coastal rain forests in Nigeria, Ghana, and the Ivory Coast. Dense forest, thick with ebony, mahogany and palm trees and drenched in water, once returned its moisture to the wind, but 90 percent of West African rain forests have been felled, and falling rain washes back into the sea. Wholesale disappearance of the West African rain forest confuses our understanding of climate, blurring distinctions between the mark of man and the rhythms of nature, the former perhaps prolonging the effects of the latter. With the cutting of rain forest, drought may persist in the Sahel, but lack of rain has desiccated this land before, when the rain forest still stood.

The disposition of tropical seas gives rise to rains watering the Sahel. Ships rerouted through South Atlantic when the 1968 Arab–Israeli War closed the Suez Canal unknowingly produced the evidence. In their logs of sea-surface temperature, as well as in data oceanographers have collected more recently, a pattern emerges. Minimal warming of tropical Atlantic, amounting to only a degree or two, accompanied by minimal cooling in waters immediately to the north, foretells death in the Sahel. When equatorial waters warm, the monsoons pause and rains fail the Sahel. When currents carry their heat a little farther north, toward the Tropic of Capricorn, monsoon winds and rains follow, and dry lands drink.

The sea holds the memory of fleeting winds and rain. Imprinted on the sea surface, where air and water touch, it is carried into deep waters and spread with the motion of currents. Eventually, it graces the shore. Once, between 1908 and 1914, and then again in the late 1960s, when westerly winds tracked south and blew warm air and moisture over the Mediterranean, strong northerly winds engorged northern Atlantic with cold, fresh polar water. The saltiness of the sea declined. In both cases, the appearance of freshwater pools coincided with drought in the Sahel. The more recent, and more studied, pool was large, hundreds of miles wide. For fifteen years it drifted through

Atlantic's icy waters, following currents across the Labrador Sea to Newfoundland, then circling back across the ocean, along the coasts of Ireland, England, and Scandinavia, up into the Arctic Circle and back to Greenland. Ripples from this pool affected the very depths of the sea. Cooler by one or two degrees, fresher by 1.4 percent, this Great Salinity Anomaly, as it was so unimaginatively called, slid over the northern waters of Atlantic, cooling sea and land.

More than three thousand years ago, an Egyptian pharaoh dreamed of a cycle of feast and famine, of lean years and fat years. Today, science can describe, if not explain, the cycle intimated in ancient literature. Computers scanning the temperature logs of ships crisscrossing Atlantic have now identified large parcels of surface water, hundreds and sometimes thousands of miles wide, some warm, some cold, trailing the currents, each winding through the sea in about twenty years. They carry the signature of yesterday's winds, and they freshen today's. They summon the rains and beckon to the deep sea. Atmosphere and ocean, sea surface and depths, are wedded. The path of prevailing westerlies, the transit of warm- and cold-water parcels through Atlantic, the birth of deep water in the Greenland and Labrador Seas synchronize in twenty- to thirty-year cycles, accounting for warming in Europe, aridity in the Middle East.

The patterns of rainfall in Europe and the Tigris–Euphrates Valley follow the cadence of the North Atlantic Oscillation. It is Atlantic's El Niño. When westerly winds blow warm, moist air over northern Europe, circulation in the Labrador Sea intensifies. Surface water throws heat to the wind, then runs cold and deep, penetrating the bottom current, spreading to the Grand Banks and on to the Bahamas. At the same time, the Greenland Sea quiets. The overturning water subsides, conveying less to the depths. The surface water warms, the sea ice retreats, and the deep current diminishes. Between 1993 and 1995, churning Labrador Sea water reached depths never before witnessed by science, while in the Greenland Sea a great deep-water pump, the Odden ice, disappeared.

During winter storms, polar winds chill and freeze the Greenland Sea, and salty brine sinks to fuel the deep current. The ice breaks apart in high waves and turbulence, and melts when storms die, only to form again in the next gale, injecting more brine into the deep. Warm temperatures in the Greenland Sea during the 1980s and 1990s kept winter sea ice at bay, preventing extensive formations of Odden ice, shutting off the deep-water pump. We cannot see water disappear into the depths of the Greenland Sea, but tags of tritium fallout mark its descent, measure its speed. When deep-water production slowed in the Greenland Sea, it declined by 80 percent. When the pump is running, surface water sinks 10,500 feet (3200 meters). In the 1990s, it barely reached 3000 feet (900 meters). In the winter of 1996–97, Odden ice suddenly reappeared in the Greenland Sea, but then it receded, ending the millennium without a further appearance.

The sea murmurs and the land responds. Europe feels the shifting wind, the changing sea temperatures. When the Odden ice returned to the Greenland Sea, a cold, harsh winter returned to Europe. In 1997, for the first time in over a decade, ice in Dutch canals thickened sufficiently to allow the Netherlands to hold its historic 125-mile (200-kilometer) skating competition. The rains, subsiding between 50 and 75 percent in northern Europe, shifted south, soothing parched lands in Spain and Africa.

Whalers have long felt these rhythms of Atlantic, noting three hundred years ago that cold in Baffin Bay foretold warmth in Europe. Scientists have begun to articulate how the movement of sea and wind is expressed in the sands of the Sahel, the snows of northern Europe. They know the dance begins with a temperature change at Atlantic's surface and a corresponding shift in the wind, but they cannot predict the end of one sea cycle and the beginning of another. They cannot say when water will turn the Sahel green, or when the Tigris and Euphrates Rivers will swell with rain. Undoubtedly, that knowledge will come, but the wonder lies not in the predictability of the model, but in the idea that life-giving rain falling on land is borne in on the pulse of a distant sea.

The natural rhythms of Atlantic measured in the droughts of the Sahel and the rains of Norway are short. We can hear the whole song in our own lifetimes. Overlying the shorter rhythms are longer melodies, playing out over generations. These voices of the sea also reach land, warming the climate and chilling it, over periods of 1000 to 1500 years. The deep sea records these songs. Year after year, century after century, millennium after millennium, the remains of tiny foraminifera accumulate in sand at the bottom of the Sargasso Sea, near Bermuda. The chemical composition of their calcium carbonate shells, measuring temperature at the sea surface, suggests that 1000 years ago, the Sargasso warmed. It was a small increase, 2 degrees Fahrenheit (1 degree Celsius), but Europe basked in the warmth. Wineries flourished in England as far north as York, Gloucester, and Hereford. Farmers, to the annoyance of Northumbrian sheepherders, expanded their fields into the Scottish uplands and hills. Mountain passes blocked by ice cleared, opening more routes across the Alps.

To us, living in more temperate climates, North Atlantic currents such as the Irminger and the East Greenland Currents have names but no faces. The pulse of these currents, running along the coasts of Iceland and Greenland, is vital to lives and cultures onshore. Nuances in the strength of a northern current, in the amount of heat it carries, can have dire or fortuitous consequences for land dwellers. Warm water carried north by oceanic currents melted the ice pack off the Icelandic and Greenland coasts, opening these once marginal and desolate lands to Norse settlement. Norse voyagers first reaching that volcanic island between Norway and Greenland in the 860s named it Iceland, for its severe winters and ice-filled fjords, but within a decade the sea warmed, the ice retreated, and the Norse came and stayed. A longer growing season yielded ample grain and grasses to support people and cattle, and the settlements thrived. Even the Arctic ice may have melted in the balmy seas. Cores of sediment drawn from the Arctic ice pack contain fossils of algae that grew in open water one thousand years ago.

The Vikings sailed on through warmed seas to Greenland and established farms on its southern and western coasts. They grew vegetables and hay in the fertile soil and raised livestock, but these colonies lasted only until the sea cooled. This history, too, is recorded in the shells of organisms buried in the Sargasso. Within a few hundred years, the warm Irminger Current, carrying heat from the Gulf Stream, gave way to the chill of the East Greenland Current. These polar waters flowed south from the Arctic, bringing pack ice to choke off the Denmark Strait between Iceland and Greenland.

The ice lingered through the year and clogged the fjords, preventing ships from landing, cutting off supplies and communication from Iceland and Norway. Norse settlements in Greenland declined and ultimately disappeared as the air chilled and the ground froze. The inhabitants starved, unable to feed their cattle on the sparse grasses of fading summers, unable to grow vegetables on the encroaching tundra. Unable or unwilling to trade their thin European clothing for warm animal skins, they froze. The nomadic Inuit, living off the bounty of the sea, following their food across the ice, survived.

Between three thousand and six thousand Norse lived in the two Greenland settlements. By 1540, all had perished. The chilling sea wrought tragedy in Iceland as well. The population, which had grown to seventy-seven thousand during the few hundred years of warmth, plummeted by half as the Little Ice Age set in and the land could no longer support such large numbers of people. A small decline in temperature, 2 degrees Fahrenheit (1 degree Celsius), cut summer short, reducing the yield of the soil. Farmers once cultivated wheat and barley throughout Iceland, but as the air chilled, production ceased, first in the north and then in the south. Livestock – sheep, horses, and cows – starved for lack of grass, and the sea ice, which once came no closer than a day's journey across the water, now encroached upon the shore, shutting in the fishermen. In the south, toward the still open water, the cod disappeared, unable to thrive in the chilly polar waters. Perhaps the Irminger veered south and ceased to temper the cold Arctic currents hugging Greenland and Iceland's shores. Perhaps a pool of polar water cloaked northern seas, slowing the deep circulation. Perhaps, as the deep current slowed, the Gulf Stream became sluggish and delivered less heat. Only now is science beginning to understand how the quiet motions of a sea ravaged a culture.

The drop in sea temperature wreaked havoc in Europe. Farmers descended from the Scottish Highlands, abandoning their fields and farms to sheepherders. Snow crowned Ben Nevis all year. Grapevines withered in England. In Germany the vines survived, but cold weather soured the grapes. For more than a century, vintners sweetened the sour wine with lead oxide, inadvertently poisoning those who imbibed, sometimes fatally. The Thames, Rhône and Guadalquiver froze in the winter, and pack ice drifted down to England. Glaciers spread down from the mountains. In Chamonix, Outzal, and the Italian frontier, Alpine glaciers engulfed farms, threatened churches, and felled trees. Meltwater streams and bursting glacial dams washed away houses and barns, roads and bridges, leaving arable fields flooded with rubble. Entire villages were destroyed and abandoned. The late springs and cool summers produced massive crop failures and widespread famine. Ireland's potatoes, stored in the ground after the harvest, froze. Mediterranean fruit and olive trees perished in the frosts and biting wind. Disease and starvation were rampant.

Atlantic's northern currents pause, and Europe plunges into a deep cold. The Little Ice Age is only the most recent cooling in a series of cooling cycles, whose sign is etched into the seafloor. The cycles reach back thousands of years, during a time when earth, released from the icy grip of glaciers, supposedly basked in a warm, stable climate. The record of the seafloor asks that we not take benign winds and salubrious seas for granted. Between Newfoundland and Ireland, pieces of volcanic glass and iron-stained silt litter the bottom, at intervals of approximately 1500 years. They were rafted down from their original sites in Iceland and Greenland on armadas of icebergs released with each pulse of cooling. They, as well as the shells of polar foraminifera layering the seafloor, track the course of cold water moving south. Polar species buried off Cap Blanc, Mauritania, suggests that even tropical seas felt a chill as the Canary Current brought cold water south.

The Little Ice Age profoundly disrupted life in Europe, but its name does not mislead. By ice age standards, the Little Ice Age was little. Europe has experienced much deeper chills, over much longer periods. For thousands of years at a time, massive ice sheets, 10,000 feet (3000 meters) thick, have buried Great Britain and Germany, forcing plant and animal life on a long retreat south. As science learns to read the deep sea more adeptly, tantalizing links between circulating currents and earth's ice ages emerge, shedding light on perplexing questions of how the sun's energy heats and cools the planet, and how earth's seas move with the dance of stars.

Earth's distance from the burning sun fashions our climate. Under a protective atmosphere, earth's temperature hovers at 59 degrees Fahrenheit (15 degrees Celsius), but over millions of years it shifts, cooler, then warmer, cooler, then warmer, as the planet glides through a delicately poised orbit. Earth's axis tilts between 22 and 25 degrees every 41,000 years, its orbit stretches to an ellipse and shrinks back to a circle every 100,000 years, and it approaches the sun at the same season every 23,000 years. As earth moves through these cycles, the intensity of sunlight striking the planet waxes and wanes, strengthening and weakening the force of seasons. Skies fill with rain, then turn to dust, the sea's heat-bearing currents swell and thin, and ice sheets advance and retreat. As earth spins through her journey, sunshine lighting the Arctic Circle increases and then diminishes, by as much as 20 percent.

Orbital variations leave their mark on earth's climate. For long stretches between 5000 and 10,000 years ago, the Sahel and the Sahara were wet and moist. In the hidden caves of Tassili n' Ajjer, a desolate sand-swept plateau in the desert, are thousands of ancient paintings portraying the herds of wild animals that once flourished there, before lush grasses turned to sand and swift rivers emptied, before brimming lakes dried up and marshes drained. In those days, earth leaned more on its axis and passed closer to the sun in her orbit. Perihelion occurred in September, rather than in January, as it does today, increasing the intensity of northern hemisphere summers. The monsoons blew stronger, reached farther north, and heated the sea. Sun, sea, and wind carried more rain, increasing precipitation in northern Africa by at least 25 percent. Bathed in fresh water, the soil turned rich and loamy and the sands grew to grass. The savanna held more moisture and released it to the sky, multiplying the rains. The Sahara shrank by almost one third. As the desert turned green in the rain, giraffe and elephant, hippopotamus and crocodile came to dwell in places that today are as dry as dust.

The relationship between earth's track through the sky and its climate is confusing. The advance and retreat of ice sheets coincide with the wobble of earth's orbit, but not always simply or directly. Layers of seafloor, holding the memory of climate change, call into question accepted truths about exactly how astronomy governs temperature on our planet. Sediment from the slopes of the Bahamas suggests that the heat of northern sunlight may not be the only cause of glacial retreat. The cores record the withdrawal of ice sheets a full 5000 or 6000 years before the northern hemisphere would have warmed from a change in orbit. Hidden in the sea are the secrets of our climate, and with each new core we come closer to understanding the weight of water, its prerogatives and persuasions.

Once we believed the ocean immutable, the deep water dark and still. Now shells of tiny animals and chemical pollutants raining from the atmosphere reveal a sea whose waters call to each other from the far ends of the planet, whose message is heard not only on the surface, but in the cold deep water sliding along the bottom. Deep currents heed the call of planetary motions, but also ebb and flow to their own rhythms as well.

The trail of CFCs sinking into the abyss of Antarctica's Weddell Sea has lightened in the last hundred years. Scientists cannot say why the deep circulation has mysteriously slowed. They suggest that perhaps deep currents counterbalance each other, Antarctic deep water thinning while northern currents swell, southern waters warming while northern seas chill. Scientists suspect that during the Little Ice Age, Antarctic deep water surged while deep circulation in Atlantic's northern waters slowed. Today's waning of Antarctic deep water may be the next phase in the 1500-year sea cycle.

The swings of deep-sea currents help explain the tumultuous end to the last ice age. The planet began to warm, but the glaciers left in fits and starts, the ice retreating and then returning, retreating and then returning. Gradual changes in solar insolation resulting from earth's steady spin though the sky do not account for these sudden and frequent bursts of glaciation. Rather, it is the motions of deep currents that jag the greater rhythm of climate prescribed by earth's orbit.

When northern currents swell, the northern hemisphere warms, and when they fade, the ice returns. The seabed records this partnership between ocean and climate, tracking the peregrinations of Atlantic's deep currents thousands of years ago when the ice ages waned.

Microscopic single-celled animals, buried in dark obscurity, illuminate the larger history of ocean circulation and climate change. The single-celled inhabitants of the deep, dwellers on the walls and floors of channels where the deep water flows, are picky, partial to one current over another, even though the water temperatures may differ only slightly. Some species find their niche with the cold current spreading south from the Arctic; others prefer icy water creeping up from Antarctica. Their shells rest in the sediment long after the animals themselves have died, long after the currents have departed.

The Vema Channel, off the coast of southern Brazil, connects two large basins of Atlantic. Cold, heavy water moves through this deep, winding passage no more than twenty feet (six meters) wide. Today, 13,000 feet (4000 meters) down in the channel, cold Antarctic water flows north. Above it, 9700 feet (3000 meters) down, deep water from the northern reaches of Atlantic moves south. Buried in the wall of the channel in this upper layer are shells from dwellers of the colder Antarctic water. They are 18,000 years old. Their presence suggests that when ice crept down onto the northern continents, the current from the north weakened and icy southern waters swelled to take its place.

Other shells, buried on the continental slope off Bermuda and in undersea mountains near the Azores, also chronicle the undulations of deep-sea currents during the ice ages. The relative amounts of cadmium and calcium in seashells reflect the fertility of surrounding waters: the less cadmium, the fewer nutrients in the sea. The northern current, having been at the surface more recently, where blooming meadows of sea grasses have depleted the sea, has fewer nutrients. Where the more impoverished current flows, shells of bottom-dwelling microscopic animals show low cadmium/calcium ratios. When the flow of deep water slows and the sea fills with more nutrient-rich water from the Pacific or from Antarctic seas, the cadmium/calcium ratios rise. Scientists, vaporizing the pinhead-sized shells with powerful currents, reducing them to individual atoms, have measured their cadmium/calcium content and discovered that during the ice ages the cold deep current flowing down from the north shrank by 50 percent.

Other pieces of the seafloor also reflect the larger circulation of water. Iron and manganese precipitate out of the sea, forming tiny particles lining the seabed. The particles contain birth certificates, isotopes of a silvery rare-earth metal that, thousands of years later, faithfully identify the waters from which the particles came, distinguishing the deep water originating in the Arctic from the cold currents of Antarctic seas. Cores of sediment taken from the southeast Atlantic seafloor tell of a weakening of the cold, northern current during the ice ages.

As the deep current grew weak, the Gulf Stream languished as well. Today, Atlantic's blue river fills the North and Norwegian Seas with warm water and warm-water animals, but warm-water species did not always dominate there. Charcoal paintings of penguins line the walls of sea caves along the Mediterranean, testament to the chill brought in from the sea. Polar shells in the seafloor bear witness to a time during the ice ages when cold water crept south and lapped against the shores of Ireland, France, and Spain, where forests gave way to Arctic grasses and tundra and animals scurried south. The Gulf Stream languished all along its course. In the Straits of Florida, where the current flows swiftly today, it grew sluggish, its flow weakening by 35 percent.

The ice ages convulsed to a close. The deep current swelled and subsided, swelled and subsided, and the ice withdrew and came back, withdrew and came back. When the deep current gained strength, the sinking water relinquished heat to the air, and the Gulf Stream, bearing even more heat, coursed in. When the deep current eased, the Gulf Stream slowed. Each time the currents weakened, the land shivered and plunged into a deep and bitter cold. The Tropics, long thought safe from the cool breath of glaciers, shuddered. Coral from waters off Barbados and shells of sea animals who once dwelled on deep underwater plateaus register a drop of 9 degrees Fahrenheit (16 degrees Celsius) throughout the depths of tropical Atlantic. As the water chilled, the land responded. Pollen and dust imprisoned in ice high in the mountains of the Andes record the descent of the tree line, the conversion of tropical rain forest to dry grassland. When Atlantic's circulation slowed, the entire basin felt the chill; no place was immune. A benign climate, which we take so much for granted, is but an impermanent gift.

Living in the present, in the moment, we believe that the latency of spring yields the bounty of autumn, and that barren winter turns to ripe summer in smooth continuous cycles that describe the past, reach into the future. The record of the sea, reaching back for hundreds and thousands of years, points to a different reality. Only a 9 degree Fahrenheit (16 degrees Celsius) difference in temperature separates the harsh ice ages from the balmy climate that nurtured the first agricultural civilizations, and earth flips easily, quickly, from one state to another.

Orphan Knoll, a lone piece of North America stranded at the very edge of the continental shelf, sits in the deep water 300 miles (480 kilometers) north of Newfoundland. Deep-sea coral embedded in its steep rock walls show the deep current languishing in as few as fifty years. Bits and pieces of Atlantic's past – solitary coral living in other parts of Atlantic, ice cores from Greenland, shells of foraminifera once drifting in icy waters – all describe currents whose paths can be quickly rerouted. The motions of deep currents determine whether large portions of earth are hospitable or hostile to human habitation. When the deep current slows, surface waters chill rapidly and the land responds. The climate we take for granted can deteriorate within our lifetime. We walk a thin edge, unseeing.

What urges, impedes the deep currents? Whose voices do deep waters heed? Global truths reside in tiny sea animals. The proportion of heavy to light oxygen atoms in their seashells mirrors that of heavy and light oxygen in the water. As earth cools, heavier oxygen rains out first, into the sea, and lighter oxygen condenses into the ice and snow of high-latitude glaciers. When earth locks into an ice age, sea and shells grow rich in heavy oxygen. When earth warms and ice melts, lighter oxygen floods the sea and materializes in seashells. Changing proportions of heavy and light oxygen in shells record pulses of fresh water pouring into the sea from melting glaciers, pulses timed to the withdrawal of deep currents.

Atlantic's deep currents are exquisitely sensitive to the levels of salt in the water. They can be weakened by heavy rains or water surging from melting glaciers and pack ice. A flood of fresh water dilutes salty Atlantic, enfeebling deep circulation. At the end of the last ice age, when the sun shone more brightly on the northern hemisphere and the warm Gulf Stream awakened, glacial meltwater pooled in southern Canada, forming a great lake reaching through Manitoba and Saskatchewan, North Dakota and Minnesota. Lake Agassiz, blocked by ice to the north and east, drained through the Mississippi River and out into the Gulf of Mexico. As earth warmed, ice to the east of the lake receded, and 11,000 years ago it uncorked the Gulf of St. Lawrence. Lake levels suddenly plunged 120 feet (37 meters), flooding Atlantic's northern regions with fresh, buoyant water. The effect was cataclysmic; Atlantic's cold, deep water current stalled, and earth's warming ceased. With the sea's heat turned off, earth cooled and once again ice draped northern latitudes. Atlantic's deep circulation shut down for 200 years; the cold lasted for 1000.

The sun continued to shine, and eventually the ice began to melt again. Lake Agassiz still existed in southern Canada; it was larger than the Great Lakes of the United States. When the last of the ice clogging Hudson Bay melted 8200 years ago, the lake suddenly drained again, leaving behind a few remnants scattered throughout Manitoba. Again the release of vast amounts of fresh water proved catastrophic. The lake poured out into Atlantic, stalling the deep circulation, precipitating another retreat into the cold.

Ice melting in the high glaciers of Atlantic may once again flood the sea with fresh water. The waters of Lake Agassiz are minuscule compared with what is frozen in the glaciers of Greenland and Antarctica. Together these glaciers hold three quarters of earth's fresh water. The Greenland glaciers, long considered a fixed, unchanging part of the landscape, insensitive to changes in temperature, are proving otherwise. Scientists studying cores of Greenland ice 100,000 years old, realize that when earth last experienced a warm period, Greenland's thawing glaciers raised sea level twelve or sixteen feet (four or five meters). They are thawing again. At higher elevations still untouched by summer sun, the ice stands firm, but along 70 percent of the coast it is thinning, in some places by as much as three feet (one meter) each year. No one can yet say why. Higher temperatures can thaw a glacier, and water melting at the surface can percolate down through the ice, lubricating it, smoothing and easing the way to the sea. These conditions explain some melting, but not all. Whatever the cause, Greenland now delivers 1.8 trillion cubic feet (5.1 cubic kilometers) of melting ice to Atlantic each year, little by little raising sea level. Lands once frozen are now losing substance to the liquid sea, and as the glaciers shrink, their fresh water may dampen Atlantic's deep-water circulation.

Melting glaciers carry fresh water into the sea, but there are other sources as well. One is the Pacific, returning water originally given by Atlantic. Trade winds evaporate water from the tropics and blow it across Central America. The salty residue left behind spawns Atlantic's deep currents, while Atlantic water taken up by wind rains into the Pacific, freshening that sea, lowering the salinity by a seemingly insignificant one fifth of a percent. This difference, too small for us to perceive without sensitive instruments, drives the circulation of an entire ocean. Buoyed by lighter, fresher water, the Pacific rides one foot (0.3 meter) higher than Atlantic. The water runs downhill through the Bering Strait, freezes into the Arctic ice pack, and eventually melts out into Atlantic's northern waters. It has been doing so for thousands of years.

Visible markers have traced the path. In 1879, Arctic pack ice trapped an American vessel, the *Jeanette*. The boat fractured under the press of the ice, and the wreckage drifted across the North Pole, washing up on the shores of southern Greenland five years later. More recently, in 1992, a rough storm swept a large container of bath toys from a cargo ship crossing the Pacific. Thousands of plastic ducks, frogs, turtles, and beavers scattered with wind and wave. Approximately ten thousand rode the currents north and floated through the Bering Strait. Frozen into the drifting pack ice, they are crossing the Arctic and melting out into waters off Greenland, where perhaps they will catch a ride on polar currents flowing south. Eventually, they may show up on the rocky coast of Newfoundland and the beaches of New England, faded from the long journey. Visible passengers crossing the Bering Strait are rare. Most often the currents glide through unseen.

Today, the Bering Strait is a narrow and shallow passage, only 50 miles (80 kilometers) wide and 150 feet (45 meters) deep. During the ice ages, when snow and rain froze into glaciers, lowering sea level, it dried up, creating a land bridge connecting Asia to North America. Some of America's early human settlers, hunter-gatherers from Asia, arrived via this land bridge over 12,000 years ago. During times of warming, when copious rains fall, the Bering Strait grows deep and wide and fills with rushing water. When earth last warmed, between 145,000 and 110,000 years ago, sea level rose fifteen feet (five meters), inundating the Bering Strait with fresh water. The water flowed through the Arctic and into Atlantic, diluting the Norwegian Sea, weakening the deep circulation, giving earth a push into the last ice age. It could do so again.

As the glaciers recede, so does the winter sea ice, further buoying the ocean with fresh water. Each winter, as the northern hemisphere tilts toward darkness, the Arctic ice pack, about 3 million square miles (7.8 million square kilometers), doubles, reaching down through the Canadian Archipelago, around Baffin Island and Greenland, and into Hudson Bay. As winter turns to spring and summer, the ice shrinks back. In recent winters, more and more open sea resists the grasp of ice. Many an explorer has sought and failed to find the fabled Northwest Passage. Iced in for hundreds, perhaps thousands of years, this shortcut from Europe to Asia is opening. In the summer of 1999, surprised Inuit fishermen watched a submarine surface in a watery hole in the ice off Baffin Island. In the summer of 2000, a cruise ship making an annual excursion to the North Pole found that the far end of earth had melted. Gulls wheeled overhead, drifting plants grew where the sun shone through thin ice. Passengers expecting to walk across the North Pole couldn't.

Data gathered by passing satellites indicate that the reach of Arctic ice has declined 3 percent in the last two decades. The ice itself has thinned as well. Multiyear ice, the thick ice that accumulates each season, has declined by 14 percent. Observers on submarines view ice from another perspective, peering up through the dark water underneath. Their data show that in the last three decades Arctic ice volume has declined by 40 percent. Some scientists, analyzing the losses and finding them too large to be accounted for by natural variations in wind, water, and temperature, ascribe the Arctic thaw to global warming. Others predict that if present trends continue, the Arctic will be free of ice during the summers within the next fifty years.

We know so little about the Arctic, this forbidding sea linking two oceans, but her waters are stirring beneath the ice. Today, this distant region is undergoing monumental change, warming to the highest temperatures seen in the last 400 years. Inuit fishermen know; for the first time they are catching coho and sockeye salmon in the Beaufort Sea. Polar bears are stressed by the heat. Melting ice floes in Hudson Bay are forcing them onto the tundra earlier and earlier in the season. Unable to fatten up on the seals they catch during their stay on the floating ice, the bears are losing weight and bearing fewer cubs. Farther south, the International Ice Patrol reported in 1999 that, for the first time in 85 years, Iceberg Alley was free of ice. Icebergs shed off the coast of Greenland drift south on the currents past Labrador. Usually five hundred or so slip into the Grand Banks shipping lanes each year, but now the water is too warm. If the *Titanic* had sailed in 1999, she would have made the crossing. Scientists clocking the speed of low-frequency sound waves travelling through the Arctic confirm the warming.

The warm water comes from Atlantic. A layer of Atlantic water flows in from the Fram Strait, off northwest Greenland. It winds around the Arctic, beneath the sea surface, around basins bounded by ridges of long-extinct volcanoes, and then exits back into Atlantic by the same route. Recently the water has warmed and shallowed, thinning the sea ice. It has also strengthened, spreading farther into the Arctic, pushing aside water circulating in from the Pacific. Scientists are quite concerned about where the displaced Pacific water is going. Fresh Pacific water escapes into Atlantic along two routes, shooting through the Fram Strait into the Greenland Sea, or winding through islands of the Canadian archipelago into the Labrador Sea. At both sites deep currents are born. At both sites deep water is suppressed by infusions of fresh water.

It is already possible that warming Atlantic water has inhibited the sea's deep currents. Salinities in deep water flowing south of the Faeroe Islands are dropping, more than they have in the last hundred years, and the water is warming. In the winter of 1996–97, water cascading over the Denmark Strait warmed to the highest temperatures ever recorded there, and the bottom layer thinned by 80 percent. We are poised on the edge of an unknown.

Years of research implicate Atlantic's northern deep current as the Achilles' heel of the sea's circulation. When the current falters, dry land shivers. The fresh water that slows deep currents may originate in melting glaciers or sea ice, or water coursing in from the Bering Strait. In a warming world, it may also come with rain. The burning of fossil fuels and the cutting down of rain forest has already warmed earth by 1 degree Fahrenheit (0.6 degree Celsius) in the last one hundred years. Scientists predict that global warming will raise the planet's temperature by another 2.7 to 10.8 degrees Fahrenheit (1.5 to 6.0 degrees Celsius) by 2100. A warm planet receives more rain. A warmer ocean evaporates more water, a warmer atmosphere holds more moisture and releases more rain. We have already begun to feel the watery effects of earth's rising temperature. Within the last hundred years, the number of intense, heavy downpours soaking and flooding areas of the United States has increased by 20 percent.

Yet the rain we see is only a tiny piece of earth's water cycle, which by and large remains veiled. The atmosphere holds very little of earth's water, only .001 percent, but where that water falls determines the balance of salt in the sea, the shape and strength of currents, the climate. Most of earth's water, 97 percent, resides in the sea. Most water entering the atmosphere evaporates from the sea, and most of earth's rain falls there, invisibly, without measure, without mark. Our very survival depends on the transport of water between sea, air, and land, but we know little about earth's water cycle. Evaporation exceeds precipitation in the equatorial Atlantic, but no one can say with certainty by how much. The best estimates differ by an amount equal to thirty times the flow of the Mississippi River. What we can't see in the ocean has monumental consequences on land. If only 1 percent of what we believe to be the rain falling in Atlantic blows west into the Mississippi watershed, the amount of water flowing through that river doubles.

Rocks strewn throughout the Mississippi River basin suggest that these calculations are not merely hypothetical. Geologists piecing together the history of flooding along the Mississippi over the last 7000 years find that only slight changes in temperature or rainfall, smaller than those occurring under global warming, triggered frequent and catastrophic flooding. Huge boulders three feet (one meter) in diameter have been swept down the river, whose floodplain has drowned in water sixteen feet (five meters) deep. This past may be reawakening in the present. In 1993, the Mississippi River, swollen with torrential rains, rose and breached its banks, flooding out homes and highways, causing millions of dollars of damage. Europe also feels the intensification of earth's water cycle. In 1995, heavy storms, which normally rain out at sea, swept in off Atlantic, drenching England, France, and the Netherlands, filling rivers to overflowing, forcing thousands of people to abandon their homes.

Scientists cannot yet measure rain falling in the ocean, but they know the effects are considerable. Pulses of fresh water, from melting glaciers and sea ice and from rising waters in the Bering Strait, have shut down Atlantic's deep circulation in the past. Those moments from Atlantic's history speak of the potential consequences today, in our time, of disturbing the atmosphere and sea with global warming. In what may prove to be a monumental imposition of man's will upon nature, we have conspired to create the conditions for earth to receive an increased abundance of fresh water. Whether it comes from rain, melting Greenland glaciers, disappearing sea ice, or some combination, there may be enough additional fresh water flowing into Atlantic to weaken and collapse the deep circulation within the next hundred years, and perhaps sooner.

And if the deep circulation of Atlantic is extinguished? Or if the current shallows, and surface waters follow ancient tracks east, to Portugal, instead of north? Water flowing up from the tropics, fuelling deep currents emanating from the Greenland and Labrador Seas, releases precious heat to blow through Europe, warming that continent by 9 degrees Fahrenheit (16 degrees Celsius). Inundated with fresh water, Atlantic's northern deep current may weaken, and shallow. Then Europe's heat pump, which has run steadily for 8000 years, may turn off, plunging mild Dublin into the icy, lonely cold of inhospitable Svalbard, 700 miles (1100 kilometers) away from the North Pole. Having no collective memory of this past, we cannot imagine its reemergence, despite the record before us. Arctic chills have hit earth suddenly, and then lingered. Climate models suggest that the sea might adjust to greater infusions of carbon dioxide, but only if the increases come slowly, not at today's frenetic pace.

A warming earth may diminish Atlantic's deep circulation. Whether the slowing or shutdown of deep currents could further accelerate warming, scientists cannot fully say. Of the six billion tons of carbon dioxide man expels into the atmosphere each year, half remains there; the rest disappears, dissolved in ocean currents, absorbed by the sea's floating meadows, inhaled by the grasses and trees of dry land. Soaking up carbon dioxide, the sea buffers earth's climate against increased emissions of greenhouse gases. Slowing the deep circulation may weaken this buffer. A portion of the sea's carbon dioxide falls to the depths with currents. More is breathed in by plants, or built into the shells of animals, before descending to the seafloor in their remains. The tiny grasses of the sea that anchor the marine food web bloom profusely in seawater bathed in nutrients. Nutrients drift toward the bottom; winds and rising currents lift them from the depths. When deep currents slacken, nutrients vital to the health of surface dwellers may stay cloistered on the dark bottom. The boundary between layers of surface water and the cold fertile sea below hardens as surface waters warm, and winds fail to penetrate the barrier. If deep-water circulation in North Atlantic waters collapses, the sea's ability to absorb carbon dioxide from the atmosphere may decline by as much as one third or one half.

The sea feels the dimming of deep currents, but science has only begun to grasp the fine workings, only begun to decipher the delicate balances between carbon dioxide drawn down by currents and that imbibed by floating plants. Climate models cannot fully describe how the biology of the sea responds to a warming earth, or how melting glaciers in Greenland, thawing sea ice in the Arctic, and rising seawaters in the Bering Strait combine to influence deep circulation. Our computers are not fast or powerful enough to capture all the complexities that make up a planet imbued with life.

It has been argued that man's contribution to the carbon dioxide content of the atmosphere pales in comparison with the immense volume pulled from the air through photosynthesis and returned through respiration and decomposition. Each year, earth and her atmosphere exchange almost 200 billion tons of carbon dioxide. However, tiny quantities can loom large. Just as barely discernible fluxes in salinity command the currents, tiny additions of carbon dioxide can raise earth's thermostat. Records of climate contained in air bubbles trapped in Antarctic ice show that for the last 160,000 years, earth's carbon dioxide levels and temperature rose and fell in concert. Temperatures ranged over approximately 12 degrees Fahrenheit (22 degrees Celsius), a fluctuation far greater than any we have known in the last 10,000 years, far greater than anything we might find comfortable. If present trends continue, man will fill the atmosphere with more carbon dioxide than earth has seen in 50 million years.

Thousands of years of climate change embedded in the fossils of the deep sea suggest that whether earth is buried in an ice age or basking in a brief interlude of warmth, the planet is feverish, heating and cooling rapidly, with deserts turning to grassland and forests to tundra and back again. The rains and waters that have sustained human civilization for so long answer the call of the sea. Small motions of Atlantic influence life and death in the Sahel, civilization and famine in the Tigris–Euphrates Valley. Larger motions open and close Greenland and Iceland to settlement, provide fields of plenty or starvation and disease in Europe. Still larger motions summon hundreds, thousands of years of ice. No matter how poorly or dimly we understand these cycles, we feel their presence.

It is likely that man's arrival on this planet was a response to the rhythms of the sea. *Australopithecus* lived on in the woodlands of Africa for one and a half million years, climbing trees to find leaves, fruit, and seeds to eat during the day and seeking refuge from hungry leopards, hyenas, and dogs at night. Nimble *Australopithecus* babies scampered among the branches at an early age; their tree-climbing mothers could not carry them. Generation after generation of *Australopithecus* survived, untouched by evolution, until, on the other side of Atlantic, the Isthmus of Panama rose from the sea. Before that time, Atlantic and Pacific waters mixed as trade winds blew water across the Caribbean into the Pacific. The rising isthmus blocked water but not wind, freshening the Pacific with rain blowing in on the trades and leaving a salty Atlantic to initiate the circulation of deep, cold currents. Warm water pulled north in the Gulf Stream, sinking in the Labrador and Greenland Seas, released heat and moisture into the air. When earth swayed in her orbit, it snowed and snowed, freezing the Arctic, beginning the ice ages.

As ice descended from the Poles, as the tundra crept into Europe and the dunes of the Sahara swept south, the African woodlands shrank, giving way to grassland, forcing *Australopithecus* down from the trees. Some *Australopithecus* mothers, not needing all their limbs to climb, carried their babies. These babies matured more slowly. They couldn't climb or clamber or fend for themselves as their tree-climbing forebears had, but while they rested in their mothers' arms their brains grew to an unprecedented size. The abrupt change in climate and habitat created an evolutionary opportunity. Slow-moving *Australopithecus* became easy prey for swift meat-eating predators; *Homo* compensated with its large brain. Within a split second of geologic time, in little more than 100,000 years, *Australopithecus* died out. *Homo* survived, filling a niche created by a change in climate wrought by the sea.

The sea brought us here. We have adjusted to its more gentle rhythms. As ocean-born rains deliver or withhold life-giving water, our civilizations and cultures flourish and fade. Now, toying with our atmosphere, we break the rhythms of the sea, nudging the climate, ignorant of whether we can adapt to the new niche we are creating. It may be an arrogant gamble.

DAVID MALOUF

Fly Away Peter

Imogen Harcourt, still carrying her equipment – camera, plates, tripod – as she had once told Jim, 'like the implements of martyrdom,' made her way down the soft sand of the dunes towards the beach.

A clear October day.

October here was spring. Sunlight and no wind.

The sea cut channels in the beach, great Vs that were delicately ridged at the edges and ribbed within, and the sunlit rippled in them, an inch, an inch and a half of shimmering gold. Further on, the surf. High walls of water were suspended a moment, held glassily aloft, then hurled themselves forward under a shower of spindrift, a white rush that ran hissing to her boots. There were gulls, dense clouds of them hanging low over the white-caps, feeding, oystercatchers darting after crabs, crested terns. A still scene that was full of intense activity and endless change.

She set down her equipment – she didn't intend to do any work; she carried all this stuff by force of habit and because she didn't like to be separated from it, it was all she had, an extension of herself that couldn't now be relinquished. She eased the strap off her shoulder, set it all down and then sat dumpily beside it, a lone figure with her hat awry, on the white sands that stretched as far as the eye could see, all the way to the Broadwater and the southern tip of Stradbroke in one direction and in the other to Point Danger and the New South Wales border. It was all untouched. Nobody came here. Before her, where she sat with her boots dug in and her knees drawn up, was the Pacific, blue to the skyline, and beyond it, Peru.

'What am I doing here?' she asked herself, putting the question for maybe the thousandth time and finding no answer, but knowing that if she were back in Norfolk there would be the same question to be put and with no answer there either.

'I am doing,' she told herself firmly, 'what those gulls are doing. Those oystercatchers. Those terns.' She pulled her old hat down hard on her curls.

The news of Jim's death had already arrived. She heard it by accident in the local store, then she heard it again from Julia Crowther, with the news that Ashley Crowther had been wounded in the same battle, though not in the same part of the field, and was convalescing in England. Then one day she ran into Jim's father.

'I lost my boy,' he told her accusingly. He had never addressed her before.

'I know,' she said. 'I'm very sorry.'

He regarded her fiercely. She had wanted to say more, to say that she understood a little of what he might feel, that for two whole days after she heard she had been unable to move; but that would have been to boast of her grief and claim for herself something she had no right to and which was too personal to be shared, though she felt, obscurely, that to share it with this man who was glaring at her so balefully and with such a deep hatred for everything he saw, might be to offer him some release from himself and to let Jim, now that he was dead, back into his life. What did he feel? What was his grief like? She couldn't tell, any more than he could have guessed at or measured hers. She said nothing. He didn't invite sympathy. It wasn't for that that he had approached her.

She sat on the beach now and watched the waves, one after another, as they rose, gathered themselves, stood poised a moment holding the sun at their crests, then toppled. There was a rhythm to it. Mathematics. It soothed, it allowed you, once you had perceived it, to breathe. Maybe she would go on from birds to waves. They were as various and as difficult to catch at their one moment.

That was it, the thought she had been reaching for. Her mind gathered and held it, on a breath, before the pull of the earth drew it apart and sent it rushing down with such energy into the flux of things. What had torn at her breast in the fact of Jim's death had been the waste of it, all those days that had been gathered towards nothing but his senseless and brutal extinction. Her pain lay in the acute vision she had had of his sitting as she had seen him on that first day, all his intense being concentrated on the picture she had taken of the sandpiper, holding it tight in his hand, but holding it also in his eye, his mind, absorbed in the uniqueness of the small creature as the camera had caught it at just that moment, with its head cocked and its fierce alert eye, and in entering that one moment of the bird's life – the bird was gone, they might never see it again – bringing up to the moment, in her vision of him, his own being that was just then so very like the birds, alert, unique, utterly present.

It was that intense focus of his whole being, it's *me*, Jim Saddler, that struck her with grief, but was also the thing – and not simply as an image either – that endured. That in itself. Not as she might have preserved it in a shot she had never in fact taken, nor even as she had held it, for so long, as an untaken image in her head, but in itself, as it for its moment was. That is what life meant, a unique presence, and it was essential in every creature. To set anything above it, birth, position, talent even, was to deny to all but a few among the infinite millions what was common and real, and what was also, in the end, most moving. A life wasn't *for* anything. It simply was.

She watched the waves build, hang and fall, one after the other in decades, in centuries, all morning and on into the early afternoon;

and was preparing, wearily, to gather up her equipment and start back – had risen in fact, and shouldered the tripod, when she saw something amazing.

A youth was walking – no, running, on the water. Moving fast over the surface. Hanging delicately balanced there with his arms raised and his knees slightly bent as if upheld by invisible strings. She had seen nothing like it. He rode rapidly towards her; then, on the crest of the wave, sharply outlined against the sky, went down fast into the darkening hollow, fell, and she saw a kind of plank flash in the sunlight and go flying up behind him.

She stood there. Fascinated. The youth, retrieving the board among the flurry of white in the shallows, knelt upon it and began paddling out against the waves. Far out, a mere dot on the sunlit water, where the waves gathered and began, she saw him paddle again, then miraculously rise, moving faster now, and the whole performance was repeated: the balance, the still dancing on the surface, the brief etching of his body against the sky at the very moment, on the wave's lip, when he would slide into its hollows and fall.

That too was an image she would hold in her mind.

Jim, she said to herself, *Jim*, *Jim*, and hugged her breast a little, raising her face to the light breeze that had come with afternoon, feeling it cold where the tears ran down. The youth, riding towards her, was blurred in the moment before the fall.

She took up her camera and set the strap to her shoulder. There was a groove. She turned her back to the sea and began climbing the heavy slope, where her boots sank and filled and the grains rolled away softly behind. At the top, among the pigweed that held the dunes together, she turned, and the youth was still there, his arms extended, riding.

It was new. So many things were new. Everything changed. The past would not hold and could not be held. One day soon, she might make a photograph of this new thing. To catch its moment, its brilliant balance up there, of movement and stillness, of tense energy and ease – that would be something.

This eager turning, for a moment, to the future, surprised and hurt her.

Jim, she moaned silently, somewhere deep inside. *Jim. Jim.* There was in there a mourning woman who rocked eternally back and forth; who would not be seen and was herself.

But before she fell below the crest of the dunes, while the ocean was still in view, she turned and looked again.

NOTES ON THE AUTHORS

Jennifer Ackerman was born in the United States in 1959. A journalist and essayist, she is the author of four works of non-fiction, most recently *Ah-Choo! The Uncommon Life of Your Common Cold*, as well as numerous articles for *National Geographic Magazine*. For more information visit jenniferackerman.net

Emily Ballou was born in the United States in 1968 and moved to Australia in 1991. An acclaimed novelist, poet and screenwriter, she is the author of two novels, *Father Lands* and *Aphelion*, and a children's book, *One Blue Sock*. 'The Beach' is taken from her award-winning book of poems about the life of Charles Darwin, *The Darwin Poems*. For more information visit emilyballou.com

William Beebe was born in Brooklyn in 1877, and developed a fascination with natural history at a young age. Appointed as the New York Zoological Society's Curator of Ornithology in 1899, he travelled widely through Mexico, Asia and the Galapagos Islands as a young man, producing a string of books and articles about his discoveries and experiences. In 1928 Otis Barton, the inventor of the bathysphere, approached him in the hope Beebe might be able to help him secure the financial and logistical support necessary to see Barton's plans to fruition. Beebe agreed, becoming an equal partner in the project. On 6 June 1930 the two men made the first manned dive in the bathysphere, descending to a depth of 245m. This dive was followed by another on 15 August, when the two men descended to the then-unthinkable depth of 923m. In the years after the dive Beebe continued to work on a variety of projects, not the least of which was the establishment of the New York Zoological Society's Tropical Research Station in Trinidad, where he died in 1962.

Judith Beveridge was born in London in 1956, and immigrated to Australia with her parents in 1960. She is the author of four books of poetry, *The Domesticity of Giraffes*, *Accidental Grace*, *Wolf Notes* and *Storm and Honey*, and has won a number of major poetry awards, including the Dame Mary Gilmore Award, the NSW Premier's Award, the Victorian Premier's Award, the Judith Wright Calanthe Poetry Prize, the Grace Leven Prize and the Josephine Ulrick Poetry Prize. She lives in Sydney.

Elizabeth Bishop was born in Massachusetts in 1911. While she received many awards and honours during her life – not the least of which was a period as Poet Laureate of the United States – in the years since her death in 1977 her reputation has continued to grow, and she is now regarded as one of the most important American poets of the 20th century. 'At the Fishhouses' is one of several poems central to her *oeuvre* that centre upon marine metaphors.

Edmund Burke was born in Ireland in 1729. One of the most significant and controversial figures of the second half of the eighteenth century, he was the author of many highly influential works of philosophy and political theory, perhaps most importantly his *Reflections on the Revolution in France* and *A Philosophical Enquiry into the Origin of Our Ideas of the Sublime and Beautiful.* He died in 1797.

Fiona Capp was born in Melbourne in 1963. She is the author of three novels, *Night Surfing*, *Last of the Sane Days* and *Musk and Byrne*, and two books of non-fiction, *Writers Defiled* and *My Blood's Country*, and a memoir, *That Oceanic Feeling*, exploring her love of the ocean and surfing. Capp lives in Brunswick, Melbourne, with her partner and son.

Rachel Carson was born in Pennsylvania in 1907 and trained as a marine biologist at Johns Hopkins University. Her first book, *Under the Sea Wind*, was published while she was still employed by the US Bureau of Fisheries, but its success, and the success of her essays and articles convinced her to begin writing full-time. This process led to the publication of her now classic exploration of the ocean and its history, *The Sea Around Us*, in 1951, and her groundbreaking study of the environmental effects of DDT, *Silent Spring*, in 1962. She died in 1964.

Owen Chase was born on Nantucket Island in 1796. In November 1820 he was first mate on the whaleship *Essex* when she was attacked by a sperm whale and sunk. His ghost-written account of the *Essex*'s sinking and the surviving crew's three month ordeal on the open ocean was published a year later, creating a small sensation, largely because of its descriptions of cannibalism in the latter stages of the crew's voyage. Despite republication in recent years it is now principally remembered as one of the sources for *Moby-Dick*. Owen Chase died on Nantucket Island in 1869.

Apsley Cherry-Garrard was born in Bedford in 1886. In 1910 he joined Robert Falcon Scott's second, and final expedition to Antarctica, an experience that formed the basis for his book *The Worst Journey in the World*. He died in 1959.

Samuel Taylor Coleridge was one of the central figures of English Romanticism. Born in Devon in 1772, his extraordinary intellectual and poetic gifts were evident from a young age, but it was the work that grew out of his friendship and creative partnership with William Wordsworth, and in particular *The Rime of the Ancient Mariner*, which established him as one of the most significant writers of his day. In composing *Mariner* Coleridge drew upon a number of sources, most obviously Cook's luminous description of his journey to the Antarctic and Bligh's account of his journey by open boat from Tonga to Timor after the *Bounty* Mutiny, but it is also clear Coleridge was familiar with Hakluyt, and more particularly the story of Captain John Davis and the crew of the *Desire*. As Bruce Chatwin points out, 'John Davis and the Mariner have these in common: a voyage to the Black South, the murder of a bird or birds, the nemesis which follows, the drift through the tropics, the rotting ship, the curses of dying men.' One of the enduring ironies of *Mariner*, a poem now seen as central to the literature of the ocean, is the fact that, save for a short trip on the Chepstow ferry (which he famously dismissed as 'exceedingly' disappointing), its author had never been to sea at the time of its composition. Coleridge died in 1834.

Joseph Conrad was born Józef Teodor Konrad Korzeniowski in Berdichev (now in the Ukraine) to Polish parents. When Conrad was four Russian authorities exiled his father – a poet, translator and Polish nationalist – for his political activities. After the premature deaths of both his parents, Conrad was raised by an uncle in Krakow, before becoming a seaman at the age of 16. For the next twelve years Conrad served on ships bound for the French Antilles and South America, before joining the British Merchant Service in 1878. In 1886, the year he became a British subject, he began writing in English, his third language, and over the next 35 years produced a string of extraordinary novels and two important autobiographical works, many of which drew heavily upon his experiences at sea. He died in England in 1924.

James Cook was born in Yorkshire in 1728. After being apprenticed to a Whitby shipowner at the age of 18, Cook became an officer in the merchant navy, before joining the Royal Navy in 1755. In 1766 the Royal Society commissioned Cook to lead an expedition to observe the Transit of Venus, the first of three voyages Cook would make in the years that followed, each of which added immeasurably to European knowledge of the Pacific and Southern Oceans. Cook kept extensive journals during his voyages. Initially these were intended for the Admiralty's use, but after reading the authorised account of his first voyage written by Dr John Hawkesworth Cook demanded – and received – permission to publish under his own name. Despite this Cook never saw his own words in print, since his account of his Second Voyage, *A Voyage to the South Pole and Round the World*, was not published until after he had departed on his third, and final voyage. Cook was killed in Hawaii in 1779. The excerpts in this book are reproduced from J.C. Beaglehole's edition of Cook's original journals, first published in 1961.

Jacques Cousteau was born in 1910 in France. He joined the French Navy in 1930, and in 1936 began the work that would lead to his invention of the aqualung. After World War II Cousteau became a celebrated filmmaker and began to explore the possibilities for underwater exploration offered by the aqualung. *The Silent World* is his account of these years. Cousteau died in 1997.

Deborah Cramer was born in Schenectady, New York, in 1951. She is the author of two books, *Great Waters: An Atlantic Passage* and *Smithsonian Ocean: Our Water Our World.* Currently a Visiting Scholar at MIT's Earth System Initiative, she lives in Massachusetts. For more information visit deborahcramer.com

Hart Crane was born in Ohio in 1899, and after an unhappy childhood moved to New York in 1917. Despite great personal and financial difficulties, over the next sixteen years he produced some of the most influential poetry of the 20th century, much of which drew upon images of water and the ocean. He committed suicide in 1932 by leaping off a steamship in the Gulf of Mexico. His body was never recovered.

Stephen Crane was born in New Jersey in 1871. After leaving school he worked as a journalist and published his first novel, *Maggie: A Girl of the Streets*, in 1893, though it was not until the publication of his second novel, *The Red Badge of Courage*, in 1895 that he found a wide audience. In 1896 while on a trip to Cuba his ship sank, and he spent several nights at sea in a lifeboat, an experience that was to provide the basis for his story, 'The Open Boat'. Despite worsening health Crane continued to travel and write until his death from tuberculosis in 1900.

Charles Darwin was born in Shropshire in 1809. After abandoning medical school he attended Christ Church College, in Cambridge as the first step towards life as an Anglican parson. However, after discovering an interest in natural history Darwin was recommended to Robert FitzRoy, captain of the *Beagle*, as a suitable candidate to accompany him on a journey to chart the coastline of South America. Despite his father's objections Darwin accepted the post, and in 1831 set off on a five year expedition with FitzRoy. Bolstered by his letters home, Darwin's reputation in scientific circles began to grow even before the *Beagle's* return to England in 1836, but his reputation was cemented by the publication of his account of his travels, *The Voyage of the Beagle*, in 1839, which was a critical and commercial success. Over the next twenty years Darwin worked steadily developing and refining the ideas that would eventually form the basis of his Theory of Evolution by Means of Natural Selection. Darwin died in 1882.

Luke Davies was born in Sydney in 1962. He is the multi-award-winning author of three novels, *Candy*, *Isabelle the Navigator*, and *God of Speed*, four volumes of poetry, a children's book, *Magpie*, and co-wrote the script for the feature film *Candy*. Since 2008 he has been living and working in the United States, in which time he has written and directed two short films, *Air* and *The Imbecile*. 'Diving the SS Coolidge' is taken from his second volume of poetry, *Running with Light*.

Daniel Duane was born in California in 1967. He is the author of numerous books, including the novel *Looking for Mo*, the memoir *Lighting Out: A Vision of California and the Mountains*, and the contemporary classic, *Caught Inside: A Surfer's Year on the California Coast*. He lives and surfs in San Francisco, California. For more information visit danielduane.com

William Falconer was born the son of a barber in Edinburgh in 1732. He left home to become a sailor, and after escaping drowning in a shipwreck in Greece, wrote and published his first poem, *The Shipwreck*, in 1762. With the patronage of the Duke of York, he commenced work on his *Universal Dictionary of the Marine*, which was the first sea-dictionary and remains one of our principal sources of understanding of life at sea in the 18th century. Shortly after its publication in 1769 the warship *Aurora*, in which he was travelling, left Cape Town and was lost with all hands, presumably while rounding the Cape of Good Hope.

Thomas Farber was born in Boston in 1944. The author of two novels, *The Beholder* and *Curves of Pursuit*, four volumes of short fiction, and a number of highly acclaimed works of non-fiction and critical writing, including the now-classic *On Water*. He divides his time between California and Hawaii. For more information visit thomasfarber.org

Richard Hakluyt was born in 1552 or 1553. The son of a merchant, he was educated at Westminster School and Christ Church, Oxford, and became fascinated by navigation and exploration as a young man. As well as working as a diplomat and spy, he edited and translated many volumes of first-hand narratives of adventure and discovery, including his *Divers Voyages*, *Principal Navigations, Voyages, and Discoveries of the English Nation*, and *The Principal Navigations, Voiages, Traffiques and Discoueries of the English Nation*, both of which were instrumental in promoting and supporting the exploration of the Americas by the English. The excerpt from 'The Last Voyage of Richard Cavendish' reproduced in this book is an account of the experience of Captain John Davis and his crew after they were separated from Cavendish's ship near the Straits of Magellan in 1592. Richard Hakluyt died in 1616.

James Hamilton-Paterson was born in London in 1941. He is the author of ten novels, the most recent of which is *Rancid Pansies*, several children's books and two volumes of poetry, as well as six highly acclaimed works of non-fiction, including the now-classic *Seven Tenths: the sea and its thresholds*. He lives in Italy.

Kevin Hart is a poet, literary critic, philosopher and theologian. Born in England in 1954 he immigrated to Australia with his family in 1965. His most recent books of poems are *Young Rain* and *Flame Tree: Selected Poems*. He has just completed a new collection of poems entitled *Morning Knowledge*. Other recent books include *The Exorbitant: Emmanuel Levinas between Jews and Christians*, *Clandestine Encounters: Philosophy in the Narratives of Maurice Blanchot*, and *Counter-Experiences: Reading Jean-Luc Marion*. He holds a Chair in the Department of Religious Studies at the University of Virginia.

Ernest Hemingway was born in Illinois in 1899. He served as an ambulance driver on the Italian Front in World War I, where he was seriously injured. In 1922 he moved to Paris, where he began to publish and became friends with figures such as F. Scott Fitzgerald and Gertrude Stein. Over the next 35 years he was present for many of the pivotal moments in European history, spending time in Spain during the Civil War and attaching himself to the 22nd Regiment of the US Army for the D-Day landings. *The Old Man and the Sea* was published in 1952 and is believed to have played an important part in securing him the Nobel Prize for Literature in 1954. Plagued by declining health and debilitating bouts of depression and paranoia, Hemingway committed suicide in 1961.

Thor Heyerdahl was born in Norway in 1914, and studied Zoology at the University of Oslo. In 1947 he set out to prove his theory that the Polynesian people originated in South America by sailing a raft from Peru to the Tuamotu Islands. While Heyerdahl's theories are now discredited, his account of that journey, *The Kon-Tiki Expedition*, remains one of the great depictions of ocean voyaging. Heyerdahl died in 2002.

Sebastian Junger is an American journalist, author and filmmaker. Born in 1962 he is the author of three books, *The Perfect Storm*, *A Death in Belmont* and *Fire*. He lives in New York City.

William Langewiesche was born in 1955. The author of four books, the most recent of which is *Fly by Wire: The Geese, The Glide and the Miracle on the Hudson*, and a number of highly celebrated works of journalism, he is currently *Vanity Fair's* International Correspondent.

Nam Le was born in Vietnam and raised in Australia. His first book, *The Boat*, won a string of major literary awards and was translated into thirteen languages. Le is the fiction editor of the *Harvard Review*.

Wayne Levin is an American photographer. Born in Los Angeles in 1945 he developed a fascination with photography after his father gave him a Brownie camera for his 12th birthday. After leaving school he enrolled in the Brooks Institute of Photography in Santa Barbara, but left in 1964 to become part of the Civil Rights Movement. After his discharge from the US Navy he joined his family in Hawaii, and later studied photography at the San Francisco Art Institute and the Pratt Institute. In 1982 he began a photographic study of surfers, and in the years since then has become recognized around the world for his studies of the ocean and its inhabitants. He has published three books of his photography, *Through a Liquid Mirror*, *Other Oceans* and *Akule*, and his photographs are held by many major galleries, including the Dayton Institute in Ohio and the Museum of Modern Art in New York. He lives in Hawaii. For more information visit waynelevinimages.com

David Malouf was born in Brisbane in 1934. The author of ten novels, most recently *Ransom*, several books of short fiction and a number of collections of poetry and non-fiction, he is one of Australia's most celebrated writers. His work draws heavily upon the landscapes of his childhood in Queensland.

Matthew Fontaine Maury was an American oceanographer and geographer. Born in Virginia in 1806, he joined the US Navy as a midshipman in 1825, where he became fascinated by navigation and oceanography. After breaking his hip in a stagecoach accident in 1839 he took up a position at the United States Naval Observatory, where he remained until he resigned his commission as a United States Naval Officer to join the Confederacy in 1861. *The Physical Geography of the Sea and its Meteorology*, first published in 1855, is widely regarded as having established the modern science of oceanography. He died in 1873.

Herman Melville was born in New York City in 1819. Despite coming from a well-established Boston family, Melville's father's business suffered continuous setbacks. In 1830 the family moved to Albany in a last-ditch attempt to rescue their fortune, but within two years Melville's father had been declared bankrupt and committed suicide. At the age of 15 Melville left school and began working in a bank to support his family, before signing on to a merchant ship bound for Liverpool. Over the next five years he travelled extensively on board merchant ships and whalers, before quitting the sea to write in 1844. Early books such as *Typee* and *Omoo*, which drew heavily upon his experiences in the Pacific, brought Melville considerable success, but the work that was to be his most enduring creation, *Moby-Dick*, was greeted with bafflement and derision, and failed to find a readership, beginning a process of decline that would lead to Melville being almost forgotten by the time of his death in 1891. In the years after Melville's death his reputation began to grow once more, leading to the 'rediscovery' of *Moby-Dick* in the 1920s. Despite its chequered history, *Moby-Dick* is now regarded as not just one of the central works of American literature, but one of the central works of world literature.

Kem Nunn is an American surfer, novelist and television writer. He is the author of five novels, including *Tapping the Source* and *The Dogs of Winter*, collaborated with David Milch on the television series *Deadwood*, and wrote and co-produced the television series *John from Cincinnati*.

Edgar Allan Poe was born in Boston in 1809, before being sent to school in England in 1815. After returning to the United States in 1820 he enrolled at the University of Virginia in 1827, but left soon after to pursue a literary career. Over the next two decades he produced a body of poetry, fiction and non-fiction as notable for its originality and energy as its vivid and unsettling imagery, and which has served as a source of ongoing inspiration to writers and filmmakers for more than 150 years. 'A Descent into the Maelström' is based upon descriptions of the *Mosktraumen*, in Norway, a system of tidal eddies and whirlpools which is sometimes described as a permanent whirlpool. Poe died under mysterious circumstances in 1849.

Jonathan Raban was born in 1942 in Norfolk. He is the author of several novels, most recently *Surveillance*, and a number of highly acclaimed works of non-fiction, many drawing upon his extensive experience as a sailor, and was the editor of *The Oxford Book of the Sea*. Since 1990 he has lived in Seattle.

Ernest Shackleton was born in Ireland in 1874. At 16 he was apprenticed on a sailing ship, and by 1898 had been certified as a

Master Mariner. In 1901 he accompanied Robert Falcon Scott south on Scott's first expedition to the Antarctic, but returned to England in 1903 after falling ill. Once his health was restored Shackleton began preparations for his own expedition to Antarctica. His first expedition, aboard the *Nimrod* in 1908, was a success, but his second, aboard the *Endurance* in 1914, ran into difficulties when the ship became trapped in pack ice. After surviving the Antarctic winter the *Endurance* was crushed by shifting ice in the spring and sank in November 1915, forcing Shackleton and his men to set up camp on an ice floe. Over the next few months they drifted slowly north, until on 9 April they made landfall on Elephant Island, just off the tip of the Antarctic Peninsula. Realising they would not survive another Antarctic winter, Shackleton decided to attempt to reach the whaling stations on South Georgia Island in one of the lifeboats, and on 24 April he and five others set sail in the *James Caird*. Despite ferocious storms and freezing conditions the boat landed on South Georgia just over two weeks later. In the years after the *Endurance* Expedition, Shackleton, dogged by ill-health, lectured and wrote, but in 1921 he set off on one last expedition to Antarctica. He reached South Georgia Island on 4 January 1922 and died of a heart attack the next day.

Joshua Slocum was born in Nova Scotia in 1844, and before running away to sea at the age of 14 spent much of his childhood living on an island near a lighthouse operated by his grandfather. After several years working as a merchant seaman, he became an American citizen in 1865 and a captain four years later, working routes from San Francisco to China and Australia. After meeting his future wife in Sydney in 1870 he began a series of voyages which culminated in his solo circumnavigation of the world in his boat, *Spray*. His account of that trip, *Sailing Alone Around the World*, was published in 1899, and made Slocum an international celebrity. In his later years Slocum's health declined and he experienced bouts of mental illness and legal troubles (including charges of raping a 12 year-old girl) but he continued to travel alone up and down the American coast until he and *Spray* disappeared in 1909 in the Atlantic. Despite a life spent at sea Slocum never learned to swim, dismissing it as a useless skill.

John Steinbeck was born in California in 1902. He studied literature and writing at Stanford University between 1919 and 1925, before leaving to work as a journalist and farm labourer. His first novel, *Cup of Gold*, was published in 1929, but it was not until the publication of his fourth novel, *Tortilla Flat*, in 1935, that he found popular success. In the years that followed he wrote a series of acclaimed novels, most notably *The Grapes of Wrath*, *Of Mice and Men* and *East of Eden*, as well as plays and screenplays. Steinbeck was fascinated by the ocean, and spent much of his childhood in towns around Monterey Bay, and even studied marine biology for a time in the 1920s. But it was his friendship with the marine biologist Ed Ricketts (himself the model for the character of Doc in *Cannery Row*) that led to Steinbeck's enduring interest in the subject, and more particularly to *The Log from the Sea of Cortez*, Steinbeck's account of the 6500km journey down the Californian coast and into the Sea of Cortez he and Ricketts took in 1940. Steinbeck received the Nobel Prize for Literature in 1962 and died in 1968.

Wallace Stevens was one of the most significant poets of the 20th century. Born in Pennsylvania in 1879, he studied at Harvard and worked briefly as a journalist, before going on to study law. In 1908 he began working for the legal department of the Hartford Accident and Indemnity insurance company in Connecticut, of which he became vice-president in 1934. He died in Hartford in 1955.

Henry David Thoreau was born in Massachusetts in 1817. One of the seminal figures of American literature, he is most famous for his essay *Civil Disobedience*, and *Walden*, his account of two years living alone in a cabin in the woods. Based on a series of short trips Thoreau made to the Cape between 1849 and 1857, *Cape Cod* was first published in book form in 1865, three years after Thoreau's death.

Derek Walcott was born in St Lucia, in the West Indies, in 1930. The author of many plays and books of poetry, most recently *White Egrets*, he was awarded the Queen's Medal for Poetry in 1988, and the Nobel Prize for Literature in 1992. He now divides his time between homes in St Lucia and New York.

Tim Winton was born in Perth in 1960. He has published twenty books for adults and children, and his work has been translated into twenty-five languages. Since his first novel, *An Open Swimmer*, won the Australian/Vogel Award in 1981, he has won the Miles Franklin Award four times (for *Shallows*, *Cloudstreet*, *Dirt Music* and *Breath*) and twice been shortlisted for the Booker Prize (for *The Riders* and *Dirt Music*). He lives in Western Australia.

ACKNOWLEDGEMENTS

Reflecting upon the writing of her luminous portrait of the oceans, *The Sea Around Us*, Rachel Carson wrote that 'to cope alone and unaided with a subject so vast, so complex, and so infinitely mysterious as the sea would be a task not only cheerless but impossible,' a sentiment I can only echo. This book would be a far poorer one without the input of the many friends and colleagues who provided ideas and suggestions. While I have lost count of the people who suggested passages or books of personal significance, I am particularly grateful to Jennifer Ackerman, Don Anderson, Delia Falconer, Thomas Farber, Rebecca Giggs, Jane Gleeson-White, Ross Gibson, Malcolm Knox, David Miller, Peter Rose, Geordie Williamson, Ron Wright and my father, Michael Bradley, all of whom pointed me to significant works which I might not otherwise have found, and whose counsel helped me refine and sharpen my sense of the purpose and structure of this book. I would also like to thank the many people who offered suggestions via my website, at least one of which has found its way into the final volume.

I must also acknowledge my debt to the many writers and editors who have preceded me, in particular Jonathan Raban, whose capacious *Oxford Book of the Sea* charted this territory before me, and Philip Edwards, whose marvellous *The Story of the Voyage* helped shape my thinking about eighteenth century travel narratives.

And finally I'd like to thank the team at Penguin Australia for their enthusiastic support of this project, and in particular my editor,

Cate Blake, and publisher, Ben Ball, whose thoughts helped spark this project.

Penguin Australia and I are grateful to the following writers and publishers for permission to reproduce material in this anthology:

'The Gray Beginnings' from *The Sea Around Us* by Rachel Carson, first published Oxford University Press, 1989, © Rachel L. Carson, reproduced by permission of Pollinger Limited and Fran Collin.

'The Sea is History' by Derek Walcott, from Derek Walcott, *Selected Poems*, reproduced by permission of Faber and Faber Ltd, faber.co.uk.

'Diving the *Coolidge*' by Luke Davies, from Luke Davies, *Running with Light*, first published by Allen and Unwin, Sydney, Australia, 1999, reproduced by permission of Allen and Unwin, allenandunwin.com.

'The Idea of Order at Key West' by Wallace Stevens, from Wallace Stevens, *Selected Poems*, reproduced by permission of Faber and Faber Ltd, faber.co.uk.

'At the Fishhouses' from *The Complete Poems 1927-1979* by Elizabeth Bishop, © 1979, 1983 by Alice Helen Methfessel. Reprinted by permission of Farrar, Straus and Giroux, LLC.

Excerpt from *Passage to Juneau* by Jonathan Raban © Jonathan Raban 1999, reproduced by permission of Pan Macmillan, London.

'The Beach' by Emily Ballou, from *The Darwin Poems* reproduced by permission of UWA Publishing.

Excerpt from *Notes from the Shore* by Jennifer Ackerman, reproduced by permission of Jennifer Ackerman.

Excerpt from *The Old Man and the Sea* by Ernest Hemingway, published by Jonathan Cape, reproduced by permission of The Random House Group Ltd.

Excerpt from *The Log from the Sea of Cortez* by John Steinbeck, reproduced by permission of Penguin Books.

Excerpt from *The Kon-Tiki Expedition* by Thor Heyerdahl, translated by F.H. Lyon and reproduced by permission of the Kon-Tiki Museum – Thor Heyerdahl's Research Foundation.

Excerpt from *On Water* by Thomas Farber reproduced with permission Thomas Farber.

Excerpt from *The Silent World* by Jacques Cousteau reproduced with permission of the Cousteau Society, cousteau.org.

Excerpt from *Seven-Tenths: The Sea and its Thresholds* by James Hamilton-Paterson, reproduced by permission of Faber and Faber Ltd, faber.co.uk.

Photographs in Wayne Levin's *Resident Spirits* reproduced by permission of Wayne Levin.

Excerpt from *Caught Inside: A Surfer's Year on the California Coast,* by Daniel Duane, reproduced by permission of Daniel Duane.

'Facing the Pacific at Night' by Kevin Hart reproduced by permission of Kevin Hart.

Excerpt from *The Perfect Storm* by Sebastian Junger reproduced by permission of Harper Collins Publishers Pty Ltd © 1997 Sebastian Junger.

Excerpt from *The Worst Journey in the World* by Apsley Cherry-Garrard appears with the permission of the Scott Polar Research Institute, University of Cambridge.

'Whale' by Judith Beveridge reproduced by permission of Judith Beveridge.

Excerpt from *The Dogs of Winter* by Kem Nunn, reproduced by permission of No Exit Press, noexit.co.uk.

Excerpt from *Breath* by Tim Winton reproduced by permission of Penguin Books Australia.

Excerpt from *That Oceanic Feeling* by Fiona Capp, first published by Allen and Unwin, Sydney, Australia, 2003, reproduced by permission of Allen and Unwin, allenandunwin.com.

'The Boat' by Nam Le from Nam Le, *The Boat*, reproduced by permission of Penguin Books Australia.

Excerpt from *The Outlaw Sea* by William Langewiesche reproduced by permission of Granta Books.

'Climate and Atlantic' from *Great Waters An Atlantic Passage* by Deborah Cramer, © Deborah Cramer 2001, used by permission of W.W. Norton & Company Inc.

Excerpt from *Fly Away Peter* by David Malouf, published by Chatto and Windus. Reproduced with permission of The Random House Group Ltd.

While all efforts have been made to locate the appropriate rights holders, any persons believing there is material contained in this collection for which appropriate permission has not been obtained should contact Penguin Books Australia.

A note on the text

Because of the difficulties inherent in standardising spelling and style across texts that span more than 400 years, I have, as much as possible, conformed to the conventions adopted in the editions upon which I have relied. While this has created some inconsistencies, it allows the flavour of the originals to be preserved better than they might otherwise have been.

ALSO FROM PENGUIN

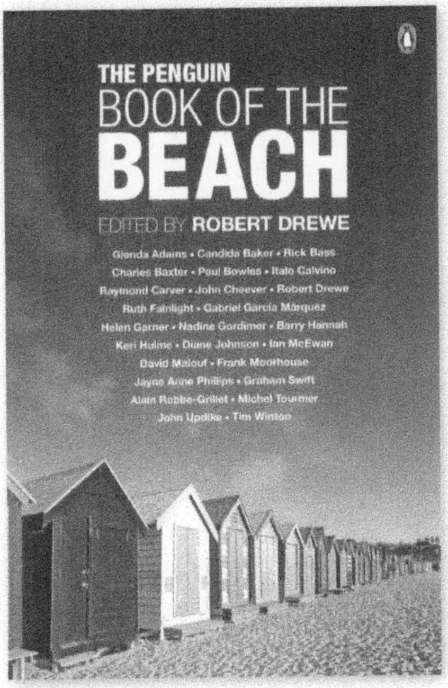

The average Australian has conducted a lifelong love affair with the beach and the ocean shores, bays, dunes, lagoons and rivers of the coast. Until now, however, no one has attempted to match the ancient sensual and artistic preoccupation with the sea to the intuitive appreciation of the coast felt by modern beachgoers.

In this illustrious international selection, Robert Drewe has drawn together twenty-five of the finest contemporary writers whose stories represent the most stimulating, startling, humorous and deeply moving writing about the beach.

Australia is a nation of drivers. We spend more time behind the wheel than almost anyone else, on fast highways, lonely bush tracks, jammed city lanes and suburban streets. The road is the place where the great dramas of our lives unfold, the route to our greatest pleasures as well as our worst nightmares. It is sexy, dangerous and unnerving.

In this landmark collection, acclaimed novelist and essayist Delia Falconer brings together some of our very best writing on every aspect of the road.

Lovers, lost children, bushrangers, killers. From the classic to the modern, from the outback to the beach, *The Penguin Book of the Road* is an entertaining ride into the heart of Australia.